LONDON MATHEMATICAL SOCIETY LECTURE NOTE SERIES

Managing Editor:
Professor N.J. Hitchin,
Mathematical Institute, 24–29 St. Giles, Oxford OX1 3DP, UK

All the titles listed below can be obtained from good booksellers or from Cambridge University Press. For a complete series listing visit http://publishing.cambridge.org/stm/mathematics/lmsn/

T0269030

London Mathematical Society Lecture Note Series. 298

Higher Operads, Higher Categories

Tom Leinster

Institut des Hautes Etudes Scientifiques

CAMBRIDGE
UNIVERSITY PRESS

CAMBRIDGE UNIVERSITY PRESS
Cambridge, New York, Melbourne, Madrid, Cape Town, Singapore, São Paulo, Delhi

Cambridge University Press
The Edinburgh Building, Cambridge CB2 8RU, UK

Published in the United States of America by Cambridge University Press, New York

www.cambridge.org
Information on this title: www.cambridge.org/9780521532150

© Tom Leinster 2004

First published 2004

A catalogue record for this publication is available from the British Library

ISBN 978-0-521-53215-0 paperback

Transferred to digital printing 2009

To
Martin Hyland
Wilson Sutherland
and
Mr Bull

Contents

Appendices

Diagram of interdependence

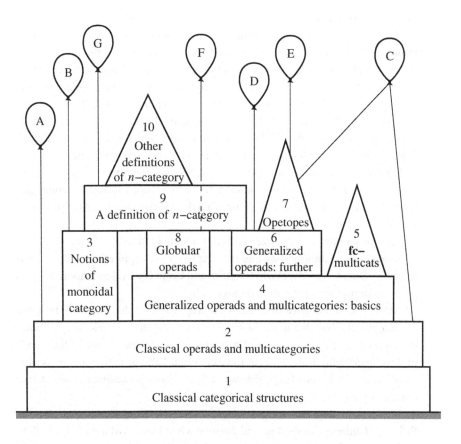

Balloons are appendices, giving support in the form of proofs. They can be omitted if some results are taken on trust.

Acknowledgements

The ideas presented here have been at the centre of my research for most of the last six years, during which time I have been influenced – for the better, I think – by more people than I can thank formally here. Above all I feel moulded by the people who were around me during my time in Cambridge. I want to thank Martin Hyland in particular: his ways of thought have been a continual inspiration, and I feel very lucky to have been his student and to have shared a department with him for so long. I have also gained greatly from conversations with Eugenia Cheng, Peter Johnstone, and Craig Snydal. Ivan Smith and Dick Thomas have cheerfully acted as consultants for daft geometry questions.

Ian Grojnowski suggested this Lecture Notes series to me, and I am very glad he did: I could not have wished for a more open-minded, patient and helpful editor than Jonathan Walthoe at CUP. I would also like to thank Nigel Hitchin, the series editor.

I am very grateful to Andrea Hesketh for sound strategic advice.

Some of the quotations starting the chapters were supplied, directly or indirectly, by Sean Carmody, David Corfield, and Colin Davey. Paul-André Melliès gave me important information on Swiss cheese.

Almost all of the software used in the preparation of this book was free, not just financially, but also in the libertarian sense: freedom to take it apart, alter it, and propagate it, like a piece of mathematics. I am grateful to the thousands of developers who brought about the truly remarkable situation that made this possible. I also acknowledge the use of Paul Taylor's excellent commutative diagrams package.

I started writing this when I was Laurence Goddard Research Fellow at St John's College, Cambridge, and finished when I was William Hodge Fellow at the Institut des Hautes Etudes Scientifiques. The index was compiled while I was visiting the University of Chicago at the invitation of Peter May. I am immensely grateful to St John's and the IHES for the opportunities they have given me, and for their unwavering dedication to research. In particular, I thank the IHES for giving me the chance to live in Paris and absorb some Russian culture.

Introduction

It must be admitted that the use of geometric intuition has no logical necessity in mathematics, and is often left out of the formal presentation of results. If one had to construct a mathematical brain, one would probably use resources more efficiently than creating a visual system. But the system is there already, it is used to great advantage by human mathematicians, and it gives a special flavor to human mathematics.

Ruelle (1999)

Higher-dimensional category theory is the study of a zoo of exotic structures: operads, n-categories, multicategories, monoidal categories, braided monoidal categories, and more. It is intertwined with the study of structures such as homotopy algebras (A_∞-categories, L_∞-algebras, Γ-spaces, ...), n-stacks, and n-vector spaces, and draws it inspiration from areas as diverse as topology, quantum algebra, mathematical physics, logic, and theoretical computer science.

No surprise, then, that the subject has developed chaotically. The rush towards formalizing certain commonly-imagined concepts has resulted in an extraordinary mass of ideas, employing diverse techniques from most of the subject areas mentioned. What is needed is a transparent, natural, and practical language in which to express these ideas.

The main aim of this book is to present one. It is the language of generalized operads. It is introduced carefully, then used to give simple descriptions of a variety of higher categorical structures.

I hope that by the end, the reader will be convinced that generalized operads provide as appropriate a language for higher-dimensional category theory as vector spaces do for linear algebra, or sheaves for algebraic geometry. Indeed, the reader may also come to share the feeling that generalized operads are as

1

applicable and pervasive in mathematics at large as are n-categories, the usual focus of higher-dimensional category theorists.

Here are some of the structures that we will study, presented informally.

Let $n \in \mathbb{N}$. An n-**category** consists of 0-**cells** (objects) $a, b, \ldots$, 1-**cells** (arrows) $f, g, \ldots$, 2-**cells** (arrows between arrows) $\alpha, \beta, \ldots$, 3-**cells** (arrows between arrows between arrows) $\Gamma, \Delta, \ldots$, and so on, all the way up to n-**cells**, together with various composition operations. The cells are usually drawn like this:

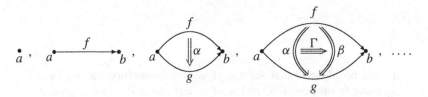

Typical example: for any topological space X there is an n-category whose k-cells are maps from the closed k-dimensional ball into X. A 0-category is just a set, and a 1-category just an ordinary category.

A **multicategory** consists of objects $a, b, \ldots$, arrows $\theta, \phi, \ldots$, a composition operation, and identities, just like an ordinary category, the difference being that the domain of an arrow is not just a single object but a finite sequence of them. An arrow is therefore drawn as

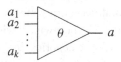

(where $k \in \mathbb{N}$), and composition turns a tree of arrows into a single arrow. Vector spaces and linear maps form a category; vector spaces and multilinear maps form a multicategory.

An **operad** is a multicategory with only one object. Explicitly, an operad consists of a set $P(k)$ for each $k \in \mathbb{N}$, whose elements are thought of as 'k-ary operations' and drawn as

with k input wires on the left, together with a rule for composing the operations and an identity operation. Example: for any vector space V there is an operad whose k-ary operations are the linear maps $V^{\otimes k} \longrightarrow V$.

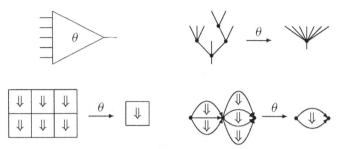

Fig. 0-A. Operations θ in four different types of generalized operad

Operads describe operations that take a finite sequence of things as input and produce a single thing as output. A finite sequence is a 1-dimensional entity, so operads can be used, for example, to describe the operation of composing a (1-dimensional) string of arrows in a (1-)category. But if we are interested in higher-dimensional structures such as n-categories then we need a more general notion of operad, one where the inputs of an operation can form a higher-dimensional shape – a grid, perhaps, or a tree, or a so-called pasting diagram. For each choice of 'input type' T, there is a class of T-**operads**. A T-operad consists of a family of operations whose inputs are of the specified type, together with a rule for composition; for instance, if the input type T is 'finite sequences' then a T-operad is an ordinary operad. Fig. 0-A shows typical operations θ in a T-operad, for four different choices of T. Similarly, there are T-**multicategories**, where the shapes at the domain and codomain of arrows are labelled with the names of objects. These are the 'generalized operads' and 'generalized multicategories' at the heart of this book.

The uniting feature of all these structures is that they are purely algebraic in definition, yet near-impossible to understand without drawing or visualizing pictures. They are inherently geometrical.

A notorious problem in this subject is the multiplicity of definitions of n-category. Something like a dozen different definitions have been proposed, and there are still very few precise results stating equivalence between any of them. This is not quite the scandal it may seem: it is hard to say what 'equivalence' should even mean. Suppose that Professors X and Y each propose a definition of n-category. To compare their definitions, you find a way of taking one of X's n-categories and deriving from it one of Y's n-categories, and *vice versa*, then you try to show that doing one process then the other gets you back to where you started. It is, however, highly unrealistic to expect that you will get back to *exactly* where you started. For most types of mathematical structure, getting back to somewhere isomorphic to your starting point would be a reasonable expectation. But for n-categories, as we shall see, this is still unrealistic: the canonical notion of equivalence of n-categories is much weaker than isomorphism. Finding a precise definition of equivalence for a given definition of n-category can be difficult. Indeed, many of the proposed definitions of n-category did not come with accompanying proposed definitions of equivalence, and this gap must be almost certainly be filled before any comparison results can be proved.

Is this all 'just language'? There would be no shame if it were: language can have the most profound effect. New language can make new concepts thinkable, and make old, apparently obscure, concepts suddenly seem natural and obvious. But there is no clear line between mathematical language and 'real' mathematics. For example, we will see that a 3-category with only one 0-cell and one 1-cell is precisely a braided monoidal category, and that the free braided monoidal category on one object is the sequence $(B_n)_{n \in \mathbb{N}}$ of braid groups. So if n-categories are just language, not 'real' mathematical objects, then the same is true of the braid groups, which describe configurations of knotted string. The distinction begins to look meaningless.

Here is a summary of the contents.

Motivation for topologists

Topology and higher-dimensional category theory are intimately related. The diagrams that one cannot help drawing when thinking about higher categorical structures can very often be taken literally as pieces of topology. We start with an informal discussion of the connections between the two subjects. This includes various topological examples of n-categories, and an account of how the world of n-categories is a mirror of the world of homotopy groups of spheres.

Part I. Background

We will build on various 'classical' notions. Those traditionally considered the domain of category theorists are in Chapter 1: ordinary categories, bicategories, strict n-categories, and enrichment. Classical operads and multicategories have Chapter 2 to themselves. They should be viewed as categorical structures too, although, anomalously, operads are best known to homotopy theorists and multicategories to categorical logicians.

The familiar concept of monoidal (tensor) category can be formulated in a remarkable number of different ways. We look at several in Chapter 3, and prove them equivalent. Monoidal categories can be identified with one-object 2-categories, so this is a microcosm of the comparison of different definitions of n-category.

Part II. Operads

This introduces the central idea of the text: that of generalized ('higher') operad and multicategory. The definitions – of generalized operad and multicategory, and of algebra for a generalized operad or multicategory – are stated and explained in Chapter 4, and some further theory is developed in Chapter 6.

There is a truly surprising theory of enrichment for generalized multicategories – it is not at all the routine extension of traditional enriched category theory that one might expect. This was to have formed Part IV of the book, but for reasons of space it was (reluctantly) dropped. A summary of the theory, with pointers to the original papers, is in Section 6.8.

The rest of Part II is made up of examples and applications. Chapter 5 is devoted to so-called **fc**-multicategories, which are generalized multicategories for a certain choice of input shape. They turn out to provide a clean setting for some familiar categorical constructions that have previously been encumbered by technical restrictions. In Chapter 7 we look at opetopic sets, structures analogous to simplicial sets and used in the definitions of n-category proposed by Baez, Dolan, and others. Again, the language of higher operads provides a very clean approach; we also find ourselves drawn inexorably into higher-dimensional topology.

Part III. n-categories

Using the language of generalized operads, some of the proposed definitions of n-category are very simple to state. We start by concentrating on one in particular, in which an n-category is defined as an algebra for a certain globular operad. A globular operad is a T-operad for a certain choice of input type T; the associated diagrams are complexes of discs, as in the last arrow θ of Fig. 0-A. Chapter 8 explains what globular operads are in pictorial terms. In Chapter 9 we choose a particular globular operad, define an n-category as an algebra for it, and explore the implications in some depth.

The many proposed definitions of n-category are not as dissimilar as they might at first appear. We go through most of them in Chapter 10, drawing together the common threads.

Appendices

This book is mostly about description: we develop language in which structures can be described simply and naturally, accurately reflecting their geometric reality. In other words, we mostly avoid the convolutions and combinatorial complexity often associated with higher-dimensional category theory. Where things run less smoothly, and in other situations where a lengthy digression threatens to disrupt the flow of the main text, the offending material is confined to an appendix. As long as a few plausible results are taken on trust, the entire main text can be read and understood without looking at any of the appendices.

A few words on terminology are needed. There is a distinction between 'weak' and 'strict' n-categories, as will soon be explained. For many years only the strict ones were considered, and they were known simply as 'n-categories'. More recently it came to be appreciated that weak n-categories are much more abundant in nature, and many authors now use 'n-category' to mean the weak version. I would happily join in, but for the following obstacle: in most parts of this book that concern n-categories, both the weak and the strict versions are involved and discussed in close proximity. It therefore seemed preferable to be absolutely clear and say either 'weak n-category' or 'strict n-category' on every occasion. The only exceptions are in this Introduction and the Motivation for topologists, where the modern convention is used.

The word 'operad' will be used in various senses. The most primitive kind of operad is an operad of sets without symmetric group action, and this is our starting point (Chapter 2). We hardly ever consider operads equipped with symmetric group actions, and when we do we call them 'symmetric operads'; see p. 58 for a more comprehensive warning.

Any finite sequence $x_1, \ldots, x_n$ of elements of a monoid has a product $x_1 \cdots x_n$. When $n = 0$, this is the unit element. Similarly, an identity arrow in a category can be regarded as the composite of a zero-length string of arrows placed end to end. I have taken the view throughout that there is nothing special about units or identities; they are merely nullary products or composites. Related to this is a small but important convention: the natural numbers, $\mathbb{N}$, start at zero.

Motivation for topologists

I'm a goddess *and* a nerd!
Bright (1999)

Higher-dimensional category theory *can* be treated as a purely algebraic subject, but that would be missing the point. It is inherently topological in nature: the diagrams that one naturally draws to illustrate higher-dimensional structures can be taken quite literally as pieces of topology. Examples of this are the braidings in a braided monoidal category and the pentagon appearing in the definitions of both monoidal category and A_∞-space.

This section is an informal description of what higher-dimensional category theory is and might be, and how it is relevant to topology. Grothendieck, for instance, suggested that tame topology should be the study of n-groupoids; others have hoped that an n-category of cobordisms between cobordisms between ... will provide a clean setting for topological quantum field theory; and there is convincing evidence that the whole world of n-categories is a mirror of the world of homotopy groups of spheres.

There are no real theorems, proofs, or definitions here. But to whet your appetite, here is a question to which we will reach an answer by the end:

Question What is the close connection between the following two facts?

A No-one ever got into trouble for leaving out the brackets in a tensor product of several objects (abelian groups, etc.). For instance, it is safe to write $A \otimes B \otimes C$ instead of $(A \otimes B) \otimes C$ or $A \otimes (B \otimes C)$.

B There exist non-trivial knots in $\mathbb{R}^3$.

The very rough idea

In ordinary category theory we have diagrams of objects and arrows such as

We can imagine more complex category-like structures in which there are diagrams such as

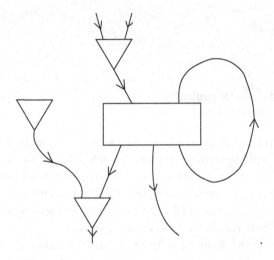

This looks like an electronic circuit diagram or a flow chart; the unifying idea is that of 'information flow'. It can be redrawn as

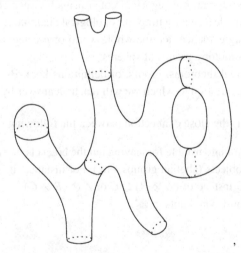

which looks like a surface or a diagram from topological quantum field theory.

We can also use diagrams like this to express algebraic laws such as commutativity:

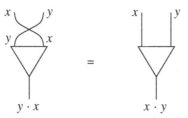

The fact that two-dimensional TQFTs are the same as commutative Frobenius algebras is an example of an explicit link between the spatial and algebraic aspects of diagrams like these.

Moreover, if we allow crossings, as in the commutativity diagram or as in

,

then we obtain pictures looking like knots; and as we shall see, there are indeed relations between knot theory and higher categorical structures.

So the idea is:

> Ordinary category theory uses 1-dimensional arrows $\longrightarrow$
> Higher-dimensional category theory uses higher-dimensional arrows

The natural topology of these higher-dimensional arrows is what makes higher-dimensional category theory an inherently topological subject.

We will be concerned with structures such as operads, generalized operads (of which the variety familiar to homotopy theorists is a basic special case), multicategories, various flavours of monoidal categories, and n-categories; in this introduction I have chosen to concentrate on n-categories. Terminology: an n-category (or 'higher-dimensional category') is not a special kind of category, but a generalization of the notion of category; compare the usage of 'quantum group'. A 1-category is the same thing as an ordinary category, and a 0-category is just a set.

n-categories

Here is a very informal

'Definition' Let $n \geq 0$. An n-**category** consists of

- **0-cells** or **objects**, A, B, ...
- **1-cells** or **morphisms**, drawn as $A \xrightarrow{\ f\ } B$

- **2-cells** 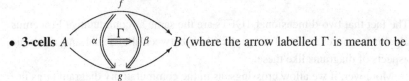 ('morphisms between morphisms')

- **3-cells** A $\alpha \left(\Longrightarrow \right) \beta$ B (where the arrow labelled Γ is meant to be

  going in a direction perpendicular to the plane of the paper)
- ...
- all the way up to n-**cells**
- various kinds of **composition**, e.g.

$$A \xrightarrow{\ f\ } B \xrightarrow{\ g\ } C \quad \text{gives} \quad A \xrightarrow{\ g \circ f\ } C$$

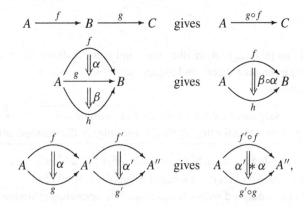

and so on in higher dimensions; and similarly **identities**.

These compositions are required to 'all fit together nicely' – a phrase hiding many subtleties. ω-**categories** (also known as ∞-**categories**) are defined similarly, by going on up the dimensions forever instead of stopping at n.

There is nothing forcing us to make the cells spherical here. We could, for instance, consider cubical structures, in which 2-cells look like

This is an inhabitant of the 'zoo of structures' mentioned earlier, but is not an n-category as such. (See Sections 1.4 and 5.2 for more on this particular structure.)

Critical example Any topological space X gives rise to an ω-category $\Pi_\omega X$ (its **fundamental ω-groupoid**), in which

- 0-cells are points of X, drawn as •
- 1-cells are paths in X (maps $[0, 1] \longrightarrow X$), drawn as •———•• – though whether that is meant to be a picture in the space X or the ω-category $\Pi_\omega X$ is deliberately ambiguous; the idea is to blur the distinction between geometry and algebra
- 2-cells are homotopies of paths (relative to endpoints), drawn as ⬤⇓⬤
- 3-cells are homotopies of homotopies of paths (that is, suitable maps $[0, 1]^3 \longrightarrow X$)
- ...
- composition is by pasting paths and homotopies.

(The word 'groupoid' means that all cells of dimension higher than zero are invertible.)

$\Pi_\omega X$ should contain all the information you want about X if your context is 'tame topology'. In particular, you should be able to compute from it the homotopy, homology and cohomology of X. You can also truncate after n steps in order to obtain $\Pi_n X$, the **fundamental n-groupoid** of X; for instance, $\Pi_1 X$ is the familiar fundamental groupoid.

Alert As you may have noticed, composition in $\Pi_\omega X$ is not genuinely associative; nor is it unital, and nor are the cells genuinely invertible (only up to homotopy). We are therefore interested in **weak** n-categories, where the 'fitting together nicely' only happens up to some kind of equivalence, rather than **strict** n-categories, where associativity and so on hold in the strict sense.

To define strict n-categories precisely turns out to be easy. To define weak n-categories, we face the same kind of challenge as algebraic topologists did in the 1960s, when they were trying to state the exact sense in which a loop space is a topological group. It is clearly not a group in the literal sense, as composition of paths is not associative; but it is associative up to homotopy, and if you pick specific homotopies to do this job then these homotopies obey laws of their own – or at least, obey them up to homotopy; and so on. At least two precise formulations of 'group up to (higher) homotopy' became popular: Stasheff's A_∞-spaces and Segal's special Δ-spaces. (More exactly, these are notions of monoid or semigroup up to homotopy; the inverses are dealt with separately.)

The situation for weak n-categories is similar but more extreme: there are something like a dozen proposed definitions and, as mentioned in the Introduction, not much has been proved about how they relate to one another. Happily, we can ignore all this here and work informally. This means that nothing in the rest of this section is true with any degree of certainty or accuracy.

At this point you might be thinking: can't we do away with this difficult theory of weak n-categories and just stick to the strict ones? The answer is: if you're interested in topology, no. The difference between the weak and strict theories is genuine and non-trivial: for while it is true that every weak 2-category is equivalent to some strict one, and so it is also true that homotopy 2-types can be modelled by strict 2-groupoids, neither of these things is true in dimensions ≥ 3. For instance, there exist spaces X (such as the 2-sphere S^2) for which the weak 3-category $\Pi_3 X$ is not equivalent to any strict 3-category.

For the rest of this section, 'n-category' will mean 'weak n-category'. The strict ones are very much the lesser-spotted species.

Some more examples of ω-categories:

Top This is very similar to the Π_ω example above. **Top** has:

- 0-cells: topological spaces
- 1-cells: continuous maps
- 2-cells X 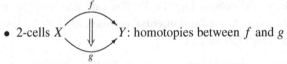 Y: homotopies between f and g
- 3-cells: homotopies between homotopies (that is, suitable maps $[0, 1]^2 \times X \longrightarrow Y$)
- ...
- composition as expected.

ChCx This ω-category has:

- 0-cells: chain complexes (of abelian groups, say)
- 1-cells: chain maps
- 2-cells: chain homotopies
- 3-cells A 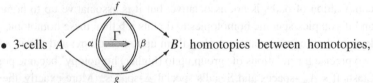 B: homotopies between homotopies, that is, maps $\Gamma : A \longrightarrow B$ of degree 2 such that $d\Gamma - \Gamma d = \beta - \alpha$
- ...

- composition: more or less as expected, but some choices are involved. For instance, if you try to write down the composite of two chain homotopies 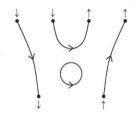 then you will find that there are two equally reasonable ways of doing it: one 'left-handed', one 'right-handed'. This is something like choosing the parametrization when deciding how to compose two loops in a space (usual choice: do everything at double speed). Somehow the fact that there is no canonical choice means that the resulting ω-category is bound to be weak.

In a reasonable world there ought to be a weak ω-functor **Chains** : **Top** $\longrightarrow$ **ChCx**.

Cobord This is an ω-category of cobordisms.

- 0-cells: 0-manifolds, where 'manifold' means 'compact, smooth, oriented manifold'. A typical 0-cell is ↑ ↓ ↑ ↑.
- 1-cells: 1-manifolds with corners, that is, cobordisms between 0-manifolds, such as

(this being a 1-cell from a 4-point 0-manifold to a 2-point 0-manifold). Atiyah–Segal-style TQFT stops here and takes *isomorphism classes* of the 1-cells just described, to make a category. We avoid this (unnatural?) quotienting out and carry on up the dimensions.

- 2-cells: 2-manifolds with corners, such as

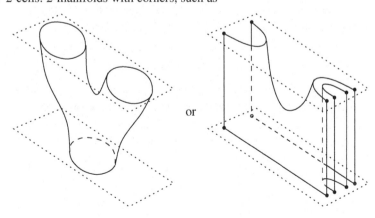

or

(leaving all the orientations off). The right-hand diagram shows a 2-

cell L 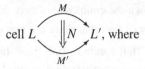 L', where

$$L = \bullet\,\bullet\,, \qquad L' = \bullet\,\bullet\,\bullet\,\bullet\,, \qquad M = \overset{\smile}{\underset{\frown}{\bullet\,\bullet}}, \qquad M' = \Big|_{\bullet}\Big|_{\bullet},$$

and N is the disjoint union of the two 2-dimensional sheets. Khovanov (2001) discusses TQFTs with corners in the language of 2-categories; his approach stops here and takes isomorphism classes of the 2-cells just described, to make a 2-category. Again, we do not quotient out but keep going up the dimensions.

- 3-cells, 4-cells, ... are defined similarly
- composition is gluing of manifolds.

Some authors discuss 'extended TQFTs' using the notion of n-vector space. A 0-vector space is a complex number, a 1-vector space is an ordinary complex vector space, and n-vector spaces for higher n are something more sophisticated. See the Notes below for references.

Stabilization

So far we have seen that topological structures provide various good examples of n-categories, and that alone might be enough to convince you that n-categories are interesting from a topological point of view. But the relationship between topology and higher-dimensional category theory is actually much more intimate than that. To see how, we analyse certain types of degenerate n-categories. It will seem at first as if this is a purely formal exercise, but before long the intrinsic topology will begin to shine through.

Some degeneracies

- A category $\mathcal{C}$ with only one object is the same thing as a monoid (semigroup with unit) M. For if the single object of $\mathcal{C}$ is called $\star$, say, then $\mathcal{C}$ just consists of the set $\mathrm{Hom}(\star, \star)$ together with a binary operation of composition and a unit element 1, obeying the usual axioms. So we have:

$$\text{morphism in } \mathcal{C} = \text{element of } M$$
$$\circ \text{ in } \mathcal{C} = \cdot \text{ in } M.$$

- A 2-category $\mathcal{C}$ with only one 0-cell is the same thing as a monoidal category $\mathcal{M}$. (Private thought: if $\mathcal{C}$ has only one 0-cell then there are only interesting things happening in the top two dimensions, so it must be *some* kind of one-dimensional structure.) This works as follows:

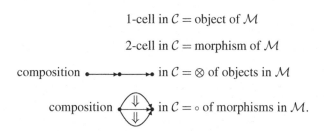

$$\text{1-cell in } \mathcal{C} = \text{object of } \mathcal{M}$$

$$\text{2-cell in } \mathcal{C} = \text{morphism of } \mathcal{M}$$

$$\text{composition} \quad \bullet \longrightarrow \bullet \longrightarrow \bullet \quad \text{in } \mathcal{C} = \otimes \text{ of objects in } \mathcal{M}$$

$$\text{composition} \quad \text{in } \mathcal{C} = \circ \text{ of morphisms in } \mathcal{M}.$$

- A monoidal category $\mathcal{C}$ with only one object is ... well, if we forget the monoidal structure for a moment then, as we have just seen, it is a monoid whose elements are the morphisms of $\mathcal{C}$ and whose multiplication is the composition in $\mathcal{C}$. Now, the monoidal structure on $\mathcal{C}$ provides not only a tensor product for objects, but also a tensor product for morphisms: so the set of morphisms of $\mathcal{C}$ has a second multiplication on it, $\otimes$. So a one-object monoidal category is a set M equipped with two monoid structures that are in some sense compatible (because of the axioms on a monoidal category). A well-known result (Lemma 1.2.4) says that in this situation, the two multiplications are in fact equal and commutative. So, a one-object monoidal category is a commutative monoid.

  This is, essentially, the argument often used to prove that the higher homotopy groups are abelian, or that the fundamental group of a topological group is abelian. In fact, we can deduce that π_2 is abelian from our 'results' so far:

  Corollary $\pi_2(X, x_0)$ is abelian, for any space X with basepoint x_0.

  Proof The 2-category $\Pi_2 X$ has a sub-2-category whose only 0-cell is x_0, whose only 1-cell is the constant path at x_0, and whose 2-cells are all the possible ones from $\Pi_2 X$ – that is, are the homotopies from the constant path to itself, that is, are the elements of $\pi_2(X, x_0)$. This sub-2-category is a 2-category with only one 0-cell and one 1-cell, that is, a monoidal category with only one object, that is, a commutative monoid.

- Next consider a 3-category with only one 0-cell and one 1-cell. We have not looked at (weak) 3-categories in enough detail to work this out properly, but it turns out that such a 3-category is the same thing as a braided monoidal category. By definition, a **braided monoidal category** is a monoidal category equipped with a map (a **braiding**)

$$A \otimes B \xrightarrow{\beta_{A,B}} B \otimes A$$

for each pair (A, B) of objects, satisfying axioms *not* including that

$$(A \otimes B \xrightarrow{\beta_{A,B}} B \otimes A \xrightarrow{\beta_{B,A}} A \otimes B) = 1.$$

The canonical example of a braided monoidal category (in fact, the braided monoidal category freely generated by a single object) is **Braid**. This has:

– objects: natural numbers 0, 1, …
– morphisms: braids, for instance

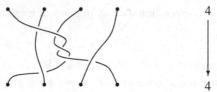

(taken up to homotopy); there are no morphisms $m \longrightarrow n$ when $m \neq n$
– tensor: placing side-by-side (which on objects means addition)
– braiding: left over right, for instance

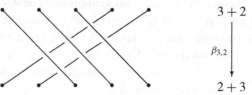

(Note that $\beta_{n,m} \circ \beta_{m,n}$ is not the identity braid.)

- We are rapidly getting out of our depth, but nevertheless: we have already considered n-categories that are only interesting in the top two dimensions for $n = 1, 2$, and 3. These are categories, monoidal categories, and braided monoidal categories respectively. What next? For $r \geq 4$, an r-category with only one i-cell for each $i < r - 1$ is, people believe, the same as a symmetric monoidal category (that is, a braided monoidal category in which $\beta_{B,A} \circ \beta_{A,B} = 1$ for all A, B). So the situation has stabilized…and this is meant to make you start thinking of stabilization phenomena in homotopy.

The big picture

Let us assemble this information on degeneracies systematically. Define an *m*-**monoidal *n*-category** to be an $(m + n)$-category with only one i-cell for each $i < m$. (This is certainly some kind of n-dimensional structure, as there are only interesting cells in the top $(n + 1)$ dimensions.) Here is what m-monoidal

n-categories are for some low values of m and n, laid out in the so-called 'periodic table'. Explanation follows.

	n			
m	0	1	2	3
0	set	category	2-category	3-category
1	monoid	monoidal category	monoidal 2-category	monoidal 3-category
2	commutative monoid	braided mon cat	braided mon 2-cat	braided mon 3-cat
3	"	symmetric mon cat	rhubarb	rhubarb
4	"	"	symmetric mon 2-cat	rhubarb
5	"	"	"	symmetric mon 3-cat
6	"	"	"	"

: take just the one-object structures

In the first row ($m = 0$), a 0-monoidal n-category is simply an n-category: it is not monoidal at all.

In the next row ($m = 1$), a 1-monoidal n-category is a monoidal n-category, in other words, an n-category equipped with a tensor product that is associative and unital up to equivalence of a suitable kind. For instance, a 1-monoidal 0-category is a one-object category (a monoid), and a 1-monoidal 1-category is a one-object 2-category (a monoidal category). A monoidal 2-category can be *defined* as a one-object 3-category, or can be defined directly as a 2-category with tensor.

We see from these examples, or from the general definition of m-monoidal n-category, that going in the direction ∕ means restricting to the one-object structures.

Now look at the third row ($m = 2$). We have already seen that a degenerate monoidal category is a commutative monoid and a doubly-degenerate 3-category is a braided monoidal category. It is customary to keep writing 'braided monoidal n-category' all along the row, but you can regard this as nothing more than name-calling.

Next consider the first *column* ($n = 0$). A one-object braided monoidal category is going to be a commutative monoid together with a little extra data (for the braiding) satisfying some axioms, but in fact this is trivial and we do not get anything new: in some sense, 'you can't get better than a commutative monoid'. This gives the entry for $m = 3, n = 0$, and the same applies all the way down the rest of the column.

A similar story can be told for the second column ($n = 1$). We saw – or rather, I claimed – that for $m \geq 3$, an m-monoidal 1-category is just a symmetric monoidal category. So again the column stabilizes, and again the point of stabilization is 'the most symmetric thing possible'.

The same goes in subsequent columns. The 'rhubarbs' could be replaced by more terminology – for instance, the first would become 'sylleptic monoidal 2-category' – but the details are not important here.

The main point is that the table stabilizes for $m \geq n + 2$ – just like $\pi_{m+n}(S^m)$. So if you overlaid a table of the homotopy groups of spheres onto the table above then they would stabilize at the same points. There are arguments to see why this should be so (and I remind you that this is all very informal and by no means completely understood). Roughly, the fact that the archetypal braided monoidal category **Braid** is not symmetric comes down to the fact that you cannot usually translate two 1-dimensional affine subspaces of 3-dimensional space past each other, and this is the same kind of dimensional calculation as you make when proving that the homotopy groups of spheres stabilize.

Answer to the initial question:

A Every weak 2-category is equivalent to a strict one. In particular, every (weak) monoidal category is equivalent to a strict one. So, for instance, we can pretend that the monoidal category of abelian groups is strict, making $\otimes$ strictly associative.

B *Not* every weak 3-category is equivalent to a strict one. We can construct a counterexample from what we have just done (details aside). Facts:

- a weak 3-category with one 0-cell and one 1-cell is a braided monoidal category
- a strict 3-category with one 0-cell and one 1-cell is a strict symmetric monoidal category
- any braided monoidal category equivalent to a symmetric monoidal category is itself symmetric.

It follows that any non-symmetric braided monoidal category is a weak 3-category not equivalent to a strict one. The canonical example is **Braid** itself, which is non-symmetric precisely because the overpass $\diagup\!\!\!\!\diagdown$

cannot be deformed to the underpass $\diagdown\!\!\!\!\diagup$ in $\mathbb{R}^3$.

Notes

Much has been written on the various interfaces between topology and higher category theory. I will just mention a few texts that I happen to have come across.

Grothendieck puts the case that tame topology is really the study of ω-groupoids in his epic (1983) letter to Quillen. A seminal paper of Shum (1994) establishes connections between higher categorical structures and knot theory; see also Yetter (2001).

Another introduction to higher categories from a topological viewpoint, with many similar themes to this one, is the first half of Baez (1997).

Specifically 2-categorical approaches to topological quantum field theory can be found in Tillmann (1998) and Khovanov (2001). n-vector spaces are explained in Kapranov and Voevodsky (1994), and their possible role in topological field theory is discussed in Lawrence (1996).

The periodic table is an absolutely fundamental object of mathematics, only discovered quite recently (Baez and Dolan, 1995), although foreshadowed in the work of Breen and of Street and the Australian school. That the table stabilizes for $m \geq n + 2$ is the 'Stabilization Hypothesis'. To state it precisely one needs to set up all the appropriate definitions first. A form of it has been proved by Simpson (1998), and a heuristic argument in the semistrict case has been given by Crans (2000, 3.8).

I have made one economy with the truth: a (weak) monoidal category with only one object is not exactly a commutative monoid, but rather a commutative monoid equipped with a distinguished invertible element; see Leinster (1999a, 1.6(vii)) for the reason.

One interesting idea not mentioned above is a higher categorical approach to non-abelian cohomology: specifically, nth cohomology should have coefficients in an n-category. This is explained in Street (1987, Introduction).

A serious and, of course, highly recommended survey of the proposed definitions of weak n-category, including ten such definitions, is my own (2001b).

PART I

Background

Chapter 1

Classical categorical structures

We will need to use some very simple notions of category theory, an esoteric subject noted for its difficulty and irrelevance.

Moore and Seiberg (1989)

You might imagine that you would need to be on top of the whole of ordinary category theory before beginning to attempt the higher-dimensional version. Happily, this is not the case. The main prerequisite for this book is basic categorical language, such as may be found in most introductory texts on the subject. Except in the appendices, we will need few actual theorems.

The purpose of this chapter is to recall some familiar categorical ideas and to explain some less familiar ones. Where the boundary lies depends, of course, on the reader, but very little here is genuinely new. Section 1.1 is on ordinary, '1-dimensional', category theory, and is a digest of the concepts that will be used later on. Impatient readers will want to skip immediately to Section 1.2, monoidal categories. This covers the basic concepts and two kinds of coherence theorem. Section 1.3 is a brief introduction to categories enriched in monoidal categories. We need enrichment in the next section, 1.4, on strict n-categories and strict ω-categories. This sets the scene for later chapters, where we consider the much more profound and interesting *weak n-categories*. Finally, in Section 1.5, we discuss bicategories, the best-known notion of weak 2-category, including coherence and their (not completely straightforward) relation to monoidal categories.

Examples of all these structures are given. Topological spaces and chain complexes are, as foreshadowed in the Motivation for topologists, a recurring theme.

1.1 Categories

This section is a sketch of the category theory on which the rest of the text is built. I have also taken the opportunity to state some notation and some small results that will eventually be needed.

In later chapters I will assume familiarity with the language of categories, functors, and natural transformations, and the basics of limits and adjunctions. I will also use the basic language of monads (sometimes still called 'triples'). As monads are less well known than the other concepts, and as they will be central to this text, I have included a short introduction to them below.

Given objects A and B of a category $\mathcal{A}$, I write $\mathcal{A}(A, B)$ for the class of maps (or morphisms, or arrows) from A to B, in preference to the less informative $\text{Hom}(A, B)$. The opposite or dual of $\mathcal{A}$ is written $\mathcal{A}^{\text{op}}$. Isomorphism between objects in a category is written $\cong$. For any two categories $\mathcal{A}$ and $\mathcal{B}$, there is a category $[\mathcal{A}, \mathcal{B}]$ whose objects are functors from $\mathcal{A}$ to $\mathcal{B}$ and whose maps are natural transformations.

My set-theoretic morals are lax; I have avoided questions of 'size' whenever possible. Where the issue is unavoidable, I have used the words **small** and **large** to mean 'forming a set' and 'forming a proper class' respectively. Some readers may prefer to re-interpret this using universes. A category is **small** if its collection of arrows is small.

The category of sets is written as **Set** and the category of small categories as **Cat**; occasionally I refer to **CAT**, the (huge) category of all categories. There are functors

$$\mathbf{Cat} \underset{I}{\overset{D}{\underset{\longleftarrow}{\overset{\longleftarrow}{\text{—ob}\longrightarrow}}}} \mathbf{Set},$$

where

- D sends a set A to the **discrete** category on A, whose object-set is A and all of whose maps are identities
- ob sends a category to its set of objects
- I sends a set A to the **indiscrete** category on A, whose object-set is A and which has precisely one map $a \longrightarrow b$ for each $a, b \in A$; all maps are necessarily isomorphisms.

We use **limits** ('inverse limits', 'projective limits') and **colimits** ('direct limits', 'inductive limits'). Binary product is written as $\times$ and arbitrary product as $\prod$; dually, binary coproduct (sum) is written as $+$ and arbitrary coproduct as $\coprod$. Nullary products are terminal (final) objects, written as 1; in particular, 1 often denotes a one-element set. The unique map from an object A of a category to a terminal object 1 of that category is written $! : A \longrightarrow 1$.

We will make particular use of **pullbacks** (fibred products). Pullback squares are indicated by right-angle marks:

$$
\begin{array}{ccc}
P & \longrightarrow & B \\
\downarrow & & \downarrow \\
A & \longrightarrow & C.
\end{array}
$$

We also write $P = A \times_C B$. In later chapters dozens of elementary manipulations of diagrams involving pullback squares are left to the hypothetical conscientious reader; almost all are made easy by the following invaluable lemma.

Lemma 1.1.1 (Pasting Lemma) *Take a commutative diagram of shape*

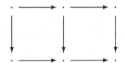

in some category, and suppose that the right-hand square is a pullback. Then the left-hand square is a pullback if and only if the outer rectangle is a pullback. □

An **adjunction** is a pair of functors $A \underset{G}{\overset{F}{\rightleftarrows}} B$ together with an isomorphism

$$
B(FA, B) \overset{\sim}{\longrightarrow} A(A, GB) \tag{1.1}
$$

natural in $A \in A$ and $B \in B$. Then F is **left adjoint** to G, G is **right adjoint** to F, and I write $F \dashv G$. In most of the examples that we meet, G is a forgetful functor and F the corresponding free functor. A typical example is that A is the category of sets, B is the category of **monoids** (sets equipped with an associative binary multiplication and a two-sided unit), G forgets the monoid structure, and F sends a set A to the monoid

$$
FA = \coprod_{n \in \mathbb{N}} A^n
$$

of finite sequences of elements of A, whose multiplication is concatenation of sequences.

Take an adjunction as above, and let $A \in A$: then applying isomorphism (1.1) to $1_{FA} \in B(FA, FA)$ yields a map $\eta_A : A \longrightarrow GFA$. The resulting natural transformation $\eta : 1_A \longrightarrow GF$ is the **unit** of the adjunction.

Dually, there is a **counit** $\varepsilon : FG \longrightarrow 1_{\mathcal{B}}$, and the unit and counit satisfy the so-called triangle identities (Mac Lane, 1971, IV.1(9)). In fact, an adjunction can equivalently be defined as a quadruple $(F, G, \eta, \varepsilon)$ where

$$\mathcal{A} \xrightarrow{F} \mathcal{B}, \qquad \mathcal{B} \xrightarrow{G} \mathcal{A}, \qquad 1_{\mathcal{A}} \xrightarrow{\eta} GF, \qquad FG \xrightarrow{\varepsilon} 1_{\mathcal{B}}$$

$$(1.2)$$

and η and ε satisfy the triangle identities.

Equivalence of categories can be formulated in several ways. By definition, an **equivalence** of categories $\mathcal{A}$ and $\mathcal{B}$ consists of functors and natural transformations (1.2) such that η and ε are isomorphisms. An **adjoint equivalence** is an adjunction $(F, G, \eta, \varepsilon)$ that is also an equivalence. A functor $F : \mathcal{A} \longrightarrow \mathcal{B}$ is **essentially surjective on objects** if for all $B \in \mathcal{B}$ there exists $A \in \mathcal{A}$ such that $FA \cong B$.

Proposition 1.1.2 *The following conditions on a functor* $F : \mathcal{A} \longrightarrow \mathcal{B}$ *are equivalent:*

a. *there exist* G, η *and* ε *such that* $(F, G, \eta, \varepsilon)$ *is an adjoint equivalence*
b. *there exist* G, η *and* ε *such that* $(F, G, \eta, \varepsilon)$ *is an equivalence*
c. *F is full, faithful, and essentially surjective on objects.*

Proof See Mac Lane (1971, IV.4.1). □

If the conditions of the proposition are satisfied then the functor F is called an **equivalence**. If the categories $\mathcal{A}$ and $\mathcal{B}$ are equivalent then we write $\mathcal{A} \simeq \mathcal{B}$ (in contrast to $\mathcal{A} \cong \mathcal{B}$, which denotes isomorphism).

Monads are a remarkably economical formalization of the notion of 'algebraic theory', traditionally formalized by universal algebraists in various rather concrete and inflexible ways. For example, there is a monad corresponding to the theory of rings, another monad for the theory of complex Lie algebras, another for the theory of topological groups, another for the theory of strict 10-categories, and so on, as we shall see.

A monad on a category $\mathcal{A}$ can be defined as a monoid in the monoidal category $([\mathcal{A}, \mathcal{A}], \circ, 1_{\mathcal{A}})$ of endofunctors on $\mathcal{A}$. Explicitly:

Definition 1.1.3 A **monad** on a category $\mathcal{A}$ consists of a functor $T : \mathcal{A} \longrightarrow \mathcal{A}$ together with natural transformations

$$\mu : T \circ T \longrightarrow T, \qquad \eta : 1_{\mathcal{A}} \longrightarrow T,$$

called the **multiplication** and **unit** respectively, such that the diagrams

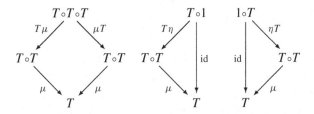

commute (the **associativity** and **unit** laws).

Any adjunction

$$
\begin{array}{c}
\mathcal{B} \\
F \uparrow \dashv \downarrow G \\
\mathcal{A}
\end{array}
$$

induces a monad (T, μ, η) on $\mathcal{A}$: take $T = G{\circ}F$, η to be the unit of the adjunction, and

$$\mu = G\varepsilon F : GFGF \longrightarrow GF$$

where ε is the counit. Often $\mathcal{A}$ is the category of sets and $\mathcal{B}$ is a category of 'algebras' of some kind, as in the following examples.

Example 1.1.4 Take the free-forgetful adjunction between the category of monoids and the category of sets, as above. Let (T, μ, η) be the induced monad. Then TA is $\coprod_{n \in \mathbb{N}} A^n$, the set of finite sequences of elements of A, for any set A. The multiplication μ strips inner brackets from double sequences:

$$T(TA) \xrightarrow{\ \mu_A\ } TA,$$
$$((a_1^1, \ldots, a_1^{k_1}), \ldots, (a_n^1, \ldots, a_n^{k_n})) \longmapsto (a_1^1, \ldots, a_1^{k_1}, \ldots, a_n^1, \ldots, a_n^{k_n})$$

$(n, k_i \in \mathbb{N}, a_i^j \in A)$. The unit η forms sequences of length 1:

$$A \xrightarrow{\ \eta_A\ } TA,$$
$$a \longmapsto (a).$$

Example 1.1.5 Fix a ring R. One can form the free R-module on any given set, and conversely one can take the underlying set of any R-module, giving an

adjunction

$$R\text{-}\mathbf{Mod}$$
$$F \dashv G$$
$$\mathbf{Set}$$

hence a monad (T, μ, η) on **Set**. Explicitly, if A is a set then TA is the set of formal R-linear combinations of elements of A. The multiplication μ realizes a formal linear combination of formal linear combinations as a single formal linear combination, and the unit η realizes an element of a set A as a trivial linear combination of elements of A.

Example 1.1.6 The same goes for all other 'algebraic theories': groups, Lie algebras, Boolean algebras, ... The functor T sends a set A to the set of formal words in the set A (which in some cases, such as that of groups, is cumbersome to describe). There is no need for the ambient category $\mathcal{A}$ to be **Set**: the theory of topological groups, for instance, gives a monad on the category **Top** of topological spaces.

A monad is meant to be an algebraic theory, so if we are handed a monad then we ought to be able to say what its 'models' are. For instance, if we are handed the monad of 1.1.5 then its 'models' should be exactly R-modules. Formally, if $T = (T, \mu, \eta)$ is a monad on a category $\mathcal{A}$ then a T**-algebra** is an object $A \in \mathcal{A}$ together with a map $h : TA \longrightarrow A$ compatible with the multiplication and unit of the monad: see Mac Lane (1971, VI.2) for the axioms. In the case of 1.1.5, a T-algebra is a set A equipped with a function

$$h : \{\text{formal } R\text{-linear combinations of elements of } A\} \longrightarrow A$$

satisfying some axioms, and this does indeed amount exactly to an R-module.

The category of algebras for a monad $T = (T, \mu, \eta)$ on a category $\mathcal{A}$ is written $\mathcal{A}^T$. There is an evident forgetful functor $\mathcal{A}^T \longrightarrow \mathcal{A}$, this has a left adjoint (forming 'free T-algebras'), and the monad on $\mathcal{A}$ induced by this adjunction is just the original T. So every monad arises from an adjunction, and informally we have

$$\{\text{monads on } \mathcal{A}\} \subset \{\text{adjunctions based on } \mathcal{A}\}.$$

The inclusion is proper: not every adjunction is of the form $\mathcal{A}^T \xrightarrow{\ \ \ T\ \ \ } \mathcal{A}$ just described. For instance, the forgetful functor **Top** $\longrightarrow$ **Set** has a left adjoint (forming discrete spaces); the induced monad on **Set** is the identity, whose category of algebras is merely **Set**, and **Set** $\not\simeq$ **Top**. The adjunctions that do arise

from monads are called **monadic**. All of the adjunctions in Examples 1.1.4–
1.1.6 are monadic, and the non-monadicity of the adjunction **Top** $\underset{\longleftarrow}{\overset{\longrightarrow}{\top}}$ **Set**
expresses the thought that topology is not algebra.

Presheaves will be important. A **presheaf** on a category $\mathcal{A}$ is a functor
$\mathcal{A}^{\mathrm{op}} \longrightarrow$ **Set**. Any object $A \in \mathcal{A}$ gives rise to a presheaf $\mathcal{A}(-, A)$ on $\mathcal{A}$,
and this defines a functor

$$
\begin{array}{ccc}
\mathcal{A} & \longrightarrow & [\mathcal{A}^{\mathrm{op}}, \mathbf{Set}] \\
A & \longmapsto & \mathcal{A}(-, A),
\end{array}
$$

the **Yoneda embedding**. It is full and faithful. This follows from the Yoneda
Lemma, which states that if $A \in \mathcal{A}$ and X is a presheaf on $\mathcal{A}$ then natural
transformations $\mathcal{A}(-, A) \longrightarrow X$ correspond one-to-one with elements of
XA.

If $\mathcal{E}$ is a category and S a set then there is a category $\mathcal{E}^S$, a power of $\mathcal{E}$,
whose objects are S-indexed families of objects of $\mathcal{E}$. On the other hand, if $\mathcal{E}$
is a category and E an object of $\mathcal{E}$ then there is a **slice category** $\mathcal{E}/E$, whose
objects are maps $D \xrightarrow{p} E$ into E and whose maps are commutative triangles.
If S is a set then there is an **S** equivalence of categories

$$
\mathbf{Set}^S \simeq \mathbf{Set}/S, \tag{1.3}
$$

given in one direction by taking the disjoint union of an S-indexed family of
sets, and in the other by taking fibres of a set over S.

There is an analogue of (1.3) in which the set S is replaced by a category.
Fix a small category $\mathbb{A}$. The replacement for $\mathbf{Set}^S$ is $[\mathbb{A}^{\mathrm{op}}, \mathbf{Set}]$, but what should
replace the slice category $\mathbf{Set}/S$? First note that any presheaf X on $\mathbb{A}$ gives rise
to a category $\mathbb{A}/X$, the **category of elements** of X, whose objects are pairs
(A, x) with $A \in \mathbb{A}$ and $x \in XA$ and whose maps

$$
(A', x') \longrightarrow (A, x)
$$

are maps $f : A' \longrightarrow A$ in $\mathbb{A}$ such that $x' = (Xf)(x)$. There is an evident
forgetful functor $\mathbb{A}/X \longrightarrow \mathbb{A}$, the **Grothendieck fibration** of X. This is an
example of a **discrete fibration**, that is, a functor $G : \mathcal{D} \longrightarrow \mathcal{C}$ such that

for any object $D \in \mathcal{D}$ and map $C' \xrightarrow{p} GD$ in $\mathcal{C}$, there is a unique map
$D' \xrightarrow{q} D$ in $\mathcal{D}$ such that $Gq = p$.

Discrete fibrations over $\mathcal{C}$ (that is, with codomain $\mathcal{C}$) can be made into a category **DFib**$(\mathcal{C})$ in a natural way, and this is the desired generalization of slice
category. We then have an equivalence

$$
[\mathbb{A}^{\mathrm{op}}, \mathbf{Set}] \simeq \mathbf{DFib}(\mathbb{A}),
$$

given in one direction by taking categories of elements, and in the other by taking fibres in a suitable sense. There is also a dual notion of **discrete opfibration**, and an equivalence

$$[\mathbb{A}, \mathbf{Set}] \simeq \mathbf{DOpfib}(\mathbb{A}).$$

A **presheaf category** is a category equivalent to $[\mathbb{A}^{op}, \mathbf{Set}]$ for some small $\mathbb{A}$. The class of presheaf categories is closed under slicing:

Proposition 1.1.7 *Let $\mathbb{A}$ be a small category and X a presheaf on $\mathbb{A}$. Then there is an equivalence of categories*

$$[\mathbb{A}^{op}, \mathbf{Set}]/X \simeq [(\mathbb{A}/X)^{op}, \mathbf{Set}]. \qquad \square$$

Finally, we will need just a whisper of internal category theory. If $\mathcal{A}$ is any category with finite products then an **(internal) group** in $\mathcal{A}$ consists of an object $A \in \mathcal{A}$ together with maps $m : A \times A \longrightarrow A$ (multiplication), $e : 1 \longrightarrow A$ (unit), and $i : A \longrightarrow A$ (inverses), such that certain diagrams expressing the group axioms commute. Thus, a group in **Set** is an ordinary group, a group in the category of smooth manifolds is a Lie group, and so on. A similar definition pertains for algebraic structures other than groups. Categories themselves can be defined in this way: if $\mathcal{A}$ is any category with pullbacks then an **(internal) category** C in $\mathcal{A}$ is a diagram

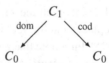

in $\mathcal{A}$ together with maps

$$C_1 \times_{C_0} C_1 \xrightarrow{\ \text{comp}\ } C_1, \qquad C_0 \xrightarrow{\ \text{ids}\ } C_1$$

in $\mathcal{A}$, satisfying certain axioms. Here $C_1 \times_{C_0} C_1$ is the pullback

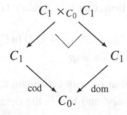

When $\mathcal{A} = \mathbf{Set}$ we recover the usual notion of small category: C_0 and C_1 are the sets of objects and of arrows, dom and cod are the domain and codomain

functions, $C_1 \times_{C_0} C_1$ is the set of composable pairs of arrows, comp and ids are the functions determining binary composition and identity maps, and the axioms specify the domain and codomain of composites and identities and express associativity and identity laws. When $\mathcal{A} = \textbf{Top}$ we obtain a notion of 'topological category', in which both the set of objects and the set of arrows carry a topology. For instance, given any space X, there is a topological category $C = \Pi_1 X$ in which $C_0 = X$ and C_1 is $X^{[0,1]}/\sim$, the space of all paths in X factored out by path homotopy relative to endpoints.

1.2 Monoidal categories

Monoidal categories come in a variety of flavours: strict, weak, plain, braided, symmetric. We look briefly at strict monoidal categories but spend most time on the more important weak case and on the coherence theorem: every weak monoidal category is equivalent to a strict one.

In the terminology of the previous section, a **strict monoidal category** is an internal monoid in **Cat**, that is, a category $\mathcal{A}$ equipped with a functor

$$\begin{array}{rlcl} \otimes: & \mathcal{A} \times \mathcal{A} & \longrightarrow & \mathcal{A}, \\ & (A, B) & \longmapsto & A \otimes B \end{array}$$

and an object $I \in \mathcal{A}$, obeying strict associativity and unit laws:

$$(A \otimes B) \otimes C = A \otimes (B \otimes C), \qquad I \otimes A = A, \qquad A \otimes I = A$$

for all objects $A, B, C \in \mathcal{A}$, and similarly for morphisms. Functoriality of $\otimes$ encodes the 'interchange laws':

$$(g' \circ f') \otimes (g \circ f) = (g' \otimes g) \circ (f' \otimes f) \tag{1.4}$$

for all maps g', f', f, f for which these composites make sense, and $1_A \otimes 1_B = 1_{A \otimes B}$ for all objects A and B.

Since one is usually not interested in equality of objects in a category, only in isomorphism, strict monoidal categories are quite rare.

Example 1.2.1 The category $[\mathcal{C}, \mathcal{C}]$ of endofunctors on a given category $\mathcal{C}$ has a strict monoidal structure given by composition (as $\otimes$) and $1_{\mathcal{C}}$ (as I).

Example 1.2.2 Given a natural number n (possibly 0), let $\textbf{n}$ denote the n-element set $\{1, \ldots, n\}$ with its usual total order. Let $\mathbb{D}$ be the category whose objects are the natural numbers and whose maps $m \longrightarrow n$ are the order-preserving functions $\textbf{m} \longrightarrow \textbf{n}$. This is the 'augmented simplex category', one object bigger than the standard topologists' Δ, and is equivalent to the category of (possibly empty) finite totally ordered sets. It has a strict monoidal structure given by addition and **0**.

Example 1.2.3 A category with only one object is just a monoid (p. 14): if the category is called $\mathcal{A}$ and its single object is called $\star$ then the monoid is the set $M = \mathcal{A}(\star, \star)$ with composition $\circ$ as multiplication and the identity $1 = 1_\star$ as unit. A one-object strict monoidal category therefore consists of a set M with monoid structures $(\circ, 1)$ and $(\otimes, 1)$ (the latter being tensor of arrows in the monoidal category), such that (1.4) holds for all $g', f', g, f \in M$. Lemma 1.2.4 below tells us that this forces the binary operations $\circ$ and $\otimes$ to be equal and commutative. So a one-object strict monoidal category is just a commutative monoid.

Lemma 1.2.4 (Eckmann–Hilton, 1962) *Suppose that $\circ$ and $\otimes$ are binary operations on a set M, satisfying (1.4) for all $g', f', g, f \in M$, and suppose that $\circ$ and $\otimes$ share a two-sided unit. Then $\circ = \otimes$ and $\circ$ is commutative.*

Proof Write 1 for the unit. Then for $g, f \in M$,

$$g \circ f = (g \otimes 1) \circ (1 \otimes f) = (g \circ 1) \otimes (1 \circ f) = g \otimes f,$$

so $\circ = \otimes$, and

$$g \circ f = (1 \otimes g) \circ (f \otimes 1) = (1 \circ f) \otimes (g \circ 1) = f \otimes g,$$

so $\circ$ is commutative. $\square$

Much more common are weak monoidal categories, usually just called 'monoidal categories'.

Definition 1.2.5 A **(weak) monoidal category** is a category $\mathcal{A}$ together with a functor $\otimes : \mathcal{A} \times \mathcal{A} \longrightarrow \mathcal{A}$, an object $I \in \mathcal{A}$ (the **unit**), and isomorphisms

$$(A \otimes B) \otimes C \xrightarrow[\sim]{\alpha_{A,B,C}} A \otimes (B \otimes C), \quad I \otimes A \xrightarrow[\sim]{\lambda_A} A, \quad A \otimes I \xrightarrow[\sim]{\rho_A} A$$

natural in $A, B, C \in \mathcal{A}$ (**coherence isomorphisms**), such that the following diagrams commute for all $A, B, C, D \in \mathcal{A}$:

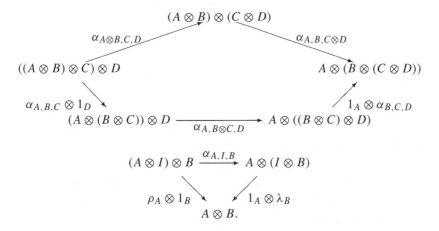

The pentagon and triangle axioms ensure that 'all diagrams' constructed out of coherence isomorphisms commute. This is one form of the coherence theorem, discussed below.

Example 1.2.6 Strict monoidal categories can be identified with monoidal categories in which all the components of α, λ and ρ are identities.

Example 1.2.7 Let $\mathcal{A}$ be a category in which all finite products exist. Choose a particular terminal object 1, and for each $A, B \in \mathcal{A}$ a particular product diagram $A \xleftarrow{\text{pr}_1} A \times B \xrightarrow{\text{pr}_2} B$. Then $\mathcal{A}$ acquires a monoidal structure with $\otimes = \times$ and $I = 1$; the maps α, λ and ρ are the canonical ones.

Example 1.2.8 For any commutative ring R, the category of R-modules is monoidal with respect to the usual tensor $\otimes_R$ and unit object R.

Example 1.2.9 Take a topological space with basepoint. There is a monoidal category whose objects are loops on the basepoint and whose maps are homotopy classes of loop homotopies (relative to the basepoint). We have to take homotopy classes so that the ordinary categorical composition obeys associativity and identity laws. Tensor is concatenation of loops (on objects) and gluing of homotopies (on maps). The coherence isomorphisms are the evident re-parametrizations.

Earlier we met the notion of (internal) algebraic structures, such as groups, in a category with finite products. There is no clear way to extend this to arbitrary monoidal categories, since to express an axiom such as $x \cdot x^{-1} = 1$ diagrammatically requires the product-projections. We can, however, define a **monoid** in a monoidal category $\mathcal{A}$ as an object A together with maps

$$m : A \otimes A \longrightarrow A, \qquad e : I \longrightarrow A$$

such that associativity and unit diagrams similar to those in Definition 1.1.3 commute. With the obvious notion of map, this gives a category **Mon**($\mathcal{A}$) of monoids in $\mathcal{A}$. When $\mathcal{A}$ is the category of sets, with product as monoidal structure, this is the usual category of monoids.

There are various notions of map between monoidal categories. In what follows we use $\otimes$, I, α, λ, and ρ to denote the monoidal structure of both the categories concerned.

Definition 1.2.10 Let $\mathcal{A}$ and $\mathcal{A}'$ be monoidal categories. A **lax monoidal functor** $F = (F, \phi) : \mathcal{A} \longrightarrow \mathcal{A}'$ is a functor $F : \mathcal{A} \longrightarrow \mathcal{A}'$ together with **coherence maps**

$$\phi_{A,B} : FA \otimes FB \longrightarrow F(A \otimes B), \qquad \phi_{\cdot} : I \longrightarrow FI$$

in $\mathcal{A}'$, the former natural in $A, B \in \mathcal{A}$, such that the following diagrams commute for all $A, B, C \in \mathcal{A}$:

$$
\begin{array}{ccc}
(FA \otimes FB) \otimes FC & \xrightarrow{\phi_{A,B} \otimes 1_{FC}} F(A \otimes B) \otimes FC \xrightarrow{\phi_{A \otimes B, C}} & F((A \otimes B) \otimes C) \\
\downarrow{\alpha_{FA,FB,FC}} & & \downarrow{F\alpha_{A,B,C}} \\
FA \otimes (FB \otimes FC) & \xrightarrow[1_{FA} \otimes \phi_{B,C}]{} FA \otimes F(B \otimes C) \xrightarrow[\phi_{A,B \otimes C}]{} & F(A \otimes (B \otimes C))
\end{array}
$$

$$
\begin{array}{ccc}
FA \otimes I & \xrightarrow{1_{FA} \otimes \phi_{\cdot}} FA \otimes FI \xrightarrow{\phi_{A,I}} & F(A \otimes I) \\
\downarrow{\rho_{FA}} & & \downarrow{F\rho_A} \\
FA & = \!\!=\!\!=\!\!=\!\!=\!\!=\!\!=\!\!=\!\!=\!\!=\!\!=\!\!=\!\!=\!\!= & FA
\end{array}
$$

$$
\begin{array}{ccc}
I \otimes FA & \xrightarrow{\phi_{\cdot} \otimes 1_{FA}} FI \otimes FA \xrightarrow{\phi_{I,A}} & F(I \otimes A) \\
\downarrow{\lambda_{FA}} & & \downarrow{F\lambda_A} \\
FA & = \!\!=\!\!=\!\!=\!\!=\!\!=\!\!=\!\!=\!\!=\!\!=\!\!=\!\!=\!\!=\!\!= & FA.
\end{array}
$$

A **colax monoidal functor** $F = (F, \phi) : \mathcal{A} \longrightarrow \mathcal{A}'$ is a functor $F : \mathcal{A} \longrightarrow \mathcal{A}'$ together with maps

$$\phi_{A,B} : F(A \otimes B) \longrightarrow FA \otimes FB, \qquad \phi. : FI \longrightarrow I$$

satisfying axioms dual to those above. A **weak** (respectively, **strict**) **monoidal functor** is a lax monoidal functor (F, ϕ) in which all the maps $\phi_{A,B}$ and $\phi.$ are isomorphisms (respectively, identities).

We write **MonCat**$_{\text{lax}}$ for the category of monoidal categories and lax maps, and similarly **MonCat**$_{\text{colax}}$, **MonCat**$_{\text{wk}}$, and **MonCat**$_{\text{str}}$. There are various alternative systems of terminology; in particular, what we call weak monoidal functors are sometimes called 'strong monoidal functors' or just 'monoidal functors'.

Example 1.2.11 The forgetful functor $U : \mathbf{Ab} \longrightarrow \mathbf{Set}$ from abelian groups to sets has a lax monoidal structure with respect to the usual monoidal structures on **Ab** and **Set**, given by the canonical maps

$$UA \times UB \longrightarrow U(A \otimes B)$$

$(A, B \in \mathbf{Ab})$ and the map $1 \longrightarrow U\mathbb{Z}$ picking out $0 \in \mathbb{Z}$.

Example 1.2.12 Let $\mathcal{C}$ be a category and $\mathcal{A}$ a monoidal category. A weak monoidal functor from $\mathcal{A}$ to the monoidal category $[\mathcal{C}, \mathcal{C}]$ of endofunctors on $\mathcal{C}$ (1.2.1) is called an **action** of $\mathcal{A}$ on $\mathcal{C}$, and amounts to a functor $\mathcal{A} \times \mathcal{C} \longrightarrow \mathcal{C}$ together with coherence isomorphisms satisfying axioms.

To state one of the forms of the coherence theorem we will need a notion of equivalence of monoidal categories, and for this we need in turn a notion of transformation.

Definition 1.2.13 Let $(F, \phi), (G, \psi) : \mathcal{A} \longrightarrow \mathcal{A}'$ be lax monoidal functors. A **monoidal transformation** $(F, \phi) \longrightarrow (G, \psi)$ is a natural transformation $\sigma : F \longrightarrow G$ such that the following diagrams commute for all $A, B \in \mathcal{A}$:

$$
\begin{array}{ccc}
FA \otimes FB & \xrightarrow{\sigma_A \otimes \sigma_B} & GA \otimes GB \\
\phi_{A,B} \downarrow & & \downarrow \psi_{A,B} \\
F(A \otimes B) & \xrightarrow{\sigma_{A \otimes B}} & G(A \otimes B)
\end{array}
\qquad
\begin{array}{ccc}
I & = & I \\
\phi. \downarrow & & \downarrow \psi. \\
FI & \xrightarrow{\sigma_I} & GI.
\end{array}
$$

A weak monoidal functor (F, ϕ) is called an **equivalence** of monoidal categories if it satisfies the conditions of the following proposition.

Proposition 1.2.14 *The following conditions on a weak monoidal functor* $(F, \phi) : \mathcal{A} \longrightarrow \mathcal{A}'$ *are equivalent:*

a. *there exist a weak monoidal functor* $(G, \psi) : \mathcal{A}' \longrightarrow \mathcal{A}$ *and invertible monoidal transformations*

$$\eta : 1_{\mathcal{A}} \longrightarrow (G, \psi) \circ (F, \phi), \qquad \varepsilon : (F, \phi) \circ (G, \psi) \longrightarrow 1_{\mathcal{A}'}$$

b. *the functor* F *is an equivalence of categories.*

Proof If F is an equivalence of categories then by Proposition 1.1.2 there exist a functor G and transformations η and ε such that $(F, G, \eta, \varepsilon)$ is an adjoint equivalence. It is easy to verify that G acquires a weak monoidal structure and that η and ε are then invertible monoidal transformations. □

A coherence theorem is, roughly, a description of a structure that makes it more manageable. For example, one coherence theorem for monoidal categories is that all diagrams built out of the coherence isomorphisms commute. Another is that any weak monoidal category is equivalent to some strict monoidal category. All non-trivial applications of monoidal categories rely on a coherence theorem in some form; the axioms as they stand are just too unwieldy. Indeed, one might argue that the 'all diagrams commute' principle should be an explicit part of the definition of monoidal category, and we will take this approach when we come to define higher-dimensional categorical structures.

'All diagrams commute' can be made precise in various ways. A very direct statement is in Mac Lane (1971, VII.2), and a less direct (but sharper) statement is 3.2.3 below. A typical instance is that the diagram

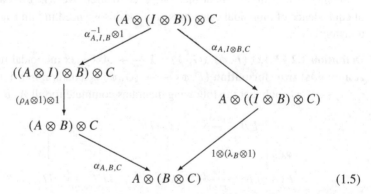

$$(1.5)$$

commutes for all objects A, B, C of a monoidal category. We will soon see how this follows from the alternative form of the coherence theorem:

Theorem 1.2.15 (Coherence for monoidal categories) *Every monoidal category is equivalent to some strict monoidal category.*

Here is how *not* to prove this: take a monoidal category $\mathcal{A}$, form a quotient strict monoidal category $\mathcal{A}'$ by turning isomorphism into equality, and show that the natural map $\mathcal{A} \longrightarrow \mathcal{A}'$ is an equivalence. To see why this fails, first note that a monoidal category may have the property that any two isomorphic objects are equal (and so, in particular, the tensor product of objects is strictly associative and unital), but even so need not be strict – the coherence isomorphisms need not be identities. An example can be found in Mac Lane (1971, VII.1). If $\mathcal{A}$ has this property then identifying isomorphic objects of $\mathcal{A}$ has no effect at all. One might attempt to go further by identifying the coherence isomorphisms with identities – for instance, identifying the two maps

$$(A \otimes B) \otimes C \overset{\alpha_{A,B,C}}{\underset{1}{\rightrightarrows}} A \otimes (B \otimes C)$$

– to make a strict monoidal category $\mathcal{A}'$; but then the quotient map $\mathcal{A} \longrightarrow \mathcal{A}'$ is not faithful, so not an equivalence.

Sketch proof of 1.2.15 This is a modification of Joyal and Street's proof (1993, 1.4). Let $\mathcal{A}$ be a monoidal category. We define a strict monoidal category $\mathcal{A}'$ and a monoidal equivalence $\mathbf{y} : \mathcal{A} \longrightarrow \mathcal{A}'$. An object of $\mathcal{A}'$ is a pair (E, δ) where E is an endofunctor of the (unadorned) category $\mathcal{A}$ and δ is a family of isomorphisms

$$\left(\delta_{A,B} : (EA) \otimes B \overset{\sim}{\longrightarrow} E(A \otimes B) \right)_{A,B \in \mathcal{A}}$$

natural in A and B and satisfying the evident coherence axioms. Tensor in $\mathcal{A}'$ is

$$(E', \delta') \otimes (E, \delta) = (E' {\circ} E, \delta'')$$

where δ'' is defined in the only sensible way; then $\mathcal{A}'$ is a strict monoidal category. The functor $\mathbf{y}$ is given by

$$\mathbf{y}(Z) = (Z \otimes -, \alpha_{Z,-,-}).$$

It is weak monoidal and full, faithful and essentially surjective on objects, so by 1.2.14 an equivalence of monoidal categories. $\square$

Joyal and Street motivate their proof as a generalization of the Cayley Theorem representing any group as a group of permutations. We find another way of looking at it when we come to bicategories (Section 1.5).

Now let us deduce that 'all diagrams commute', or at least, by way of example, that diagram (1.5) commutes. For any objects A, B and C of any monoidal category, write

$$(A \otimes (I \otimes B)) \otimes C \underset{\omega_{A,B,C}}{\overset{\chi_{A,B,C}}{\rightrightarrows}} A \otimes (B \otimes C)$$

for the two composite coherence maps shown in (1.5). Now take a particular monoidal category $\mathcal{A}$ and objects $A, B, C \in \mathcal{A}$; we want to conclude that $\chi_{A,B,C} = \omega_{A,B,C}$. We have a monoidal equivalence (F, ϕ) from $\mathcal{A}$ to a strict monoidal category $\mathcal{A}'$, and a serially commutative diagram

$$
\begin{array}{ccc}
(FA \otimes (I \otimes FB)) \otimes FC & \underset{\omega_{FA,FB,FC}}{\overset{\chi_{FA,FB,FC}}{\rightrightarrows}} & FA \otimes (FB \otimes FC) \\
\phi \downarrow & & \downarrow \phi \\
F((A \otimes (I \otimes B)) \otimes C) & \underset{F\omega_{A,B,C}}{\overset{F\chi_{A,B,C}}{\rightrightarrows}} & F(A \otimes (B \otimes C))
\end{array}
$$

where the maps labelled ϕ are built out of ϕ. and various $\phi_{D,E}$s. ('Serially commutative' means that both the top and the bottom square commute.) Since $\mathcal{A}'$ is strict, $\chi_{FA,FB,FC} = \omega_{FA,FB,FC}$; then since the ϕs are isomorphisms, $F\chi_{A,B,C} = F\omega_{A,B,C}$; then since F is faithful, $\chi_{A,B,C} = \omega_{A,B,C}$, as required.

There are similar diagrammatic coherence theorems for monoidal functors, saying, for instance, that if (F, ϕ) is a lax monoidal functor then any two maps

$$((FA \otimes FB) \otimes I) \otimes (FC \otimes FD) \rightrightarrows F(A \otimes ((B \otimes C) \otimes D))$$

built out of copies of the coherence maps are equal. The form of the codomain is important, being F applied to a product of objects; in contrast, the coherence maps can be assembled to give two maps

$$FI \rightrightarrows FI \otimes FI$$

that are in general not equal. See Lewis (1972) for both this counterexample and a precise statement of coherence for monoidal functors.

Many everyday monoidal categories have a natural symmetric structure. Formally, a symmetric monoidal category is a monoidal category $\mathcal{A}$ together with a specified isomorphism $\gamma_{A,B} : A \otimes B \longrightarrow B \otimes A$ for each pair (A, B) of objects, satisfying coherence axioms. There are various coherence theorems for symmetric monoidal categories (Mac Lane, 1963, Joyal and Street, 1993). Beware, however, that the symmetry isomorphisms $\gamma_{A,B}$ cannot be turned into identities: every symmetric monoidal category is equivalent to some symmetric strict monoidal category, but not usually to any strict symmetric monoidal category. The latter structures – commutative monoids in **Cat** – are rare.

More general than symmetric monoidal categories are braided monoidal categories, mentioned in the Motivation for topologists. See Joyal and Street (1993) for the definitions and coherence theorems, and Gordon, Power and Street (1995) for a 3-dimensional perspective.

1.3 Enrichment

In many basic examples of categories C, the hom-sets $C(A, B)$ are richer than mere sets. For instance, if C is a category of chain complexes then $C(A, B)$ is an abelian group, and if C is a suitable category of topological spaces then $C(A, B)$ is itself a space.

This idea is called 'enrichment' and can be formalized in various ways. The best-known, enrichment in a monoidal category, is presented here. We will see later (Section 6.8) that it is not the most natural or general formalization, but it serves a purpose before we reach that point.

Definition 1.3.1 Let V be a category. A V**-graph** X is a set X_0 together with a family $(X(x, x'))_{x,x' \in X_0}$ of objects of V. A **map of** V**-graphs** $f : X \longrightarrow Y$ is a function $f_0 : X_0 \longrightarrow Y_0$ together with a family of maps

$$\left(X(x, x') \xrightarrow{f_{x,x'}} Y(f_0 x, f_0 x') \right)_{x,x' \in X_0}.$$

We usually write both f_0 and $f_{x,x'}$ as just f. The category of V-graphs is written V-**Gph**.

A **Set**-graph is, then, an ordinary directed graph. A category is a directed graph equipped with composition and identities, suggesting the following definition.

Definition 1.3.2 Let $(V, \otimes, I)$ be a monoidal category. A **category enriched in** V, or V**-enriched category**, is a V-graph A together with families of maps

$$\left(A(b, c) \otimes A(a, b) \xrightarrow{\text{comp}_{a,b,c}} A(a, c) \right)_{a,b,c \in A_0} , \quad \left(I \xrightarrow{\text{ids}_a} A(a, a) \right)_{a \in A_0}$$

in V satisfying associativity and identity axioms (expressed as commutative diagrams). A V**-enriched functor** $F : A \longrightarrow B$ is a map of the underlying V-graphs commuting with the composition maps $\text{comp}_{a,b,c}$ and identity maps ids_a. This defines a category V-**Cat**.

A (**Set**, $\times$, 1)-enriched category is, of course, an ordinary (small) category, and an (**Ab**, $\otimes$, $\mathbb{Z}$)-enriched category is an **Ab**-category in the sense of homological algebra. A one-object V-enriched category is a monoid in the monoidal category $\mathcal{V}$. Compare and contrast internal and enriched categories: in the case of topological spaces, for instance, an internal category in **Top** is an ordinary category equipped with a topology on the set of objects and a topology on the set of all arrows, whereas a category enriched in **Top** is an ordinary category equipped with a topology on each hom-set.

Any lax monoidal functor $Q = (Q, \phi) : \mathcal{V} \longrightarrow \mathcal{W}$ induces a functor

$$Q_* : \mathcal{V}\text{-}\mathbf{Cat} \longrightarrow \mathcal{W}\text{-}\mathbf{Cat}.$$

In particular, if $\mathcal{V}$ is any monoidal category then the functor $\mathcal{V}(I, -) :$ $\mathcal{V} \longrightarrow \mathbf{Set}$ has a natural lax monoidal structure, and the induced functor defines the **underlying category** of a $\mathcal{V}$-enriched category. This does exactly what we would expect in the familiar cases of $\mathcal{V}$.

In the next section we will enrich in categories $\mathcal{V}$ whose monoidal structure is ordinary (cartesian) product. The following result will be useful; its proof is straightforward.

Proposition 1.3.3

a. *If $\mathcal{V}$ is a category with finite products then the category $\mathcal{V}$-**Cat** also has finite products.*

b. *If $Q : \mathcal{V} \longrightarrow \mathcal{W}$ is a finite-product-preserving functor between categories with finite products then the induced functor $Q_* : \mathcal{V}$-**Cat** $\longrightarrow \mathcal{W}$-**Cat** also preserves finite products.* $\square$

The theory of categories enriched in a monoidal category can be taken much further: see Kelly (1982), for instance. Under the assumption that $\mathcal{V}$ is symmetric monoidal closed and has all limits and colimits, very large parts of ordinary category theory can be extended to the $\mathcal{V}$-enriched context.

1.4 Strict *n*-categories

Strict *n*-categories are not encountered nearly as often as their weak cousins, but there are nevertheless some significant examples.

We start with a very short definition of strict *n*-category, and some examples. This definition, being iterative, can seem opaque, so we then formulate a much longer but equivalent definition providing a complementary viewpoint. We then look briefly at the infinite-dimensional case, strict ω-categories, and at the cubical analogue of strict *n*-categories.

The definition of strict *n*-category uses enrichment in a category with finite products.

Definition 1.4.1 Let $(\mathbf{Str}\text{-}n\text{-}\mathbf{Cat})_{n \in \mathbb{N}}$ be the sequence of categories given inductively by

$$\mathbf{Str}\text{-}0\text{-}\mathbf{Cat} = \mathbf{Set}, \qquad \mathbf{Str}\text{-}(n+1)\text{-}\mathbf{Cat} = (\mathbf{Str}\text{-}n\text{-}\mathbf{Cat})\text{-}\mathbf{Cat}.$$

A **strict n-category** is an object of $\mathbf{Str}\text{-}n\text{-}\mathbf{Cat}$, and a **strict n-functor** is a map in $\mathbf{Str}\text{-}n\text{-}\mathbf{Cat}$.

This makes sense by Proposition 1.3.3(a).

Strict 0-categories are sets and strict 1-categories are categories. A strict 2-category A consists of a set A_0, a category $A(a, b)$ for each $a, b \in A_0$, composition functors as in 1.3.2, and an identity object of $A(a, a)$ for each $a \in A_0$, all obeying associativity and identity laws.

Example 1.4.2 There is a (large) strict 2-category $\mathcal{A}$ in which $\mathcal{A}_0$ is the class of topological spaces and, for spaces X and Y, $\mathcal{A}(X, Y)$ is the category whose objects are continuous maps $X \longrightarrow Y$ and whose arrows are homotopy classes of homotopies. (We need to take homotopy classes so that composition in $\mathcal{A}(X, Y)$ is associative and unital.) The composition functors

$$\mathcal{A}(Y, Z) \times \mathcal{A}(X, Y) \longrightarrow \mathcal{A}(X, Z) \tag{1.6}$$

and the identity objects $1_X \in \mathcal{A}(X, X)$ are the obvious ones.

Example 1.4.3 Similarly, there is a strict 2-category $\mathcal{A}$ in which $\mathcal{A}_0$ is the class of chain complexes and $\mathcal{A}(X, Y)$ is the category whose objects are chain maps $X \longrightarrow Y$ and whose arrows are homotopy classes of chain homotopies, in the sense of p. 12. (This time we need to take homotopy classes in order that the composition functors (1.6) really are functorial.)

Example 1.4.4 There is, self-referentially, a strict 2-category $\mathbf{Cat}$ of categories. Here $\mathbf{Cat}_0$ is the class of small categories and $\mathbf{Cat}(C, D)$ is the functor category $[C, D]$. In fact, there is for each $n \in \mathbb{N}$ a strict $(n+1)$-category of strict n-categories: it can be proved by induction that for each n the category $\mathbf{Str}\text{-}n\text{-}\mathbf{Cat}$ is cartesian closed, which implies that it is naturally enriched in itself, in other words, forms a strict $(n+1)$-category. Sensitive readers may find this shocking: the entire $(n+1)$-category of strict n-categories can be extracted from the mere 1-category.

We now build up to an alternative and more explicit definition of strict n-category.

A category can be regarded as a directed graph with structure. The most obvious n-dimensional analogue of a directed graph uses spherical or 'globular' shapes, as in the following definition.

Definition 1.4.5 Let $n \in \mathbb{N}$. An n-**globular set** X is a diagram

$$X(n) \underset{t}{\overset{s}{\rightrightarrows}} X(n-1) \underset{t}{\overset{s}{\rightrightarrows}} \quad \cdots \quad \underset{t}{\overset{s}{\rightrightarrows}} X(0)$$

of sets and functions, such that

$$s(s(x)) = s(t(x)), \qquad t(s(x)) = t(t(x)) \tag{1.7}$$

for all $m \in \{2, \ldots, n\}$ and $x \in X(m)$.

Alternatively, an n-globular set is a presheaf on the category $\mathbb{G}_n$ generated by objects and arrows

$$n \underset{\tau_n}{\overset{\sigma_n}{\leftrightarrows}} n-1 \underset{\tau_{n-1}}{\overset{\sigma_{n-1}}{\leftrightarrows}} \quad \cdots \quad \underset{\tau_1}{\overset{\sigma_1}{\leftrightarrows}} 0$$

subject to equations

$$\sigma_m \circ \sigma_{m-1} = \tau_m \circ \sigma_{m-1}, \qquad \sigma_m \circ \tau_{m-1} = \tau_m \circ \tau_{m-1}$$

($m \in \{2, \ldots, n\}$). The category of n-globular sets can then be defined as the presheaf category $[\mathbb{G}_n^{\mathrm{op}}, \mathbf{Set}]$.

Let X be an n-globular set. Elements of $X(m)$ are called m-**cells** of X and drawn as labels on an m-dimensional disc. Thus, $a \in X(0)$ is drawn as

$$a$$

(and sometimes called an **object** rather than a 0-cell), and $f \in X(1)$ is drawn as

$$a \overset{f}{\underset{}{\bullet\!\!\longrightarrow\!\!\bullet}} b$$

where $a = s(f)$ and $b = t(f)$. We call $s(x)$ the **source** of x, and $t(x)$ the **target**; these are alternative names for 'domain' and 'codomain'. A 2-cell $\alpha \in X(2)$ is drawn as

where

$$f = s(\alpha), \qquad g = t(\alpha), \qquad a = s(f) = s(g), \qquad b = t(f) = t(g).$$

That $s(f) = s(g)$ and $t(f) = t(g)$ follows from the globularity equations (1.7). A 3-cell $x \in X(3)$ is drawn as

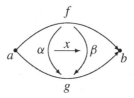

where $\alpha = s(x)$, $\beta = t(x)$, and so on. Sometimes, as in the Motivation for topologists, I have used double-shafted arrows for 2-cells, triple-shafted arrows for 3-cells, and so on, and sometimes, as here, I have stuck to single-shafted arrows for cells of all dimensions; this is a purely visual choice.

Example 1.4.6 Let $n \in \mathbb{N}$ and let S be a topological space: then there is an n-globular set in which an m-cell is a labelled m-dimensional disc in S. Formally, let D^m be the closed m-dimensional Euclidean disc (ball), and consider the diagram

$$ D^n \rightleftarrows D^{n-1} \rightleftarrows \quad \cdots \quad \rightleftarrows D^0 = 1 $$

formed by embedding D^{m-1} as the upper or lower cap of D^m. This is a functor $\mathbb{G}_n \longrightarrow \mathbf{Top}$ (an 'n-coglobular space'), and so induces a functor $\mathbb{G}_n^{\mathrm{op}} \longrightarrow \mathbf{Set}$, the n-globular set

$$ \mathbf{Top}(D^n, S) \rightrightarrows \mathbf{Top}(D^{n-1}, S) \rightrightarrows \quad \cdots \quad \rightrightarrows \mathbf{Top}(D^0, S). $$

Example 1.4.7 Analogously, any non-negatively graded chain complex C of abelian groups give rise to an n-globular set X for each $n \in \mathbb{N}$. An m-cell of X is not quite just an element of C_m; for instance, we regard an element f of C_1 as a 1-cell $a \longrightarrow b$ for any $a, b \in C_0$ such that $d(f) = b - a$. In general, an element of $X(m)$ is a $(2m + 1)$-tuple

$$ \mathbf{c} = (c_m, c_{m-1}^-, c_{m-1}^+, c_{m-2}^-, c_{m-2}^+, \ldots, c_0^-, c_0^+) $$

where $c_m \in C_m, c_p^-, c_p^+ \in C_p$, and

$$ d(c_m) = c_{m-1}^+ - c_{m-1}^-, \qquad d(c_p^-) = d(c_p^+) = c_{p-1}^+ - c_{p-1}^- $$

for all $p \in \{1, \ldots, m-1\}$. We then put

$$ s(\mathbf{c}) = (c_{m-1}^-, c_{m-2}^-, c_{m-2}^+, \ldots, c_0^-, c_0^+) $$

and dually the target.

In the following alternative definition, a strict n-category is an n-globular set equipped with identities and various binary composition operations, satisfying various axioms. Identities are simple: every p-cell x has an identity $(p+1)$-cell 1_x on it $(0 \le p < n)$, as in

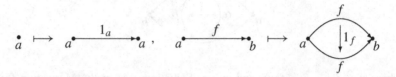

$(p = 0, 1)$. There are m different binary composition operations for m-cells. When $m = 1$ this is ordinary categorical composition. We saw the two possibilities for composing 2-cells on p. 10. The three ways of composing 3-cells are drawn as

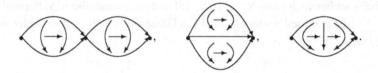

To express this formally we write, for any n-globular set A and $0 \le p \le m \le n$,

$$A(m) \times_{A(p)} A(m) = \{(x', x) \in A(m) \times A(m) \mid t^{m-p}(x) = s^{m-p}(x')\},$$

the set of pairs of m-cells with the potential to be joined along p-cells.

Definition 1.4.8 Let $n \in \mathbb{N}$. A **strict n-category** is an n-globular set A equipped with

- a function $\circ_p : A(m) \times_{A(p)} A(m) \longrightarrow A(m)$ for each $0 \le p < m \le n$; we write $\circ_p (x', x)$ as $x' \circ_p x$ and call it a **composite** of x and x'
- a function $i : A(p) \longrightarrow A(p+1)$ for each $0 \le p < n$; we write $i(x)$ as 1_x and call it the **identity** on x,

satisfying the following axioms:

a. (sources and targets of composites) if $0 \le p < m \le n$ and $(x', x) \in A(m) \times_{A(p)} A(m)$ then

$$s(x' \circ_p x) = s(x) \qquad \text{and } t(x' \circ_p x) = t(x') \qquad \text{if } p = m - 1$$
$$s(x' \circ_p x) = s(x') \circ_p s(x) \text{ and } t(x' \circ_p x) = t(x') \circ_p t(x) \text{ if } p \le m - 2$$

b. (sources and targets of identities) if $0 \le p < n$ and $x \in A(p)$ then $s(1_x) = x = t(1_x)$

c. (associativity) if $0 \le p < m \le n$ and $x, x', x'' \in A(m)$ with (x'', x'), $(x', x) \in A(m) \times_{A(p)} A(m)$ then

$$(x'' \circ_p x') \circ_p x = x'' \circ_p (x' \circ_p x)$$

d. (identities) if $0 \le p < m \le n$ and $x \in A(m)$ then

$$i^{m-p}(t^{m-p}(x)) \circ_p x = x = x \circ_p i^{m-p}(s^{m-p}(x))$$

e. (binary interchange) if $0 \le q < p < m \le n$ and $x, x', y, y' \in A(m)$ with

$$(y', y), (x', x) \in A(m) \times_{A(p)} A(m),$$
$$(y', x'), (y, x) \in A(m) \times_{A(q)} A(m)$$

then

$$(y' \circ_p y) \circ_q (x' \circ_p x) = (y' \circ_q x') \circ_p (y \circ_q x)$$

f. (nullary interchange) if $0 \le q < p < n$ and $(x', x) \in A(p) \times_{A(q)} A(p)$ then $1_{x'} \circ_q 1_x = 1_{x' \circ_q x}$.

If A and B are strict n-categories then a **strict n-functor** is a map $f : A \longrightarrow B$ of the underlying n-globular sets commuting with composition and identities. This defines a category **Str-n-Cat** of strict n-categories.

Proposition 1.4.9 *The categories* **Str-n-Cat** *defined in 1.4.1 and 1.4.8 are equivalent.*

Sketch proof We first compare the underlying graph structures. Define for each $n \in \mathbb{N}$ the category n-**Gph** of n-**graphs** by

$$0\text{-}\mathbf{Gph} = \mathbf{Set}, \qquad (n+1)\text{-}\mathbf{Gph} = (n\text{-}\mathbf{Gph})\text{-}\mathbf{Gph}.$$

An $(n+1)$-globular set amounts to a graph of n-globular sets: precisely, an $(n+1)$-globular set X corresponds to the graph $(X(a, b))_{a,b \in X(0)}$ where $X(a, b)$ is the n-globular set defined by

$$(X(a, b))(m) = \{x \in X(m+1) \mid s^{m+1}(x) = a, \ t^{m+1}(x) = b\}.$$

So by induction, n-**Gph** $\simeq [\mathbb{G}_n^{\mathrm{op}}, \mathbf{Set}]$.

Now we bring in the algebra. Given a strict $(n+1)$-category A in the sense of 1.4.8, the functions $\circ_p$ and $i : A(p) \longrightarrow A(p+1)$ taken over $1 \le p < n$ give a strict n-category structure on $A(a, b)$ for each $a, b \in A(0)$. This determines a graph $(A(a, b))_{a,b \in A(0)}$ of strict n-categories. Moreover, the functions $\circ_0$ and $i : A(0) \longrightarrow A(1)$ give this graph the structure of a category enriched in **Str-n-Cat**. With a little work we find that a strict $(n+1)$-category in the sense of 1.4.8 is, in fact, *exactly* a category enriched in **Str-n-Cat**, and by induction we are done.   □

Some examples of n-categories are more easily described using one definition than the other. It also helps to have the terminology of both at hand. We can now say that the strict 2-category **Cat** of 1.4.4 has categories as 0-cells, functors as 1-cells, and natural transformations as 2-cells. There are two kinds of composition of natural transformations. We usually write the ('vertical') composite of natural transformations

as $\beta{\circ}\alpha$ (rather than the standard $\beta \circ_1 \alpha$) and the ('horizontal') composite of natural transformations

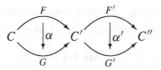

as $\alpha' * \alpha$ (rather than $\alpha' \circ_0 \alpha$).

Example 1.4.10 Let $n \in \mathbb{N}$ and let C be a chain complex, as in 1.4.7. Then the n-globular set X arising from C has the structure of a strict n-category: if $(\mathbf{d}, \mathbf{c}) \in X(m) \times_{X(p)} X(m)$ then $\mathbf{d} \circ_p \mathbf{c} = \mathbf{e}$ where, for $0 \le r \le m$ and $\sigma \in \{-, +\}$,

$$
e_r^\sigma = \begin{cases}
d_r^\sigma + c_r^\sigma & \text{if } r > p \\
c_p^- & \text{if } r = p \text{ and } \sigma = - \\
d_p^+ & \text{if } r = p \text{ and } \sigma = + \\
c_r^\sigma = d_r^\sigma & \text{if } r < p.
\end{cases}
$$

Strict ω-categories can also be defined in both styles. The globular or 'global' definition is obvious. Let $\mathbb{G}$ be the category generated by objects and arrows

$$
\cdots \; \underset{\tau_{m+1}}{\overset{\sigma_{m+1}}{\rightleftarrows}} \; m \; \underset{\tau_m}{\overset{\sigma_m}{\rightleftarrows}} \; m-1 \; \underset{\tau_{m-1}}{\overset{\sigma_{m-1}}{\rightleftarrows}} \; \cdots \; \underset{\tau_1}{\overset{\sigma_1}{\rightleftarrows}} \; 0
$$

subject to the usual globularity equations. Then $[\mathbb{G}^{op}, \mathbf{Set}]$ is the category of **globular sets**, and the category **Str-ω-Cat** of **strict ω-categories** is defined just as in 1.4.8 but without the upper limit of n.

Example 1.4.11 Any chain complex C gives rise to a strict ω-category X, just as in the previous example. In fact, X is an abelian group in **Str-ω-Cat**, and in this way the category of abelian groups in **Str-ω-Cat** is equivalent to the category of non-negatively graded chain complexes of abelian groups.

For the enriched or 'local' definition, we first define a sequence

$$\cdots \xrightarrow{S_{n+1}} \mathbf{Str\text{-}}(n+1)\mathbf{\text{-}Cat} \xrightarrow{S_n} \mathbf{Str\text{-}}n\mathbf{\text{-}Cat} \xrightarrow{S_{n-1}} \cdots \mathbf{Cat} \xrightarrow{S_0} \mathbf{Set}$$

of finite-product-preserving functors by $S_0 = \mathrm{ob}$ and $S_{n+1} = (S_n)_*$, which is possible by Proposition 1.3.3(b). We then take **Str-ω-Cat** to be the limit of this diagram in **CAT**. It is easy to prove

Proposition 1.4.12 *The two categories* **Str-ω-Cat** *just defined are equivalent.*

□

Strict n-categories use globular shapes; there is also a cubical analogue. Again, there is a short inductive definition and a longer explicit version. The short form uses internal rather than enriched categories. If $\mathcal{V}$ is a category with pullbacks then we write **Cat($\mathcal{V}$)** for the category of internal categories in $\mathcal{V}$, which, it can be shown, also has pullbacks.

Definition 1.4.13 Let $(\mathbf{Str\text{-}}n\mathbf{\text{-}tuple\text{-}Cat})_{n\in\mathbb{N}}$ be the sequence of categories given inductively by

$$\mathbf{Str\text{-}0\text{-}tuple\text{-}Cat} = \mathbf{Set},$$

$$\mathbf{Str\text{-}}(n+1)\mathbf{\text{-}tuple\text{-}Cat} = \mathbf{Cat}(\mathbf{Str\text{-}}n\mathbf{\text{-}tuple\text{-}Cat}).$$

A **strict n-tuple category** (or 'strict cubical n-category') is an object of **Str-n-tuple-Cat**.

A strict single (= 1-tuple) category is just a category. A strict double (= 2-tuple) category D is a diagram

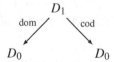

of categories and functors, with extra structure. The objects of D_0 are called the **0-cells** or **objects** of D, the maps in D_0 are the **vertical 1-cells** of D, the objects of D_1 are the **horizontal 1-cells** of D, and the maps in D_1 are the

2-cells of D, as in the picture

$$
\begin{array}{ccc}
a & \xrightarrow{\ m\ } & a' \\
f \downarrow & \Downarrow \theta & \downarrow f' \\
b & \xrightarrow{\ p\ } & b'
\end{array}
\tag{1.8}
$$

where $a \xrightarrow{\ f\ } b, a' \xrightarrow{\ f'\ } b'$ are maps in D_0 and $m \xrightarrow{\ \theta\ } p$ is a map in D_1, with
$\mathrm{dom}(m) = a$, $\mathrm{dom}(\theta) = f$, and so on. The 'extra structure' consists of various
kinds of composition and identities, so that vertical 1-cells can be composed
vertically, horizontal 1-cells can be composed horizontally, and 2-cells can be
composed both vertically and horizontally. Thus, any $p \times q$ grid of 2-cells has
a unique 2-cell composite, for any $p, q \in \mathbb{N}$.

Example 1.4.14 A strict double category in which all vertical 1-cells (or all
horizontal 1-cells) are identities is just a strict 2-category.

Example 1.4.15 A strict 2-category A gives rise to a strict double category D
in two other ways: take the 0-cells of D to be those of A, both the vertical and
the horizontal 1-cells of D to be the 1-cells of A, and the 2-cells (1.8) of D to
be the 2-cells

$$
\begin{array}{ccc}
a & \xrightarrow{\ m\ } & a' \\
f \downarrow & \nearrow\!\!\!\!\nearrow \theta & \downarrow f' \\
b & \xrightarrow{\ p\ } & b'
\end{array}
$$

in A, or the same with the arrow for θ reversed.

The longer definition of strict n-tuple category is omitted since we will not
need it, but goes roughly as follows. As for the globular case, an n-tuple category is a presheaf with extra algebraic structure.

Let $\mathbb{H} = \mathbb{G}_1 = (1 \underset{\tau}{\overset{\sigma}{\rightleftarrows}} 0)$, so that a presheaf on $\mathbb{H}$ is a directed graph. If
$n \in \mathbb{N}$ then an n-**cubical set** is a presheaf on $\mathbb{H}^n$. An n-cubical set consists of a
set $X(M)$ for each $M \subseteq \{1, \ldots, n\}$, and a function $X(\xi) : X(M) \longrightarrow X(P)$
for each $P \subseteq M$ and function $\xi : M \backslash P \longrightarrow \{-, +\}$, satisfying functoriality
axioms. For instance, a 2-cubical set X consists of sets

$$
\begin{aligned}
X(\emptyset) &= \{\text{0-cells}\} \\
X(\{1\}) &= \{\text{vertical 1-cells}\} \\
X(\{2\}) &= \{\text{horizontal 1-cells}\} \\
X(\{1, 2\}) &= \{\text{2-cells}\}
\end{aligned}
$$

with various source and target functions between them.

A strict n-tuple category can then be defined as an n-cubical set A together with a composition function

$$A(P \cup \{m\}) \times_{A(P)} A(P \cup \{m\}) \longrightarrow A(P \cup \{m\})$$

and an identity function

$$A(P) \longrightarrow A(P \cup \{m\})$$

for each $P \subseteq \{1, \ldots, n\}$ and $m \in \{1, \ldots, n\} \backslash P$, satisfying axioms. Strict n-categories can be identified with strict n-tuple categories whose underlying n-cubical set is degenerate in a certain way, generalizing Example 1.4.14.

1.5 Bicategories

Bicategories are to strict 2-categories as weak monoidal categories are to strict monoidal categories. There are many other formalizations of the idea of weak 2-category (see Section 3.4), but bicategories are the oldest and best-known.

We define a bicategory as a category 'weakly enriched' in **Cat**. An alternative definition as a 2-globular set with structure can be found in Bénabou (1967), where bicategories were introduced.

Definition 1.5.1 A **bicategory** $\mathcal{B}$ consists of

- a class $\mathcal{B}_0$, whose elements are called the **objects** or **0-cells** of $\mathcal{B}$
- for each $A, B \in \mathcal{B}_0$, a category $\mathcal{B}(A, B)$, whose objects f are called **1-cells** of $\mathcal{B}$ and written $A \xrightarrow{\ f\ } B$ and whose arrows γ are called **2-cells** of $\mathcal{B}$ and written

- for each $A, B, C \in \mathcal{B}_0$, a functor

$$\mathcal{B}(B, C) \times \mathcal{B}(A, B) \longrightarrow \mathcal{B}(A, C)$$

(**composition**), written

$$\begin{aligned} (g, f) &\longmapsto g \circ f, \\ (\delta, \gamma) &\longmapsto \delta * \gamma \end{aligned}$$

on 1-cells f, g and 2-cells γ, δ

- for each $A \in \mathcal{B}_0$, an object $1_A \in \mathcal{B}(A, A)$ (the **identity** on A)

- for each triple of 1-cells, an isomorphism

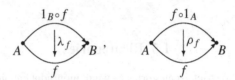

in $\mathcal{B}(A, D)$ (the **associativity coherence isomorphism**)
- for each 1-cell $A \xrightarrow{f} B$, isomorphisms

in $\mathcal{B}(A, B)$ (the **unit coherence isomorphisms**)

such that the coherence isomorphisms $\alpha_{f,g,h}$, λ_f, and ρ_f are natural in f, g, and h and satisfy pentagon and triangle axioms like those in 1.2.5 (replacing $((A \otimes B) \otimes C) \otimes D$ with $((k \circ h) \circ g) \circ f$, etc.).

Functoriality of composition encodes 'interchange laws', just as for monoidal categories (equation (1.4), p. 31): the two evident derived composites of a diagram of shape

are equal, and similarly for 2-cells formed from diagrams $\bullet \longrightarrow \bullet \longrightarrow \bullet$.

Example 1.5.2 A bicategory with only one object is just a monoidal category.

Example 1.5.3 Strict 2-categories can be identified with bicategories in which all the components of α, λ and ρ are identities.

Example 1.5.4 Any topological space S gives rise to a bicategory $\Pi_2 S$, the **fundamental 2-groupoid** of S. ('2-groupoid' means that all the 2-cells are isomorphisms and all the 1-cells are equivalences, in the sense defined below.) The objects of $\Pi_2 S$ are the points of S. The 1-cells $a \longrightarrow b$ are the paths from a to b, that is, the maps $f : [0, 1] \longrightarrow S$ satisfying $f(0) = a$ and $f(1) = b$. The 2-cells are the homotopy classes of path homotopies, relative to endpoints. Any particular point $s \in S$ determines a one-object full sub-bicategory of $\Pi_2 S$

(that is, the sub-bicategory whose only object is s and with all possible 1- and 2-cells), and this is the monoidal category of 1.2.9.

Notice, incidentally, that a chain complex gives rise to a strict n-category for each n (1.4.10, 1.4.11) but a space gives rise to only *weak* structures. This reflects the difference in difficulty between homology and homotopy.

The fundamental ω-groupoid of a space is constructed in 9.2.7.

Example 1.5.5 There is a bicategory $\mathcal{B}$ in which objects are rings, 1-cells $A \longrightarrow B$ are (B, A)-bimodules, and 2-cells are maps of bimodules. (A (B, A)-**bimodule** is an abelian group M equipped with a left B-module structure and a right A-module structure satisfying $(b \cdot m) \cdot a = b \cdot (m \cdot a)$ for all $b \in B, m \in M, a \in A$.) Composition is tensor: if M is a (B, A)-bimodule and N a (C, B)-bimodule then $N \otimes_B M$ is a (C, A)-bimodule.

Example 1.5.6 An n-category has 2^n duals, including the original article. For a bicategory $\mathcal{B}$, the dual obtained by reversing the 1-cells is traditionally called $\mathcal{B}^{\mathrm{op}}$ and that obtained by reversing the 2-cells is called $\mathcal{B}^{\mathrm{co}}$. So the following four pictures show corresponding 2-cells in $\mathcal{B}$, $\mathcal{B}^{\mathrm{op}}$, $\mathcal{B}^{\mathrm{co}}$ and $\mathcal{B}^{\mathrm{co\,op}} = \mathcal{B}^{\mathrm{op\,co}}$:

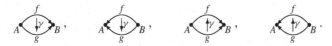

Since **Cat** is a bicategory (1.4.4), we may take definitions from category theory and try to imitate them in an arbitrary bicategory $\mathcal{B}$. For example, a **monad** in $\mathcal{B}$ is an object A of $\mathcal{B}$ together with a 1-cell $A \xrightarrow{t} A$ and 2-cells

$$\mu : t{\circ}t \longrightarrow t, \qquad \eta : 1_A \longrightarrow t,$$

rendering commutative the diagrams of 1.1.3 (with Ts changed to ts, $T\mu$ changed to $1_t * \mu$, and so on). An **adjunction** in $\mathcal{B}$ is a pair (A, B) of objects together with 1- and 2-cells

$$A \xrightarrow{f} B, \qquad B \xrightarrow{g} A, \qquad 1_A \xrightarrow{\eta} g{\circ}f, \qquad f{\circ}g \xrightarrow{\varepsilon} 1_B$$

$$(1.9)$$

satisfying the triangle identities,

$$(\varepsilon * 1_f){\circ}(1_f * \eta) = 1_f, \qquad (1_g * \varepsilon){\circ}(1_g * \eta) = 1_g.$$

An **equivalence** between objects A and B of $\mathcal{B}$ is a quadruple $(f, g, \eta, \varepsilon)$ of cells as in (1.9) such that η and ε are isomorphisms (in their respective hom-categories). An **adjoint equivalence** is a quadruple that is both an adjunction and an equivalence. A 1-cell f in a bicategory is called an **equivalence** if it

satisfies the equivalent conditions of the following result, which generalizes
Proposition 1.1.2(a–b):

Proposition 1.5.7 *Let B be a bicategory. The following conditions on a 1-cell
$f : A \longrightarrow B$ in B are equivalent:*

a. *there exist g, η and ε such that $(f, g, \eta, \varepsilon)$ is an adjoint equivalence*
b. *there exist g, η and ε such that $(f, g, \eta, \varepsilon)$ is an equivalence.*

Proof More is true: given an equivalence $(f, g, \eta, \varepsilon)$, there is a unique ε' such
that $(f, g, \eta, \varepsilon')$ is an adjoint equivalence, namely

$$f \circ g \xrightarrow{\varepsilon^{-1}*1_{f \circ g}} f \circ g \circ f \circ g \xrightarrow{1_f * \eta^{-1} * 1_g} f \circ g \xrightarrow{\varepsilon} 1_B.$$

Brackets have been omitted, as if B were a strict 2-category; the conscientious
reader can both fill in the missing coherence isomorphisms and verify that η
and ε' satisfy the triangle identities (a long but elementary exercise). $\square$

We write $A \simeq B$ if there exists an equivalence $A \longrightarrow B$.

Strict 2-categories form a strict 3-category (1.4.4), so we would expect
there to be notions of functor between bicategories and transformation between functors, and then a further notion of map between transformations. On
the other hand, one-object bicategories – monoidal categories – only form a
2-category: there are various notions of functor between monoidal categories,
and a notion of transformation between monoidal functors, but nothing more
after that. We will soon see how this apparent paradox is resolved.

For functors everything goes smoothly.

Definition 1.5.8 Let B and B' be bicategories. A **lax functor** $F = (F, \phi)$:
$B \longrightarrow B'$ consists of

- a function $F_0 : B_0 \longrightarrow B'_0$, usually just written F
- for each $A, B \in B_0$, a functor $F_{A,B} : B(A, B) \longrightarrow B'(FA, FB)$, usually
  also written F
- for each composable pair (f, g) of 1-cells in B, a 2-cell $\phi_{g,f}$:
  $Fg \circ Ff \longrightarrow F(g \circ f)$
- for each $A \in B_0$, a 2-cell $\phi_A : 1_{FA} \longrightarrow F1_A$

satisfying naturality and coherence axioms analogous to those of Definition 1.2.10. **Colax**, **weak** and **strict functors** are also defined analogously.

Example 1.5.9 Let $\mathcal{A}$ and $\mathcal{A}'$ be monoidal categories, and $\Sigma\mathcal{A}$ and $\Sigma\mathcal{A}'$ the
corresponding one-object bicategories. Then lax monoidal functors $\mathcal{A} \longrightarrow$
$\mathcal{A}'$ are exactly lax functors $\Sigma\mathcal{A} \longrightarrow \Sigma\mathcal{A}'$, and similarly for colax, weak and
strict functors.

The evident composition of lax functors between bicategories is, perhaps surprisingly, strictly associative and unital. This means that we have a category **Bicat**$_{\text{lax}}$ of bicategories and lax functors, and subcategories **Bicat**$_{\text{wk}}$ and **Bicat**$_{\text{str}}$. The same goes for colax functors, but we concentrate on the lax (and above all, the weak) case.

Definition 1.5.10 Let $(F, \phi), (G, \psi) : \mathcal{B} \longrightarrow \mathcal{B}'$ be lax functors between bicategories. A **lax transformation** $\sigma : F \longrightarrow G$ consists of

- for each $A \in \mathcal{B}_0$, a 1-cell $\sigma_A : FA \longrightarrow GA$
- for each 1-cell $f : A \longrightarrow B$ in $\mathcal{B}$, a 2-cell

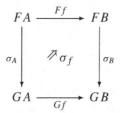

such that σ_f is natural in f and satisfies coherence axioms as in Street (1996, p. 568) or Leinster (1998b, 1.2). A **colax transformation** is the same but with the direction of σ_f reversed. **Weak** and **strict transformations** are lax transformations in which all the σ_fs are, respectively, isomorphisms or identities.

Example 1.5.11 Let $F, G : \mathcal{A} \longrightarrow \mathcal{A}'$ be lax monoidal functors between monoidal categories. As in 1.5.9, these correspond to lax functors $\Sigma F, \Sigma G : \Sigma \mathcal{A} \longrightarrow \Sigma \mathcal{A}'$ between bicategories. A lax transformation $\Sigma F \longrightarrow \Sigma G$ is an object S of $\mathcal{A}'$ together with a map

$$\sigma_X : GX \otimes S \longrightarrow S \otimes FX$$

for each object X of $\mathcal{A}$, satisfying coherence axioms. (We are writing $S = \sigma_\star$, where $\star$ is the unique object of $\Sigma \mathcal{A}$.)

Transformations of one-object bicategories are, therefore, more general than transformations of monoidal categories. Monoidal transformations can be identified with colax bicategorical transformations σ for which $\sigma_\star = I$. Even putting aside the reversal of direction, we see that what seems appropriate for bicategories in general seems inappropriate in the one-object case. We find out more in 1.5.14 and 1.5.16 below.

Definition 1.5.12 Let

be lax transformations between lax functors between bicategories. A **modifica-
tion** $\Gamma : \sigma \longrightarrow \tilde{\sigma}$ consists of a 2-cell $\Gamma_A : \sigma_A \longrightarrow \tilde{\sigma}_A$ for each $A \in \mathcal{B}_0$, such
that Γ_A is natural in A and satisfies a coherence axiom (Street, 1996, p. 569 or
Leinster, 1998b, 1.3).

Functors, transformations, and modifications can be composed in various
ways. Let us consider just weak functors and transformations from now on. It
is straightforward to show that for any two bicategories $\mathcal{B}$ and $\mathcal{B}'$, there is a
functor bicategory $[\mathcal{B}, \mathcal{B}']$ whose objects are the weak functors from $\mathcal{B}$ to $\mathcal{B}'$,
whose 1-cells are weak transformations, and whose 2-cells are modifications.
This is not usually a strict 2-category, because composing 1-cells in $[\mathcal{B}, \mathcal{B}']$ involves composing 1-cells in $\mathcal{B}'$, but it is a strict 2-category if $\mathcal{B}'$ is. In particular,
there is a strict 2-category $[\mathcal{B}^{\mathrm{op}}, \mathbf{Cat}]$ of 'presheaves' on any bicategory $\mathcal{B}$.

We might expect bicategories to form a weak 3-category, if we knew what
one was. Gordon, Power and Street gave an explicit definition of tricategory
(weak 3-category) in (1995), and showed that bicategories form a tricategory. It
is worth noting, however, that there are two equally sensible and symmetrically
opposite ways of making bicategories into a tricategory, because there are two
ways of horizontally composing weak transformations, as is easily verified.

There are coherence theorems for bicategories analogous to those for monoidal categories. 'All diagrams commute' is handled in exactly the same way.
For the other statement of coherence we need the right notion of equivalence
of bicategories.

Let $(F, \phi) : \mathcal{B} \longrightarrow \mathcal{B}'$ be a weak functor between bicategories. We
call F a **local equivalence** if for each $A, B \in \mathcal{B}_0$, the functor $F_{A,B} :$
$\mathcal{B}(A, B) \longrightarrow \mathcal{B}'(FA, FB)$ is an equivalence of categories, and **essentially
surjective on objects** if for each $A' \in \mathcal{B}'_0$, there exists $A \in \mathcal{B}_0$ such that
$FA \simeq A'$. We call F a **biequivalence** if it satisfies the equivalent conditions of
the following result, analogous to Proposition 1.1.2(b–c):

Proposition 1.5.13 *The following conditions on a weak functor* $F : \mathcal{B} \longrightarrow \mathcal{B}'$
between bicategories are equivalent:

a. *there exists a weak functor* $G : \mathcal{B}' \longrightarrow \mathcal{B}$ *such that* $1_{\mathcal{B}} \simeq G \circ F$ *in* $[\mathcal{B}, \mathcal{B}]$
   and $F \circ G \simeq 1_{\mathcal{B}'}$ *in* $[\mathcal{B}', \mathcal{B}']$
b. *F is a local equivalence and essentially surjective on objects.*

Sketch proof (a) ⇒ (b) is straightforward. For the converse, choose for each $A' \in \mathcal{B}'$ an object $GA' \in \mathcal{B}$ together with an *adjoint* equivalence between FGA' and A', which is possible by 1.5.7. Then the remaining constructions and checks are straightforward, if tedious. □

Condition (a) of the Proposition says that there is a system of functors, *transformations and modifications* relating $\mathcal{B}$ and $\mathcal{B}'$. It therefore seems quite plausible that in the one-object case, biequivalence is a looser relation than monoidal equivalence. Nevertheless,

Corollary 1.5.14 *Two monoidal categories are monoidally equivalent if and only if the corresponding one-object bicategories are biequivalent.*

Proof The monoidal and bicategorical notions of weak functor are the same (1.5.9), so this follows from conditions (b) of Propositions 1.2.14 and 1.5.13. □

Theorem 1.5.15 (Coherence for bicategories) *Every bicategory is biequivalent to some strict 2-category.*

Sketch proof Let $\mathcal{B}$ be a bicategory. There is a weak functor

$$\mathbf{y} : \mathcal{B} \longrightarrow [\mathcal{B}^{\mathrm{op}}, \mathbf{Cat}],$$

the analogue of the Yoneda embedding for categories, sending an object A of $\mathcal{B}$ to the weak functor

$$\mathcal{B}(-, A) : \mathcal{B}^{\mathrm{op}} \longrightarrow \mathbf{Cat}$$

and so on. Just as the ordinary Yoneda embedding is full and faithful, $\mathbf{y}$ is a local equivalence. So if $\mathcal{B}'$ is the sub-strict-2-category of $[\mathcal{B}^{\mathrm{op}}, \mathbf{Cat}]$ consisting of all the objects in the image of $\mathbf{y}$ and all the 1- and 2-cells between them then $\mathbf{y}$ defines a biequivalence from $\mathcal{B}$ to $\mathcal{B}'$. □

Example 1.5.16 Corollary 1.5.14 enables us to deduce coherence for monoidal categories (1.2.15) from coherence for bicategories. In fact, the proof of 1.2.15 is exactly the proof of 1.5.15 in the one-object case. Let $\mathcal{B}$ be a bicategory with single object $\star$ and let $\mathcal{A}$ be the corresponding monoidal category. Then $\mathcal{B}'$ is the sub-2-category of $[\mathcal{B}^{\mathrm{op}}, \mathbf{Cat}]$ whose single object is the representable functor $\mathcal{B}(-, \star)$, and a weak transformation from $\mathcal{B}(-, \star)$ to itself is an object (E, δ) of $\mathcal{A}'$ as in the proof of 1.2.15, and so on.

Notes

Almost everything in this chapter is very well-known (in the sense of the phrase to which mathematicians are accustomed). I have used Mac Lane (1971) as my main reference for ordinary category theory, with Borceux (1994a, 1994b) as backup. The standard text on enriched category theory is Kelly (1982). For higher category theory, Street (1996) provides a useful survey and reference list. Strict n-categories were introduced by Ehresmann (1965).

I thank Bill Fulton and Ross Street for an enlightening exchange on equivalence of bicategories, and Nathalie Wahl for a useful conversation on how not to prove coherence.

Chapter 2
Classical operads and multicategories

Some 'pictures' are not really pictures, but rather are windows to Plato's heaven.

Brown (1999)

Where category theory has arrows, higher-dimensional category theory has higher-dimensional arrows. One of the simplest examples of a 'higher-dimensional arrow' is one like

Think of this as a box with n input wires coming in on the left, where n is any natural number, and one output wire emerging on the right. (For instance, when $n = 1$ this is just an arrow as in an ordinary category.) With this in mind, it is easy to imagine what composition of such arrows might look like: outputs of one arrow attach to inputs of another.

A categorical structure with arrows like this is called a multicategory. (Multicategories and n-categories are not the same!) A very familiar example: the objects (drawn as labels on wires) are vector spaces, and the arrows are multilinear maps. The special case of a multicategory where there is only one object – that is, the wires are unlabelled – is particularly interesting. Such a structure is called an operad; for a basic example, fix a topological space X and define an arrow with n inputs to be a continuous map $X^n \longrightarrow X$.

There is a curiously widespread impression that operads are frighteningly complicated structures. Among users of operads, there is a curiously widespread impression that multicategories – usually known to them as 'coloured operads' – are some obscure and esoteric elaboration of the basic notion of operad. I hope this chapter will correct both impressions. Both structures are as natural as can be, and, if one draws some pictures, very simple to understand.

We start with multicategories (Section 2.1) then specialize to operads (Section 2.2). (This order of presentation may convince sceptical operad-theorists that multicategories are natural structures in their own right.) The basic definitions are given, with a broad range of examples. An assortment of further topics on operads and multicategories is covered in Section 2.3.

Warning Readers already familiar with operads may be used to them coming equipped with symmetric group actions. In this text operads *without* symmetries are the default. This is partly to fit with the convention that monoidal categories are by default non-symmetric, and rings, groups and monoids non-commutative, but is mostly for reasons that will emerge later. So:

> **'Operad' means what is sometimes called 'non-Σ operad' or 'non-symmetric operad'.**

Operads equipped with symmetries will be called 'symmetric operads'.

2.1 Classical multicategories

Let us make the description above more precise. A category consists of objects, arrows between objects, and a way of composing arrows. Precisely the same description applies to multicategories. The only difference lies in the shape of the arrows: in a category, an arrow looks like

$$a \xrightarrow{\;\theta\;} b,$$

with one object as its domain and one object as its codomain, whereas in a multicategory, an arrow looks like

$$
\begin{array}{c}
a_1 \\
a_2 \\
\vdots \\
a_n
\end{array}
\left.\rule{0pt}{2.2em}\right\}
\;\theta\;
\triangleright
\;-\; a
\tag{2.1}
$$

($n \in \mathbb{N}$), with a finite sequence of 'input' objects as its domain and one 'output' object as its codomain. Arrows can be composed when outputs are joined to inputs, which for categories means that any string of arrows

$$a_0 \xrightarrow{\;\theta_1\;} a_1 \xrightarrow{\;\theta_2\;} \quad \cdots \quad \xrightarrow{\;\theta_n\;} a_n$$

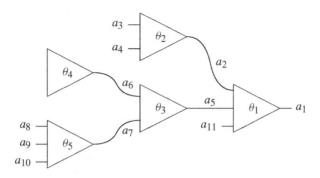

Fig. 2-A. Composable diagram of arrows in a multicategory

has a well-defined composite, and for multicategories means (more interestingly in geometrical terms) that any tree of arrows such as that in Fig. 2-A has a well-defined composite, in this case of the form

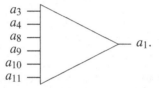

Perhaps the most familiar example is where the objects are vector spaces and the arrows are multilinear maps. Commonly the multicategory structure is obscured by the device of considering multilinear maps as linear maps out of a tensor product; but in many situations it is the multicategory, not the monoidal category, that is fundamental.

Definition 2.1.1 A **multicategory** C consists of

- a class C_0, whose elements are called the **objects** of C
- for each $n \in \mathbb{N}$ and $a_1, \ldots, a_n, a \in C_0$, a class $C(a_1, \ldots, a_n; a)$, whose elements θ are called **arrows** or **maps** and depicted as in (2.1) or as

$$a_1, \ldots, a_n \xrightarrow{\ \theta\ } a \qquad (2.2)$$

- for each $n, k_1, \ldots, k_n \in \mathbb{N}$ and $a, a_i, a_i^j \in C_0$, a function (Fig. 2-B)

$$C(a_1, \ldots, a_n; a) \times C(a_1^1, \ldots, a_1^{k_1}; a_1) \times \cdots \times C(a_n^1, \ldots, a_n^{k_n}; a_n)$$
$$\longrightarrow C(a_1^1, \ldots, a_1^{k_1}, \ldots, a_n^1, \ldots, a_n^{k_n}; a),$$

called **composition** and written

$$(\theta, \theta_1, \ldots, \theta_n) \longmapsto \theta \circ (\theta_1, \ldots, \theta_n)$$

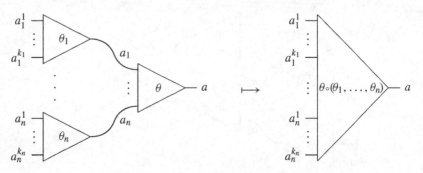

Fig. 2-B. Composition in a multicategory

- for each $a \in C_0$, an element $1_a \in C(a; a)$, called the **identity** on a

satisfying

- associativity:

$$\theta \circ \left(\theta_1 \circ (\theta_1^1, \ldots, \theta_1^{k_1}), \ldots, \theta_n \circ (\theta_n^1, \ldots, \theta_n^{k_n}) \right)$$
$$= (\theta \circ (\theta_1, \ldots, \theta_n)) \circ (\theta_1^1, \ldots, \theta_1^{k_1}, \ldots, \theta_n^1, \ldots, \theta_n^{k_n})$$

whenever $\theta, \theta_i, \theta_i^j$ are arrows for which these composites make sense

- identity:

$$\theta \circ (1_{a_1}, \ldots, 1_{a_n}) = \theta = 1_a \circ (\theta)$$

whenever $\theta : a_1, \ldots, a_n \longrightarrow a$ is an arrow.

Operads are precisely multicategories with only one object, and are the subject of the next section. For now we look at multicategories with many objects – often a proper class of them.

Example 2.1.2 A multicategory in which every arrow is **unary** (that is, of the form (2.2) with $n = 1$) is the same thing as a category.

Example 2.1.3 Any monoidal category $(A, \otimes)$ has an **underlying multicategory** C. It has the same objects as A, and a map

$$a_1, \ldots, a_n \longrightarrow a$$

in C is a map

$$a_1 \otimes \cdots \otimes a_n \longrightarrow a$$

in A. Composition in C is derived from composition and tensor in A.

If A is a non-strict monoidal category then there is some ambiguity in the meaning of '$a_1 \otimes \cdots \otimes a_n$'. We will address this properly in Section 3.3, but for now let us just choose a particular bracketing, e.g. $((a_1 \otimes a_2) \otimes a_3) \otimes a_4$ for $n = 4$.

Given a commutative ring R, the monoidal category of R-modules with their usual tensor gives rise in this way to a multicategory of R-modules and R-multilinear maps. Similarly, given a category A with finite products there is an underlying multicategory C with the same objects as A and with

$$C(a_1, \ldots, a_n; a) = A(a_1 \times \cdots \times a_n, a).$$

We will make particular use of the multicategory C coming from the category $A = \mathbf{Set}$; we write $C = \mathbf{Set}$ too.

Example 2.1.4 More generally, if A is a monoidal category and C_0 a collection of objects of A (not necessarily closed under tensor) then there is a multicategory C whose class of objects is C_0 and whose arrows are defined as in the previous example.

It is important to realize that *not* every multicategory is the underlying multicategory of a monoidal category (2.1.3). Example 2.1.4 makes this reasonably clear. Another way of seeing it is to consider one-object multicategories (operads, next section): if A is a monoidal category with single object a then in the underlying multicategory C, the cardinality of the hom-set $C(a, \ldots, a; a)$ is independent of the number of as in the domain. But this independence property certainly does not hold for all one-object multicategories, as numerous examples in Section 2.2 show.

Example 2.1.5 Sometimes tensor is irrelevant; sometimes it does not even exist. For instance, let V be a symmetric monoidal category and let $\mathbf{Ab}(V)$ be the category of abelian groups in V. In general V will not have enough colimits to define a tensor product on $\mathbf{Ab}(V)$ and so make it into a monoidal category. But it is always possible to define multilinear maps between abelian groups in V, giving $\mathbf{Ab}(V)$ the structure of a multicategory.

Example 2.1.6 If A is a category with finite coproducts then Example 2.1.3 reveals that there is a multicategory C with the same objects as A, with

$$C(a_1, \ldots, a_n; a) = A(a_1, a) \times \cdots \times A(a_n, a),$$

and with composition given by

$$(f_1, \ldots, f_n) \circ ((f_1^1, \ldots, f_1^{k_1}), \ldots, (f_n^1, \ldots, f_n^{k_n}))$$
$$= (f_1 \circ f_1^1, \ldots, f_1 \circ f_1^{k_1}, \ldots, f_n \circ f_n^1, \ldots, f_n \circ f_n^{k_n})$$

($f_i \in A(a_i, a)$, $f_i^j \in A(a_i^j, a_i)$). Indeed, these formulas make sense and define a multicategory C for any category A whatsoever. An arrow of C can be drawn as

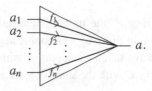

Example 2.1.7 A category in which each hom-set has at most one element is the same thing as a preordered set. Similarly, a multicategory C in which $C(a_1, \ldots, a_n; a)$ has at most one element for each $n, a_1, \ldots, a_n, a$ is some kind of generalized poset. It amounts to a class $X = C_0$ of elements together with an $(n+1)$-ary relation $\leq_n$ on X for each $n \in \mathbb{N}$, satisfying generalized reflexivity and transitivity axioms. For instance, let $d \in \mathbb{N}$, let $X = \mathbb{R}^d$, and define $\leq_n$ by $(a_1, \ldots, a_n) \leq_n a$ if and only if a is in the convex hull of $\{a_1, \ldots, a_n\}$. This gives a 'generalized poset'; the axioms express basic facts about convex hulls.

The final example is more substantial.

Example 2.1.8 The **Swiss cheese multicategory SC** has the natural numbers as its objects, with $d \in \mathbb{N}$ thought of as the d-dimensional disc. Maps are configurations of discs, half-discs, quarter-discs, and so on, as now described.

Let $\mathbb{R}^\infty$ be the set of sequences $(x_n)_{n=1}^\infty$ of real numbers. For each $d \in \mathbb{N}$, let

$$B^{[d]} = \{\mathbf{x} \in \mathbb{R}^\infty \mid \sum_{n=1}^\infty x_n^2 \leq 1, \text{ and } x_n \geq 0 \text{ for all } n > d\},$$

and define a sequence of subsets $F_1^d, F_2^d, \ldots$ of $B^{[d]}$ by

$$F_n^d = \begin{cases} \emptyset & \text{if } n \leq d \\ \{\mathbf{x} \in B^{[d]} \mid x_n = 0\} & \text{if } n > d. \end{cases}$$

(The idea is that if the hyperplane $x_n = 0$ contains one of the flat faces of $B^{[d]}$ then F_n^d is that flat face, and otherwise F_n^d is empty.) Let G be the group of bijections on $\mathbb{R}^\infty$ of the form $\mathbf{x} \longmapsto \mathbf{a} + \lambda\mathbf{x}$ for some $\mathbf{a} \in \mathbb{R}^\infty$ and $\lambda > 0$. Then there is a category A in which the objects are the natural numbers and a map $d' \longrightarrow d$ is an element $\alpha \in G$ such that

$$\alpha B^{[d']} \subseteq B^{[d]}, \qquad \alpha F_n^{d'} \subseteq F_n^d \text{ for all } n \geq 1.$$

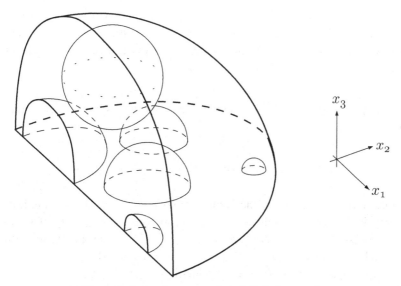

Fig. 2-C. Swiss cheese map $1, 1, 2, 2, 2, 3 \longrightarrow 1$

(So α sends $B^{[d']}$ into $B^{[d]}$, with flat faces mapping into flat faces.) By Example 2.1.6, there arises a multicategory C with object-set $\mathbb{N}$ and with

$$C(d_1, \ldots, d_k; d) = A(d_1, d) \times \cdots \times A(d_k, d).$$

The Swiss cheese multicategory **SC** is the sub-multicategory of C consisting of the same objects but only those arrows

$$(\alpha_1, \ldots, \alpha_k) \in C(d_1, \ldots, d_k; d)$$

for which the images $\alpha_i B^{[d_i]}$ and $\alpha_j B^{[d_j]}$ are disjoint whenever $i \neq j$.

A few calculations reveal that $\mathbf{SC}(d_1, \ldots, d_k; d)$ is only non-empty when $d_i \geq d$ for each i, and that the description of this hom-set is unchanged if we replace $\mathbb{R}^\infty$ by $\mathbb{R}^{\max\{d_1, \ldots, d_k\}}$ throughout. It follows that, for instance, a map

$$1, 1, 2, 2, 2, 3 \longrightarrow 1$$

(Fig. 2-C) is a configuration of two quarter-balls, three half-balls, and one whole ball inside the unit quarter-ball in $\mathbb{R}^3$, with the (straight) 1-dimensional face of each quarter-ball lying on the 1-dimensional face of the unit quarter-ball and the (flat) 2-dimensional face of each half-ball lying on the 2-dimensional face $x_3 = 0$ of the unit quarter-ball. The six little fractions of balls must be disjoint.

For each $d \in \mathbb{N}$ there is a sub-multicategory $\mathbf{SC}_d$ consisting of only the two objects d and $d+1$ (and all the maps between them). This was Voronov's original 'Swiss-cheese operad' (a '2-coloured operad'); see Voronov (1998) and Kontsevich (1999, 2.5). The one-object sub-multicategories of **SC** are more famous: for each $d \in \mathbb{N}$, the sub-multicategory consisting of d (and all maps) is exactly the little d-dimensional discs operad (2.2.16).

Many of the multicategories mentioned have a natural symmetric structure: there is a bijection

$$- \cdot \sigma : C(a_1, \ldots, a_n; a) \xrightarrow{\sim} C(a_{\sigma(1)}, \ldots, a_{\sigma(n)}; a)$$

for each $a_1, \ldots, a_n, a \in C_0$ and permutation $\sigma \in S_n$. A 'symmetric multicategory' is defined as a multicategory equipped with such a family of bijections, satisfying axioms. We give the precise definition in 2.2.21; see also Appendix A. For example, if A is a symmetric monoidal category then its underlying multicategory C is also symmetric.

Similarly, many of these examples are also naturally 'enriched': $C(a_1, \ldots, a_n; a)$ is more than a mere set. For instance, if C is vector spaces and multilinear maps then $C(a_1, \ldots, a_n; a)$ naturally has the structure of a vector space itself, and in some of the oldest examples of operads the 'hom-sets' are topological spaces (as we shall see). One way to formalize this is to allow $C(a_1, \ldots, a_n; a)$ to be an object of some chosen symmetric monoidal category $\mathcal{V}$ (generalizing from $\mathcal{V} = \mathbf{Set}$). This is actually not the most natural generalization, essentially because the tensor product on $\mathcal{V}$ is redundant; we come back to this in Section 6.8.

Definition 2.1.9 Let C and C' be multicategories. A **map of multicategories** $f : C \longrightarrow C'$ consists of a function $f_0 : C_0 \longrightarrow C'_0$ (usually just written f) together with a function

$$C(a_1, \ldots, a_n; a) \longrightarrow C'(f(a_1), \ldots, f(a_n); f(a))$$

(written $\theta \longmapsto f(\theta)$) for each $a_1, \ldots, a_n, a \in C_0$, such that composition and identities are preserved. The category **Multicat** consists of small multicategories and maps between them.

Example 2.1.10 Let A and A' be monoidal categories, with respective underlying multicategories C and C'. Then a map $C \longrightarrow C'$ of multicategories is precisely a lax monoidal functor $A \longrightarrow A'$.

Example 2.1.11 Monoids can be described as maps. A **monoid** in a multicategory C consists of an object a of C together with arrows

$$a, a \xrightarrow{\mu} a, \qquad \cdot \xrightarrow{\eta} a$$

(where the domain of η is the empty sequence) obeying associativity and identity laws. The terminal multicategory 1 consists of one object, $\star$, say, and one arrow

$$\underbrace{\star, \ldots, \star}_{n} \longrightarrow \star$$

for each $n \in \mathbb{N}$. A map from 1 into a multicategory C therefore consists of an object a of C together with an arrow

$$\underbrace{a, \ldots, a}_{n} \xrightarrow{\mu_n} a$$

for each $n \in \mathbb{N}$, obeying 'all possible laws', and this is exactly a monoid in C.

In ordinary category theory, functors into **Set** play an important role. The same goes for multicategories.

Definition 2.1.12 Let C be a multicategory. An **algebra for** C, or C**-algebra**, is a map from C into the multicategory **Set** of Example 2.1.3.

The name makes sense if a multicategory is regarded as an algebraic theory with as many sorts as there are objects; it also generalizes the standard terminology for operads. Explicitly, a C-algebra X consists of

- for each object a of C, a set $X(a)$, whose elements x can be drawn as

- for each map $\theta : a_1, \ldots, a_n \longrightarrow a$ in C, a function

$$\overline{\theta} = X(\theta) : X(a_1) \times \cdots \times X(a_n) \longrightarrow X(a)$$

(Fig. 2-D),

satisfying axioms of the same shape as the associativity and second identity axioms in Definition 2.1.1. This explicit form makes it clear that a **map of** C**-algebras**, $\alpha : X \longrightarrow Y$, should be defined as a family of functions

$$\left(X(a) \xrightarrow{\alpha_a} Y(a) \right)_{a \in C_0}$$

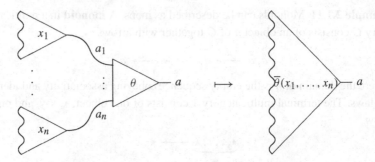

Fig. 2-D. Action of a multicategory on an algebra

satisfying the evident compatibility condition; so we have a category **Alg**(C) of C-algebras. An alternative definition uses the notion of transformation between maps between multicategories, as described in Section 2.3.

Example 2.1.13 If A is a strict monoidal category then an algebra for its underlying multicategory is, by 2.1.10, just a lax monoidal functor from A to (**Set**, ×, 1).

Example 2.1.14 If C is a multicategory in which all arrows are unary (2.1.2), and so essentially just a category, then **Alg**(C) is the ordinary functor category [C, **Set**].

Example 2.1.15 As the pictures suggest, there is for each multicategory C an algebra X defined by taking X(a) to be the set C(; a) of arrows in C from the empty sequence into a. When C is the multicategory of modules over some commutative ring R (Example 2.1.3), this is the evident forgetful map C ⟶ **Set**.

Example 2.1.16 Any family $(X(a))_{a \in S}$ of sets, indexed over any set S, gives rise to a multicategory **End**(X) with object-set S, the **endomorphism multicategory** of X. This is constructed by transporting back from the multicategory **Set**: an arrow $a_1, \ldots, a_n \longrightarrow a$ in **End**(X) is a function

$$X(a_1) \times \cdots \times X(a_n) \longrightarrow X(a),$$

and composition and identities are as in **Set**. An algebra for a multicategory C can then be defined, equivalently, as a family $(X(a))_{a \in C_0}$ of sets together with a multicategory map $f : C \longrightarrow \mathbf{End}(X)$ fixing the objects.

Set-valued functors on ordinary categories are also the same thing as discrete opfibrations. There is a parallel theory of opfibrations for multicategories

(Sections 3.3 and 6.3, and Leinster, 2003), but this will not be of central concern.

2.2 Classical operads

An operad is a multicategory with only one object. In a sense there is no more to be said: the definitions of map between operads, algebra for an operad, and so on, are just special cases of the definitions for multicategories.

On the other hand, operads have a distinctive feel to them and are worth considering in their own right. The analogy to keep in mind is monoids versus categories. A monoid is nothing but a one-object category, and many basic monoid-theoretic concepts are specializations of category-theoretic concepts; nevertheless, monoids still form a natural and interesting class of structures. The same goes for operads and multicategories. Conversely, a category can be regarded as a many-object monoid, and a multicategory as a many-object operad; this leads to the alternative name 'coloured operad' for multicategory.

Most of this section is examples. But first we give the definitions, describe equivalent 'explicit' forms, and discuss the ever-important question of terminology.

Definition 2.2.1 An **operad** is a multicategory with exactly one object. The category **Operad** is the full subcategory of **Multicat** whose objects are the operads.

Let C be an operad, with single object called $\star$. Then C consists of one set

$$P(n) = C(\underbrace{\star, \ldots, \star}_{n}; \star)$$

for each $n \in \mathbb{N}$, together with composition and identities satisfying axioms. Of course, it makes no difference (up to isomorphism) what the single object is called, so an operad can equivalently be described as consisting of

- a sequence $(P(n))_{n \in \mathbb{N}}$ of sets, whose elements θ will be called the n-**ary operations** of P and drawn as

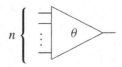

- for each $n, k_1, \ldots, k_n \in \mathbb{N}$, a function

$$P(n) \times P(k_1) \times \cdots \times P(k_n) \longrightarrow P(k_1 + \cdots + k_n)$$
$$(\theta, \theta_1, \ldots, \theta_n) \longmapsto \theta \circ (\theta_1, \ldots, \theta_n),$$

  called **composition** (as in Fig. 2-B but without the labels a, a_i, a_i^j)
- an element $1 = 1_P \in P(1)$, called the **identity**,

satisfying associativity and identity axioms as in the definition of multicategory (2.1.1). This direct description, or something like it, is what one usually sees as the definition of operad. Similarly, a map $f : P \longrightarrow Q$ of operads consists of a family

$$(f_n : P(n) \longrightarrow Q(n))_{n \in \mathbb{N}}$$

of functions, preserving composition and identities.

All of this works just as well if the $P(n)$s are allowed to be objects of some chosen symmetric monoidal category $\mathcal{V}$, rather than necessarily being sets; these are **operads in** $\mathcal{V}$, or $\mathcal{V}$**-operads**, and form a category $\mathcal{V}$**-Operad**. We noted a more general version of this for multicategories, where the appropriate terminology was 'multicategory enriched in $\mathcal{V}$'; we also said there that symmetric monoidal categories are actually not the most natural setting (6.8).

Symmetries are often present in operads. Again, this was noted in the more general context of multicategories. A symmetric structure on an operad P consists, then, of an action of the symmetric group S_n on $P(n)$ for each $n \in \mathbb{N}$, subject to axioms; the precise definition is in 2.2.21. Most authors use the term 'operad' to mean what I call a symmetric operad, and 'non-symmetric' or 'non-Σ' operad for what I call a (plain) operad. For us the non-symmetric case will be by far the more important. The generalized (multicategories and) operads that we consider for much of this book are in some sense a more advanced replacement for symmetric operads; see p. 224 for further explanation of this point of view.

Many authors use the term 'coloured operad' to mean multicategory. The 'colours' are the objects, and the idea is that a multicategory is a more complicated version of an operad. The analogous usage would make a category a 'coloured monoid' and a groupoid a 'coloured group'. I prefer the term 'multicategory' because it emphasizes that we are dealing with a categorical structure, with objects and arrows. Moreover, although thinking of the objects as colours is practical when there are only a small, finite, number of them, it becomes somewhat baroque otherwise: in the multicategory of abelian groups, for instance, one has to paint each group a different colour (the real numbers are green, the cyclic group of order 10 is pink, and so on). This is not entirely a frivolous issue: the coloured viewpoint has given rise to eyebrow-raising statements such as

> [...] it serves us well to have a subtle generalization of operad known as a bicolored operad. Still more colorful operads can be defined, but they are currently not of great importance

(Markl, Shnider and Stasheff, 2002, p. 115); of course, most of the coloured operads that one meets every day (**Set, Top,** *R*-**Mod,** ...) have not just more than two colours, but a proper class of them. The quotation above is analogous to, and indeed includes (Example 2.1.2), the statement that no important category has more than two objects.

Algebras for operads are very important. Translating the multicategorical definition (2.1.12) into explicit language, an algebra for an operad *P* is a set *X* together with a function

$$\overline{\theta} : X^n \longrightarrow X$$

for each $n \in \mathbb{N}$ and $\theta \in P(n)$, satisfying the evident axioms. If *P* is a symmetric operad then a further axiom must be satisfied. If *P* is an operad in a symmetric monoidal category $\mathcal{V}$ then *X* is not a set but an object of $\mathcal{V}$, and an algebra structure on *X* is a family $(P(n) \otimes X^{\otimes n} \longrightarrow X)_{n \in \mathbb{N}}$ of maps, satisfying axioms. (More generally, *X* could be an object of a monoidal category either acted on by $\mathcal{V}$ or enriched in $\mathcal{V}$.) It is clear in all situations what a map of algebras is.

The first group of examples follows the slogan 'an operad is an algebraic theory'.

Example 2.2.2 The terminal operad 1 has exactly one *n*-ary operation for each $n \in \mathbb{N}$. An algebra for 1 is a set *X* together with a function

$$X^n \longrightarrow X$$
$$(x_1, \ldots, x_n) \longmapsto (x_1 \cdot \ldots \cdot x_n)$$

for each $n \in \mathbb{N}$, satisfying the axioms

$$((x_1^1 \cdot \ldots \cdot x_1^{k_1}) \cdot \ldots \cdot (x_n^1 \cdot \ldots \cdot x_n^{k_n})) = (x_1^1 \cdot \ldots \cdot x_n^{k_n})$$
$$x = (x).$$

Hence **Alg**(1) is the category of monoids.

If 1 is considered as a *symmetric* operad then there is a further axiom on its algebras *X*:

$$(x_{\sigma(1)} \cdot \ldots \cdot x_{\sigma(n)}) = (x_1 \cdot \ldots \cdot x_n)$$

for all $\sigma \in S_n$ (as will follow from the definition, p. 80). This means that algebras for 1 are now *commutative* monoids.

Example 2.2.3 Various sub-operads of 1 are commonly encountered. The smallest operad P is given by $P(1) = 1$ and $P(n) = \emptyset$ for $n \neq 1$; its algebras are merely sets. The unique operad P satisfying $P(0) = \emptyset$ and $P(n) = 1$ for $n \geq 1$ has semigroups as its algebras. (A **semigroup** is a set equipped with an associative binary operation; a monoid is a semigroup with identity.) As a kind of dual, the unique operad P satisfying $P(n) = 1$ for $n \leq 1$ and $P(n) = \emptyset$ for $n > 1$ has as its algebras **pointed sets** (sets equipped with a basepoint).

Example 2.2.4 Let M be a monoid. Then there is an operad P in which $P(1) = M$, $P(n) = \emptyset$ for $n \neq 1$, and the composition and identity are the multiplication and unit of M. An algebra for P consists of a set X together with a function $\overline{\theta} : X \longrightarrow X$ for each $\theta \in M$, satisfying axioms; in other words, it is a set with a left M-action.

This example is the one-object case of a multicategorical example that we have already seen: a multicategory all of whose arrows are unary is just a category (2.1.2), and its algebras are just functors from that category into **Set** (2.1.14).

Example 2.2.5 We have seen so far that the theories of monoids, of semigroups, of pointed sets, and of M-sets (for a fixed monoid M) can all be described by operads. The natural question is: which algebraic theories can be described by operads? The answer is: the strongly regular finitary theories. Here is the definition; the proof is in C.1.

A finitary algebraic theory is **strongly regular** if it can be presented by operations and strongly regular equations. In turn, an equation (made up of variables and finitary operation symbols) is **strongly regular** if the same variables appear in the same order, without repetition, on each side. So all of

$$(x \cdot y) \cdot z = x \cdot (y \cdot z), \qquad x \cdot 1 = x, \qquad (x^y)^z = x^{y \cdot z},$$

but none of

$$x \cdot 0 = 0, \qquad x \cdot y = y \cdot x, \qquad x \cdot (y + z) = x \cdot y + x \cdot z,$$

are strongly regular. For instance, the theory of monoids is strongly regular. One would guess that the theories of commutative monoids and of groups are not strongly regular, because their usual presentations involve the equations $x \cdot y = y \cdot x$ and $x^{-1} \cdot x = 1$ respectively, neither of which is strongly regular. For the moment the possibility remains that these theories can be presented by some devious selection of operations and strongly regular equations, but in 4.1.6 we will see a method for proving that a given theory is not strongly regular, and in particular it can be applied to confirm that no such devious selections exist.

The name 'strongly regular' is due to Carboni and Johnstone (1995), and, as they explain, is something of an accident.

Example 2.2.6 A much wider range of algebraic theories is covered if symmetries are allowed and if the $P(n)$s are allowed to be objects of a symmetric monoidal category $\mathcal{V}$ instead of just sets. For instance, if $\mathcal{V}$ is the category of vector spaces over some field then there is a symmetric operad P in $\mathcal{V}$ generated by one element $\theta \in P(2)$ subject to the equations

$$\theta + \theta \cdot \tau = 0$$
$$\theta \circ (1, \theta) + (\theta \circ (1, \theta)) \cdot \sigma + (\theta \circ (1, \theta)) \cdot \sigma^2 = 0$$

where $\tau \in S_2$ is a 2-cycle, $\sigma \in S_3$ is a 3-cycle, and the action of S_n on $P(n)$ is denoted by a dot; P-algebras are exactly Lie algebras. Operads of this kind will not be of direct concern here, our generalization of the notion of operad being in a different direction, but they are very much in use.

The most familiar examples of multicategories are those underlying monoidal categories (2.1.3) or, more generally, their full sub-multicategories (2.1.4). Here is the one-object case.

Example 2.2.7 A one-object strict monoidal category is a commutative monoid (1.2.3), so any commutative monoid $(A, +, 0)$ has an 'underlying' operad P. Concretely, $P(n) = A$ for all n, composition is

$$P(n) \times P(k_1) \times \cdots \times P(k_n) \longrightarrow P(k_1 + \cdots + k_n)$$
$$(a, a_1, \ldots, a_n) \longmapsto a + a_1 + \cdots + a_n,$$

and the identity is $0 \in P(1)$.

Example 2.2.8 For any object b of a monoidal category B, there is an operad structure on the sequence of sets $(B(b^{\otimes n}, b))_{n \in \mathbb{N}}$, given by substitution. This is a special case of Example 2.1.4. We call it **End**(b), the **endomorphism operad** of b. In the case $B = \mathbf{Set}$ this is compatible with the **End** notation of 2.1.16, and specializing the observations there, an algebra for an operad P amounts to a set X together with a map $P \longrightarrow \mathbf{End}(X)$ of operads.

Example 2.2.9 The **operad of curves** P is defined by

$$P(n) = \{\text{smooth maps } \mathbb{R} \longrightarrow \mathbb{R}^n\}$$

and substitution. If B denotes the monoidal category of smooth manifolds and smooth maps, with product as monoidal structure, then P is the endomorphism operad of $\mathbb{R}$ in B^{op}.

Example 2.2.10 Fix a commutative ring k. Substituting polynomials into variables gives an operad structure on the sequence of sets $(k[X_1, \ldots, X_n])_{n \in \mathbb{N}}$; this is the endomorphism operad of the object $k[X]$ of the monoidal category (commutative k-algebras)^{op}.

Example 2.2.11 The same construction works for any algebraic theory: if T is a monad on **Set** then there is a natural operad structure on $(T(n))_{n \in \mathbb{N}}$. The first n here denotes an n-element set, so $T(n)$ is the set of words in n variables. Informally, composition is substitution of words; formally, the composition functions can be written down in terms of the monad structure on T (exercise). This is the endomorphism operad of the free algebra on one generator in the opposite of the category of algebras, where tensor is coproduct of algebras.

In particular, this gives operad structures on each of $(n)_{n \in \mathbb{N}}$ (from the theory of sets, that is, the identity monad), $(k[X_1, \ldots, X_n])_{n \in \mathbb{N}}$ (from the theory of k-algebras, the previous example), and $(k^n)_{n \in \mathbb{N}}$ (from the theory of k-modules).

Example 2.2.12 There is an operad P in which $P(n)$ is the set of isomorphism classes of Riemann surfaces whose boundaries are identified with the disjoint union of $(n + 1)$ copies of the circle S^1 (in order, with the first n thought of as inputs and the last as an output); composition is gluing. Many geometric and topological variants of this example exist. See the end of Section 6.6 for remarks on a more sophisticated version in which there is no need to quotient out by isomorphism.

Example 2.2.13 We saw in 2.1.6 that given any category A, we can pretend that A has coproducts to obtain a multicategory C with the same objects as A. The one-object version is: given any monoid M, there is an operad P with $P(n) = M^n$ and composition

$$(\alpha_1, \ldots, \alpha_n) \circ ((\alpha_1^1, \ldots, \alpha_1^{k_1}), \ldots, (\alpha_n^1, \ldots, \alpha_n^{k_n}))$$

$$= (\alpha_1 \alpha_1^1, \ldots, \alpha_1 \alpha_1^{k_1}, \ldots, \alpha_n \alpha_n^1, \ldots, \alpha_n \alpha_n^{k_n})$$

$(\alpha_i, \alpha_i^j \in M)$. Some specific instances are in 2.2.14 and 2.2.16 below.

Example 2.2.14 Let G be the group of affine automorphisms of the complex plane – maps of the form $z \longmapsto a + bz$ with $a, b \in \mathbb{C}$ and $b \neq 0$. (So G is generated by translations, rotations and dilatations $z \longmapsto \lambda z$ with $\lambda > 0$.) Let

P be the operad $(G^n)_{n \in \mathbb{N}}$ defined in the previous example. Then P has a sub-operad Q given by

$$Q(n) = \{(\alpha_1, \dots, \alpha_n) \in G^n \mid 0 = \alpha_1(0), \, \alpha_1(1) = \alpha_2(0), \, \dots,$$
$$\alpha_{n-1}(1) = \alpha_n(0), \, \alpha_n(1) = 1\}$$

$(n \geq 1)$ and $Q(0) = \emptyset$. Since G acts freely and transitively on the set of ordered pairs of distinct points in the plane, we have

$$Q(n) \cong \{(z_0, \dots, z_n) \in \mathbb{C}^{n+1} \mid 0 = z_0 \neq z_1 \neq \cdots \neq z_n = 1\}, \qquad (2.3)$$

and we call Q the **operad of finite planar sequences**. The set $\mathcal{K}$ of nonempty compact subsets of $\mathbb{C}$ is a Q-algebra, with action

$$\overline{(\alpha_1, \dots, \alpha_n)} : \qquad \begin{array}{ccc} \mathcal{K}^n & \longrightarrow & \mathcal{K} \\ (S_1, \dots, S_n) & \longmapsto & \alpha_1 S_1 \cup \cdots \cup \alpha_n S_n. \end{array}$$

Given any operad R, algebra X for R, and operation $\theta \in R(n)$, we can study the **fixed points** of θ, that is, the elements $x \in X$ satisfying $\overline{\theta}(x, \dots, x) = x$. Here this is the study of affinely self-similar planar sets, and a theorem of Hutchinson (1981) implies that any $(\alpha_1, \dots, \alpha_n) \in Q(n)$ for which each α_i is a contraction (or in terms of (2.3), any $(z_0, \dots, z_n)$ for which $|z_{i+1} - z_i| < 1$ for each i) has a unique fixed point in $\mathcal{K}$. Some of these fixed points are shown in Fig. 2-E.

Operads rose to fame for their role in loop space theory. Let $\mathbf{Top}_*$ be the category of topological spaces with a distinguished basepoint. For any $Y \in \mathbf{Top}_*$, the set $\mathbf{Top}_*(S^1, Y)$ of basepoint-preserving maps from the circle S^1 into Y, endowed with the canonical (= compact-open) topology, is called the **loop space** on Y and written $\Omega(Y)$. This defines an endofunctor Ω of $\mathbf{Top}_*$. If $d \in \mathbb{N}$ then $\Omega^d(Y) \cong \mathbf{Top}_*(S^d, Y)$; a space homeomorphic to $\Omega^d(Y)$ for some Y is called a d-**fold loop space**.

Loop spaces are nearly monoids. A binary multiplication on $\Omega(Y)$ is a rule for composing two loops in Y; the standard choice is to travel the first then the second, each at double speed. The unit is the constant loop. The associativity and unit laws are obeyed not quite exactly, but up to homotopy. Moreover, if one chooses particular homotopies to witness this, then these homotopies obey laws of their own – not quite exactly, but up to homotopy; and so *ad infinitum*. A loop space therefore admits an algebraic structure of a rather complex kind, and it is to describe this complex structure that operads are so useful. In the following group of examples we meet various operads for which loop spaces, and more generally d-fold loop spaces, are naturally algebras.

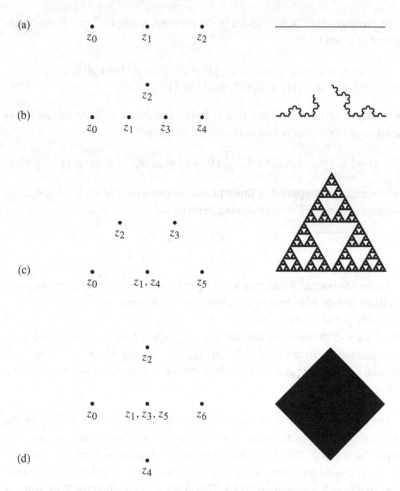

(a) z_0 z_1 z_2

(b) z_0 z_1 z_2 z_3 z_4

(c) z_0 z_1, z_4 z_2 z_3 z_5

(d) z_0 z_1, z_3, z_5 z_2 z_4 z_6

Fig. 2-E. On the left, $(z_0, \dots, z_n) \in Q(n)$, and on the right, its unique fixed point in $\mathcal{K}$: (a) interval, (b) Koch curve, (c) Sierpiński gasket, (d) Péano curve

Example 2.2.15 The coproduct in **Top**$_*$ is the wedge product $\vee$ (disjoint union with basepoints identified). This makes **Top**$_*$ into a monoidal category, so for each $d \in \mathbb{N}$ there is an endomorphism operad $\mathbf{U}_d$ given by $\mathbf{U}_d(k) = \mathbf{Top}_*(S^d, (S^d)^{\vee k})$. Any d-fold loop space is naturally a $\mathbf{U}_d$-algebra via the evident maps

$$
\begin{aligned}
\mathbf{U}_d(k) \times (\Omega^d(Y))^k &\cong \mathbf{Top}_*(S^d, (S^d)^{\vee k}) \times \mathbf{Top}_*((S^d)^{\vee k}, Y) \\
&\longrightarrow \mathbf{Top}_*(S^d, Y) \\
&\cong \Omega^d(Y).
\end{aligned}
$$

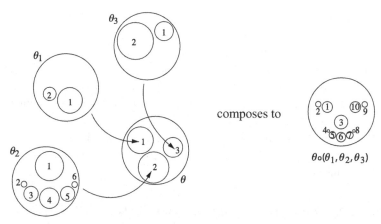

Fig. 2-F. Window to Plato's heaven

Borrowing the terminology of Salvatore (2000), U_d is the **universal operad for d-fold loop spaces**.

Example 2.2.16 Let $d \in \mathbb{N}$ and let G be the group of transformations α of $\mathbb{R}^d$ of the form $\alpha(\mathbf{x}) = \mathbf{a} + \lambda\mathbf{x}$, with $\mathbf{a} \in \mathbb{R}^d$ and $\lambda > 0$. Denote by $D(\mathbf{a}, \lambda)$ the closed disc (ball) in $\mathbb{R}^d$ with centre $\mathbf{a}$ and radius λ. Then $(G^k)_{k \in \mathbb{N}}$ is naturally an operad (Example 2.2.13), and the **little d-discs operad D_d** is the sub-operad defined by

$$D_d(k) = \{(\alpha_1, \ldots, \alpha_k) \in G^k \mid \text{the images of } D(\mathbf{0}, 1) \text{ under } \alpha_1, \ldots, \alpha_k$$
$$\text{are disjoint subsets of } D(\mathbf{0}, 1)\}.$$

Since G acts freely and transitively on the set $\{D(\mathbf{a}, \lambda) \mid \mathbf{a} \in \mathbb{R}^d, \lambda > 0\}$ of discs, $D_d(k)$ may be identified with the set of configurations of d ordered disjoint 'little' discs inside the unit disc: the ith little disc is $\alpha_i D(\mathbf{0}, 1)$. Fig. 2-F shows an example of composition

$$D_2(3) \times (D_2(2) \times D_2(6) \times D_2(2)) \longrightarrow D_2(10)$$

in the little 2-discs operad D_2.

The same can be done with cubes instead of discs; this leaves the homotopy type of $D_d(k)$ unchanged. $D_d(k)$ is also homotopy equivalent to the space of configurations of k ordered, distinct points in $\mathbb{R}^d$, but there is no obvious way to put an operadic composition on this sequence of spaces.

Any d-fold loop space $\Omega^d(Y)$ is naturally a D_d-algebra. Concretely, a configuration of k little d-discs shows how to glue together k based maps

$S^d \longrightarrow Y$ to make a single based map $S^d \longrightarrow Y$ (noting that S^d is $D(\mathbf{0}, 1)$ with its boundary collapsed to a point). Abstractly, any element $(\alpha_1, \ldots, \alpha_k)$ of $\mathbf{D}_d(k)$ defines a continuous injection from the disjoint union of k copies of $D(\mathbf{0}, 1)$ into a single copy of $D(\mathbf{0}, 1)$. Collapsing the boundaries and taking the inverse gives a based map $S^d \longrightarrow (S^d)^{\vee k}$. This process embeds $\mathbf{D}_d$ as a sub-operad of $\mathbf{U}_d$ (2.2.15), and the inclusion $\mathbf{D}_d \hookrightarrow \mathbf{U}_d$ induces a functor $\mathbf{Alg}(\mathbf{U}_d) \longrightarrow \mathbf{Alg}(\mathbf{D}_d)$ in the opposite direction. So any d-fold loop space is a $\mathbf{D}_d$-algebra. In fact, the converse is almost true – roughly, any algebra for $\mathbf{D}_d$ (as a symmetric operad in **Top**) is a d-fold loop space – see Adams (1978, Ch. 2), for instance.

Example 2.2.17 The first really important operad to be considered in topology was Stasheff's (1963a) operad K of associahedra. (The terms 'operad' and 'associahedra' came later.) This is a non-symmetric, topological operad; $K(n)$ is an $(n - 2)$-dimensional solid polyhedron whose vertices are indexed by the n-leafed, planar, binary, rooted trees. Any loop space is naturally a K-algebra. See 7.3 below and Markl, Shnider and Stasheff (2002) for more.

Example 2.2.18 Suppose we are interested in paths rather than based loops: then the appropriate replacement for the space $\mathbf{U}_1(k) = \mathbf{Top}_*(S^1, (S^1)^{\vee k})$ of 2.2.15 is the space

$$E(k) = \{\gamma \in \mathbf{Top}([0, 1], [0, k]) \mid \gamma(0) = 0 \text{ and } \gamma(1) = k\}.$$

If Y is any space, and $Y(y, y')$ denotes the space of $[0, 1]$-parametrized paths from y to y' in Y, then there is a natural map

$$E(k) \times Y(y_0, y_1) \times \cdots \times Y(y_{k-1}, y_k) \longrightarrow Y(y_0, y_k)$$

for each $y_0, \ldots, y_k \in Y$. Here we have something like an E-algebra, in that these maps satisfy axioms resembling very closely those for an algebra for an operad, but it is not an algebra as such; we will meet the appropriate language when we come to generalized operads in Part II. The operad E and its 'nearly-algebras' are used in Trimble's proposed definition of weak n-category (Section 10.1 and Leinster, 2001b).

A miscellaneous example:

Example 2.2.19 Any operad P gives rise to a 'bicoloured operad' (2-object multicategory), $\mathbf{Map}_P$, that has sometimes been found useful (e.g., Markl, Shnider and Stasheff, 2002, §2.9). Call the colours (objects) 0 and 1, and for

$\varepsilon_1, \ldots, \varepsilon_n, \varepsilon \in \{0, 1\}$, put

$$\mathbf{Map}_P(\varepsilon_1, \ldots, \varepsilon_n; \varepsilon) = \begin{cases} P(n) & \text{if } \varepsilon_i \leq \varepsilon \text{ for each } i \\ \emptyset & \text{otherwise.} \end{cases}$$

Then a $\mathbf{Map}_P$-algebra is a map $f : X \longrightarrow Y$ of P-algebras.

To place this in context, let A be any category and C any multicategory. Write $\overline{A}$ for the multicategory obtained by pretending that A has coproducts (2.1.6): then there is an isomorphism of categories

$$\mathbf{Alg}(\overline{A} \times C) \cong [A, \mathbf{Alg}(C)]. \tag{2.4}$$

This is easy to show directly; alternatively, once we have defined transformations of multicategories and hence a category $[C, D]$ for any multicategory D (see Section 2.3), it follows by taking $D = \mathbf{Set}$ in the more general isomorphism

$$[\overline{A} \times C, D] \cong [A, [C, D]].$$

Let $\mathbf{2}$ be the arrow category ($\bullet \longrightarrow \bullet$): then (2.4) with $A = \mathbf{2}$ says that an algebra for $\overline{\mathbf{2}} \times C$ is a map of C-algebras. Indeed, the 2-object multicategory $\mathbf{Map}_P$ defined originally is simply $\overline{\mathbf{2}} \times P$.

The next example provides the language for defining symmetric operads and multicategories.

Example 2.2.20 The sequence $(S_n)_{n \in \mathbb{N}}$ consisting of the underlying sets of the symmetric groups is naturally an operad. We call it the **operad of symmetries, S**. Fig. 2-G shows an example of composition

$$S_3 \times (S_2 \times S_4 \times S_5) \longrightarrow S_{11},$$
$$(\sigma, \rho_1, \rho_2, \rho_3) \longmapsto \sigma \circ (\rho_1, \rho_2, \rho_3)$$

with

$$\sigma = \begin{pmatrix} 1 & 2 & 3 \\ 2 & 3 & 1 \end{pmatrix},$$

$$\rho_1 = \begin{pmatrix} 1 & 2 \\ 2 & 1 \end{pmatrix}, \quad \rho_2 = \begin{pmatrix} 1 & 2 & 3 & 4 \\ 1 & 4 & 2 & 3 \end{pmatrix}, \quad \rho_3 = \begin{pmatrix} 1 & 2 & 3 & 4 & 5 \\ 5 & 1 & 2 & 3 & 4 \end{pmatrix}$$

(that is, $\sigma(1) = 2$, $\sigma(2) = 3$, etc.), and

$$\sigma \circ (\rho_1, \rho_2, \rho_3) = \begin{pmatrix} 1 & 2 & 3 & 4 & 5 & 6 & 7 & 8 & 9 & 10 & 11 \\ 7 & 6 & 8 & 11 & 9 & 10 & 5 & 1 & 2 & 3 & 4 \end{pmatrix}.$$

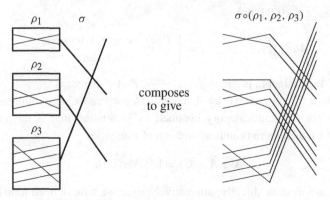

Fig. 2-G. Composition in the operad of symmetries

Formally, let $\sigma \in S_n$, $\rho_1 \in S_{k_1}, \dots, \rho_n \in S_{k_n}$: then for $1 \leq i \leq n$ and $1 \leq j \leq k_i$,

$$\sigma \circ (\rho_1, \dots, \rho_n)(k_1 + \cdots + k_{i-1} + j)$$
$$= k_{\sigma^{-1}(1)} + \cdots + k_{\sigma^{-1}(\sigma(i)-1)} + \rho_i(j).$$

This gives **S** the structure of an operad.

A different construction takes the n-ary operations to be the total orders on the set $\{1, \dots, n\}$ and composition to be lexicographic combination. In this formulation it is clear that composition is associative and unital. This operad of total orders is isomorphic to the operad of symmetries – but for the proof, beware that you need to use the right one out of the two obvious bijections between {total orders on $\{1, \dots, n\}$} and S_n. It is also homotopy equivalent, in a suitable sense, to the little intervals operad $\mathbf{D}_1$.

As discussed on p. 64, a symmetric structure on a multicategory C should consist of a map

$$- \cdot \sigma : C(a_1, \dots, a_n; b) \longrightarrow C(a_{\sigma(1)}, \dots, a_{\sigma(n)}; b) \qquad (2.5)$$

for each $a_1, \dots, a_n, b \in C_0$ and $\sigma \in S_n$. These maps should satisfy the obvious axioms

$$(\theta \cdot \sigma) \cdot \rho = \theta \cdot (\sigma\rho), \qquad \theta = \theta \cdot 1_{S_n} \qquad (2.6)$$

($\theta \in C(a_1, \dots, a_n; b)$, $\sigma, \rho \in S_n$), which guarantee that $- \cdot \sigma$ is a bijection. The symmetric action should also be compatible with composition in C, as for

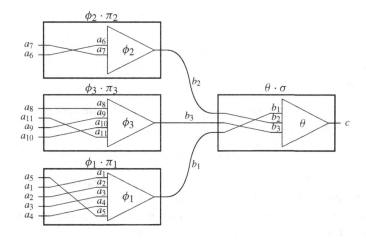

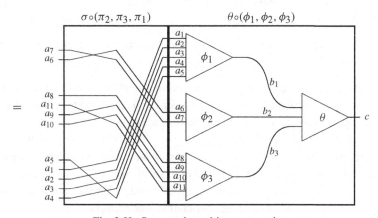

Fig. 2-H. Symmetric multicategory axiom

instance in Fig. 2-H. In general, we want

$$(\theta \cdot \sigma) \circ (\phi_{\sigma(1)} \cdot \pi_{\sigma(1)}, \ldots, \phi_{\sigma(n)} \cdot \pi_{\sigma(n)})$$

$$= (\theta \circ (\phi_1, \ldots, \phi_n)) \cdot (\sigma \circ (\pi_{\sigma(1)}, \ldots, \pi_{\sigma(n)})) \tag{2.7}$$

whenever $\theta, \phi_1, \ldots, \phi_n$ are maps in C and $\sigma, \pi_1, \ldots, \pi_n$ are permutations for which these expressions make sense. The permutation $\sigma \circ (\pi_{\sigma(1)}, \ldots, \pi_{\sigma(n)})$ on the right-hand side is the composite in **S**. (It is easier to get axiom (2.7) right for multicategories than for the special case of operads – the different objects should stop us from writing down nonsense.)

Definition 2.2.21 A **symmetric multicategory** is a multicategory C together with a map (2.5) for each $a_1, \ldots, a_n, b \in C_0$ and $\sigma \in S_n$, satisfying the axioms in (2.6) and (2.7). A **map of symmetric multicategories** is a map f of multicategories such that $f(\theta \cdot \sigma) = f(\theta) \cdot \sigma$ whenever θ is an arrow and σ a permutation for which this makes sense. The category of symmetric multicategories is written **SymMulticat**. A **symmetric operad** is a one-object symmetric multicategory.

Any symmetric monoidal category is naturally a symmetric multicategory, via the symmetry maps

$$\sigma \cdot - : a_{\sigma(1)} \otimes \cdots \otimes a_{\sigma(n)} \xrightarrow{\sim} a_1 \otimes \cdots \otimes a_n.$$

This is true in particular of the category of sets. An **algebra** for a symmetric multicategory C is a map $C \longrightarrow \mathbf{Set}$ of symmetric multicategories. In general, C has more algebras when regarded as a non-symmetric multicategory than when the symmetries are taken into account.

An equivalent definition of symmetric multicategory is given in Appendix A: 'fat symmetric multicategories', in many ways more graceful.

Example 2.2.22 The operad **S** of symmetries becomes a symmetric operad by multiplication in the symmetric groups.

Example 2.2.23 There is a symmetric multicategory $\mathcal{O}$ whose algebras (as a *symmetric* multicategory) are exactly operads. This example will be done informally; we replace it with a precise construction later.

The objects of $\mathcal{O}$ are the natural numbers. To define the arrows we use finite, rooted, planar trees in which each vertex may have any natural number of branches (including 0) coming up out of it. An element of $\mathcal{O}(m_1, \ldots, m_k; n)$ is an n-leafed tree with k vertices that are totally ordered in such a way that the ith vertex has m_i branches coming up out of it. For example,

$$\mathcal{O}(2, 2; 3) = \left\{ \raisebox{-1em}{\includegraphics{}} \right\}. \tag{2.8}$$

Composition is substitution of trees into vertices (much as in the little discs operad), the identity on n is the n-leafed tree with only one vertex, and the symmetric group action is by permutation of the order of the vertices. An $\mathcal{O}$-algebra consists of a set $P(n)$ for each $n \in \mathbb{N}$ together with a map

$$P(m_1) \times \cdots \times P(m_k) \longrightarrow P(n)$$

for each element of $\mathcal{O}(m_1, \ldots, m_k; n)$, satisfying axioms, and this is exactly an operad. For example, the first element α of $\mathcal{O}(2, 2; 3)$ listed in (2.8) induces the function

$$\begin{aligned} \bar{\alpha} : \quad P(2) \times P(2) &\longrightarrow P(3) \\ (\theta, \theta') &\longmapsto \theta \circ (\theta', 1), \end{aligned}$$

part of the operadic structure of P. If σ is the non-trivial element of S_2 then the second element of $\mathcal{O}(2, 2; 3)$ listed in (2.8) is $\alpha \cdot \sigma$, and a consequence of P being an algebra for $\mathcal{O}$ as a *symmetric* multicategory is that $\overline{\alpha \cdot \sigma}(\theta', \theta) = \overline{\alpha}(\theta, \theta')$.

Similarly, there is a symmetric multicategory $\mathcal{O}'$ whose algebras are symmetric operads; it is the same as $\mathcal{O}$ except that the trees are equipped with an ordering of the leaves as well as the vertices. And more generally, for any set S there are symmetric multicategories $\mathcal{O}_S$ and $\mathcal{O}'_S$ whose algebras are, respectively, non-symmetric and symmetric multicategories with object-set S; the object-sets of both $\mathcal{O}_S$ and $\mathcal{O}'_S$ are $(\coprod_{n \in \mathbb{N}} S^n) \times S$.

There is no symmetric multicategory whose algebras are *all* multicategories. There is, however, a *generalized* multicategory with this property, as we shall see.

2.3 Further theory

So far we have seen the basic definitions in the theory of multicategories and operads, and some examples. Here we consider a few further topics. Most are special cases of constructions for generalized multicategories that we meet later; some are generalizations of concepts familiar for ordinary categories.

We start with two alternative ways of defining (operad and) multicategory, both in use; they go by the names of 'circle-i' ($\circ_i$) and 'PROs' respectively. Then we extend three concepts of category theory to multicategories: the free category on a graph, transformations between maps between categories, and modules over categories.

Circle-i

The 'circle-i' method takes composition of diagrams of shape

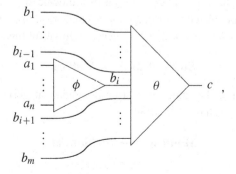

along with identities, to be the basic operations in a multicategory. Thus, a multicategory C can be defined as a set C_0 of objects together with hom-sets $C(a_1, \ldots, a_n; a)$, a function

$$
\begin{aligned}
\circ_i : \quad & C(b_1, \ldots, b_m; c) \\
& \times C(a_1, \ldots, a_n; b_i) \longrightarrow C(b_1, \ldots, b_{i-1}, a_1, \ldots, a_n, \\
& \qquad\qquad\qquad\qquad\qquad\qquad b_{i+1}, \ldots, b_m; c) \\
& (\theta, \phi) \longmapsto \theta \circ_i \phi
\end{aligned}
$$

for each $1 \leq i \leq m, n \in \mathbb{N}, a_1, \ldots, a_n, b_1, \ldots, b_m \in C_0$, and an element $1_a \in C(a; a)$ for each $a \in C_0$, satisfying certain axioms. This is an equivalent definition: given a multicategory in the usual sense, we put

$$
\theta \circ_i \phi = (1_{b_1}, \ldots, 1_{b_{i-1}}, \phi, 1_{b_{i+1}}, \ldots, 1_{b_m}),
$$

and given a multicategory in the new sense, the composite maps $\theta \circ (\phi_1, \ldots, \phi_n)$ can be built using n operations of the form $\circ_i$.

This was, in fact, Lambek's original definition of multicategory (1969, p. 103). His motivating example was that of a deductive system: objects are statements, maps $a_1, \ldots, a_n \longrightarrow b$ are deductions of b from $a_1, \ldots, a_n$, and the $\circ_i$ operation is Gentzen cut. The style of definition is also useful if for some reason one does not want one's multicategories to have identities (as in Markl, Shnider and Stasheff, 2002, p. 45): for with all the $\circ_i$s (but not identities) one can build an operation for composing diagrams in the shape of any non-trivial tree, whereas with the usual $\circ$s (but not identities) one only obtains the non-trivial trees whose leaves are at uniform height. We stick firmly to the original definition, as that is what is generalized to give the all-important definition of generalized multicategory. The $\circ_i$ definition does not generalize in the same way: consider, for instance, the **fc**-multicategories of Chapter 5.

PROs and PROPs

To reach the second alternative definition of multicategory we consider how multicategories are related to strict monoidal categories. (The weak case is left until Section 3.3.) As we saw in 2.1.3, there is a forgetful functor

$$
\textbf{StrMonCat} \xrightarrow{\;U\;} \textbf{Multicat}
$$

where the domain is the category of strict monoidal categories and strict monoidal functors. This has a left adjoint

$$
\textbf{Multicat} \xrightarrow{\;F\;} \textbf{StrMonCat}.
$$

Given a multicategory C, the objects (respectively, arrows) of $F(C)$ are finite ordered sequences of objects (respectively, arrows) of C, and the tensor product in $F(C)$ is concatenation of sequences. So a typical arrow

$$(a_1, a_2, a_3, a_4, a_5) \longrightarrow (a'_1, a'_2, a'_3)$$

in $F(C)$ looks like

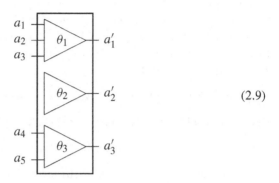

$$(2.9)$$

where $\theta_1 : a_1, a_2, a_3 \longrightarrow a'_1$ in C, etc. An arrow $(a_1, \ldots, a_n) \longrightarrow (a)$ in $F(C)$ is simply an arrow $a_1, \ldots, a_n \longrightarrow a$ in C.

The monoidal categories that arise freely from multicategories can be characterized intrinsically, and this makes it possible to redefine a multicategory as a monoidal category with certain properties. Let C be a multicategory. First, the strict monoidal category $F(C)$ has the property that its underlying monoid of objects is the free monoid on the set C_0. Second, let S be a set and let A be a strict monoidal category whose monoid of objects is the free monoid on S; then for any elements $b_1, \ldots, b_m, a_1, \ldots, a_n$ of S, tensor product in A defines a map

$$\coprod_{a_1^1, \ldots, a_n^{k_n}} \left(A((a_1^1, \ldots, a_1^{k_1}), (a_1)) \times \cdots \times A((a_n^1, \ldots, a_n^{k_n}), (a_n)) \right)$$

$$\longrightarrow A((b_1, \ldots, b_m), (a_1, \ldots, a_n))$$

$$(2.10)$$

where the union is over all $n, k_1, \ldots, k_n \in \mathbb{N}$ and $a_i^j \in S$ such that there is an equality of formal sequences

$$(a_1^1, \ldots, a_1^{k_1}, \ldots, a_n^1, \ldots, a_n^{k_n}) = (b_1, \ldots, b_m).$$

The crucial point is that when $A = F(C)$, the map (2.10) is always a bijection.

A **PRO** is a pair (S, A) where S is a set and A is a strict monoidal category, such that

- the monoid of objects of A is equal to the free monoid on S
- for all $b_1, \ldots, b_m, a_1, \ldots, a_n \in S$, the canonical map (2.10) is a bijection.

A **map of PROs** $(u, f) : (S, A) \longrightarrow (S', A')$ is a function $u : S \longrightarrow S'$ together with a strict monoidal functor $f : A \longrightarrow A'$ such that the objects-function $f_0 : A_0 \longrightarrow A'_0$ is the result of applying the free monoid functor to u. This gives a category **PRO** of PROs. There is a forgetful functor **PRO** $\longrightarrow$ **StrMonCat**, and F lifts in the obvious way to give a functor $\widetilde{F} :$ **Multicat** $\longrightarrow$ **PRO**.

Proposition 2.3.1 *The functor* $\widetilde{F} :$ **Multicat** $\longrightarrow$ **PRO** *is an equivalence.*

Proof Given a PRO (S, A), there is a multicategory C with $C_0 = S$ and

$$C(a_1, \ldots, a_n; a) = A((a_1, \ldots, a_n), (a)),$$

and the condition that the maps (2.10) are bijections implies that $\widetilde{F}(C) \cong (S, A)$. The rest of the proof is straightforward. $\square$

The same kind of equivalence can be established for symmetric multicategories and monoidal categories. The symmetric analogue of a PRO is a PROP. These structures were introduced by Adams and Mac Lane (Mac Lane, 1963) and developed by Boardman and Vogt (1973); the names stand for 'PROduct (and Permutation) category'. Boardman and Vogt called a pair (S, A) an 'S-coloured PRO(P)', and paid particular attention to the single-coloured case. Spelling it out: the category **sPRO** of single-coloured PROs has as objects those strict monoidal categories A whose underlying monoid of objects is equal to $(\mathbb{N}, +, 0)$ and for which the canonical map

$$\coprod_{k_1 + \cdots + k_n = m} A(k_1, 1) \times \cdots \times A(k_n, 1) \longrightarrow A(m, n)$$

is a bijection for all $m, n \in \mathbb{N}$, and as arrows those strict monoidal functors that are the identity on objects. We have immediately:

Corollary 2.3.2 *The functor* $\widetilde{F}$ *restricts to an equivalence of categories* **Operad** $\longrightarrow$ **sPRO**. $\square$

Free multicategories

Free structures are the formal origin of much of the geometry in this subject. A basic case is that free multicategories are made out of trees, as now explained.

A **multigraph** is a set X_0 together with a set $X(a_1, \ldots, a_n; a)$ for each $n \in \mathbb{N}$ and $a_1, \ldots, a_n, a \in X_0$. Forgetting composition and identities gives a functor $U : \mathbf{Multicat} \longrightarrow \mathbf{Multigraph}$. This has a left adjoint F, the free multicategory functor, which can be described as follows. Let X be a multigraph. The free multicategory FX on X has the same objects: $(FX)_0 = X_0$. Its arrows are formal gluings of arrows of X, that is, the hom-sets of FX are generated recursively by the clauses

- if $a \in X_0$ then $1_a \in (FX)(a; a)$
- if $\xi \in X(a_1, \ldots, a_n; a)$ and

$$\theta_1 \in (FX)(a_1^1, \ldots, a_1^{k_1}; a_1), \quad \ldots, \quad \theta_n \in (FX)(a_n^1, \ldots, a_n^{k_n}; a_n)$$

then $\xi \circ (\theta_1, \ldots, \theta_n) \in (FX)(a_1^1, \ldots, a_n^{k_n}; a)$.

Here 1_a and $\xi \circ (\theta_1, \ldots, \theta_n)$ are just formal expressions, but also make it clear how identities and composition in FX are to be defined. A typical arrow in FX is

$$\xi_1 \circ (\xi_2 \circ (1_{a_3}, 1_{a_4}), \; \xi_3 \circ (\xi_4 \circ (), \xi_5 \circ (1_{a_8}, 1_{a_9}, 1_{a_{10}})), \; 1_{a_{11}})$$

where

$$\xi_1 \in X(a_2, a_5, a_{11}; a_1), \qquad \xi_2 \in X(a_3, a_4; a_2),$$

and so on, naturally drawn as in Fig. 2-A (p. 59) with ξ_is in place of θ_is. The multigraph X is embedded in FX by sending $\xi \in X(a_1, \ldots, a_n; a)$ to

$$\xi \circ (1_{a_1}, \ldots, 1_{a_n}) \in (FX)(a_1, \ldots, a_n; a).$$

Example 2.3.3 The **operad tr of trees** is defined as $F1$, the free multicategory on the terminal multigraph. Explicitly, the sets $\mathbf{tr}(n)$ $(n \in \mathbb{N})$ are generated recursively by

- $\mathbf{tr}(1)$ has an element $|$
- if $n, k_1, \ldots, k_n \in \mathbb{N}$ and $\tau_1 \in \mathbf{tr}(k_1), \ldots, \tau_n \in \mathbf{tr}(k_n)$, then $\mathbf{tr}(k_1 + \cdots + k_n)$ has an element $(\tau_1, \ldots, \tau_n)$.

Here $|$ is a formal symbol and $(\tau_1, \ldots, \tau_n)$ a formal n-tuple. The elements of $\mathbf{tr}(n)$ are called n-**leafed trees**, and drawn as diagrams with n edges coming into the top and one edge (the **root**) emerging from the bottom, in the following way:

- | ∈ **tr**(1) is drawn as |
- if $\tau_1 \in$ **tr**$(k_1), \ldots, \tau_n \in$ **tr**(k_n), and if

represents the diagram of τ_i, then the tree $(\tau_1, \ldots, \tau_n)$ is drawn as

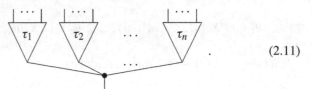

(2.11)

For example, **tr**(3) has an element $(\,|\,,\,|\,,\,|\,)$, drawn as

,

and **tr**(4) has an element $((\,|\,,\,|\,,\,|\,),\,|\,)$, drawn as

.

These diagrams are just like Fig. 2-A (p. 59), but rotated and unlabelled.

Some special cases can trap the unwary. For $n = 1$ and $n = 0$, diagram (2.11) looks like

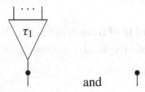

and

respectively. The tree on the right is an element of **tr**(0), that is, has 0 leaves; a leaf is an edge *without* a vertex at its upper end. Note in particular that the trees

...

are all different. Formally, the first is the element () of **tr**(0), and the rest are elements of **tr**(1), namely, $|$, $(|)$, $((|))$, ... Moral: the vertices matter.

The embedding $1 \longrightarrow F1$ picks out

$$v_n = (|, \ldots, |) = \quad\text{\Large Y}\quad \in \mathbf{tr}(n),$$

the n-**leafed corolla**, for each $n \in \mathbb{N}$. The operadic composition in **tr** is 'grafting' (gluing roots to leaves), with unit $| \in \mathbf{tr}(1)$.

As should be apparent, 'tree' is used to mean finite, rooted, planar tree. Nonplanar trees arise similarly from *symmetric* operads. We will examine planar trees in detail, including how they form a category, in Section 7.3.

Example 2.3.4 Let X be the multigraph with a single object $\star$ and in which $X(\star, \ldots, \star; \star)$ has one element if there are 0 or 2 copies of $\star$ to the left of the semicolon, and no elements otherwise. For reasons that will emerge in the next chapter, FX is called the **operad of classical trees, ctr**. It is the suboperad of **tr** containing just those trees in which each vertex has either 0 or 2 vertices coming up out of it. Explicitly, the sets $\mathbf{ctr}(n)$ ($n \in \mathbb{N}$) are generated recursively by

- **ctr**(1) has an element $|$
- **ctr**(0) has an element $\uparrow$
- if $k_1, k_2 \in \mathbb{N}$, $\tau_1 \in \mathbf{ctr}(k_1)$, and $\tau_2 \in \mathbf{ctr}(k_2)$, then $\mathbf{ctr}(k_1 + k_2)$ has an element (τ_1, τ_2).

Transformations

Mac Lane recounts that he and Eilenberg started category theory in order to enable them to talk about natural transformations. Transformations for multicategories will not be so important here but are still worth a look. The definition is suggested by the definition of a map between algebras (p. 65).

Definition 2.3.5 Let $C \overset{f}{\underset{f'}{\rightrightarrows}} D$ be a pair of maps between multicategories. A **transformation** $\alpha : f \longrightarrow f'$ is a family $\left(f(a) \xrightarrow{\alpha_a} f'(a) \right)_{a \in C_0}$ of unary maps in D, such that (Fig. 2-I)

$$\alpha_a \circ (f(\theta)) = f'(\theta) \circ (\alpha_{a_1}, \ldots, \alpha_{a_n})$$

for every map $a_1, \ldots, a_n \xrightarrow{\theta} a$ in C.

Transformations compose in the evident ways, making **Multicat** into a strict 2-category. In particular, there is a category $[C, D]$ for any multicategories C

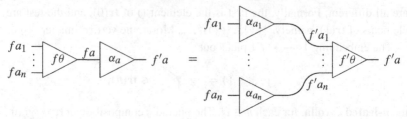

Fig. 2-I. Axiom for a transformation

and D, consisting of maps $C \longrightarrow D$ and transformations, and when $D = \mathbf{Set}$ this is $\mathbf{Alg}(C)$.

Actually, **Multicat** is not just a 2-category: in the language of Chapter 5, it is an **fc**-multicategory. One of the ingredients missing in the 2-category but present in the **fc**-multicategory is modules, which we consider next.

Modules

Recall that given categories C and D, a (D, C)-**module** (also called a bimodule, profunctor or distributor) is a functor $X : C^{\mathrm{op}} \times D \longrightarrow \mathbf{Set}$. We write $X : C \longrightarrow\!\!\!\!\!\!+\; D$. When C and D are monoids (one-object categories), X is a set with a left D-action and a compatible right C-action; in general, X is a family $(X(c, d))_{c \in C, d \in D}$ of sets 'acted on' by the arrows of C and D.

Definition 2.3.6 Let C and D be multicategories. A (D, C)-**module** X, written $X : C \longrightarrow\!\!\!\!\!\!+\; D$, consists of

- for each $a_1, \ldots, a_n \in C$ and $b \in D$, a set $X(a_1, \ldots, a_n; b)$ (Fig. 2-J(a))
- for each $a_i^j \in C$ and $b_i, b \in D$, a function

$$D(b_1, \ldots, b_n; b) \times X(a_1^1, \ldots, a_1^{k_1}; b_1) \times \cdots$$
$$\times X(a_n^1, \ldots, a_n^{k_n}; b_n) \longrightarrow X(a_1^1, \ldots, a_n^{k_n}; b),$$
$$(\phi, \xi_1, \ldots, \xi_n) \longmapsto \phi \cdot (\xi_1, \ldots, \xi_n)$$

- for each $a_i^j, a_i \in C$ and $b \in D$, a function

$$X(a_1, \ldots, a_n; b) \times C(a_1^1, \ldots, a_1^{k_1}; a_1) \times \cdots$$
$$\times C(a_n^1, \ldots, a_n^{k_n}; a_n) \longrightarrow X(a_1^1, \ldots, a_n^{k_n}; b),$$
$$(\xi, \theta_1, \ldots, \theta_n) \longmapsto \xi \cdot (\theta_1, \ldots, \theta_n),$$

satisfying the evident axioms for compatibility of the two actions with composition and identities in D and C, together with a further axiom stating

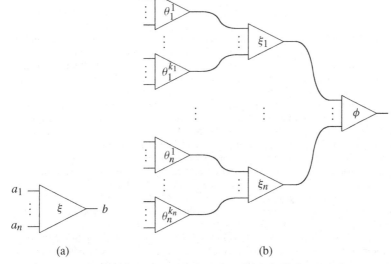

Fig. 2-J. (a) 'Element' of a module, (b) compatibility of left and right actions

compatibility with each other (Fig. 2-J(b)):

$$(\phi \cdot (\xi_1, \ldots, \xi_n)) \cdot (\theta_1^1, \ldots, \theta_n^{k_n})$$
$$= \phi \cdot (\xi_1 \cdot (\theta_1^1, \ldots, \theta_1^{k_1}), \ldots, \xi_n \cdot (\theta_n^1, \ldots, \theta_n^{k_n}))$$

whenever these expressions make sense.

When C and D have only unary arrows, this is the usual definition of module between categories. When C and D are operads, X is a sequence $(X(n))_{n \in \mathbb{N}}$ of sets with left D-action and right C-action. When $C = D$, taking $X(a_1, \ldots, a_n; b) = C(a_1, \ldots, a_n; b)$ gives a canonical module $C \longrightarrow C$. For any C and D, there is an obvious notion of **map between** (D, C)**-modules**, making (D, C)-modules into a category.

Just as for rings, it is fruitful to consider one-sided modules. Thus, when D is a multicategory, a **left D-module** is a family $(X(b))_{b \in D}$ of sets together with a left D-action – nothing other than a D-algebra. When C is a multicategory, a **right C-module** is a family $(X(a_1, \ldots, a_n))_{a_1, \ldots, a_n \in C}$ of sets together with a right C-action; these structures have sometimes been considered in the special case of operads (see Voronov, 1998, §1, for instance).

Notes

The story of operads and multicategories is a typical one in mathematics, strewn with failures of communication between specialists in different areas. Lazard seems to have been the first person to have published the basic idea, in work on formal group laws (1955); his 'analyseurs' are close to what were later dubbed operads. Lambek (1969) introduced multicategories, in the context of logic and linguistics. He says that Bénabou and Cartier had both considered multicategories previously; indeed, the idea might have occurred to anyone who knew what both a category and a multi-linear map were. Perhaps a year or two later (but effectively in a parallel universe), the special case of operads became very important in homotopy theory, with the work of Boardman and Vogt (1973) (who approached them via PROPs) and May (1972) (who gave operads their name). They in turn were building on the work of Stasheff (1963a), who defined the operad of associahedra without isolating the operad concept explicitly.

As far as I can tell, all three parties (Lazard, Lambek, and the homotopy theorists) were unaware of the work of the others until much later. Users of multicategories did not realize how interesting the one-object case was; users of operads were slow to see how simple and natural was the many-object case. It seems to have been more than twenty years before the appearance of the first paper containing both Lambek and Boardman–Vogt or May in its bibliography: Beilinson and Drinfeld (c.1997).

Activity in operads and multicategories surged in the mid-1990s with, among other things, the emergence of mirror symmetry and higher category theory. See, for example, Loday, Stasheff and Voronov (1997) and Baez and Dolan (1997). Some important structures that I have not included here are cyclic and modular operads (Getzler and Kapranov, 1995, 1994); cyclic and modular multicategories can be defined similarly.

Some authors on operads make a special case of nullary operations. May (1972, 1.1) insisted in the case of topological operads P that $P(0)$ should have only one element. Markl, Shnider and Stasheff (2002) do not have a $P(0)$ at all; they start at $P(1)$. They also consider pseudo-operads, which do not have an identity $1_P \in P(1)$: see p. 82 above.

I thank Paolo Salvatore for illumination on the various operads for which loop spaces are algebras.

Example 2.1.8 is probably a misnomer: the Swiss Cheese Board has recently been running an advertising campaign under the banner 'Si c'est troué, c'est pas Suisse'.

Chapter 3
Notions of monoidal category

In one sense all he ever wanted to be was someone with many nicknames.

Marcus (2002)

The concept of monoidal category is in such widespread use that one might expect – or hope, at least – that its formalization would be thoroughly understood. Nevertheless, it is not. Here we look at five different possible definitions, plus one infinite family of definitions, of monoidal category. We prove equivalence results between almost all of them.

Apart from the wish to understand a common mathematical structure, there is a reason for doing this motivated by higher-dimensional category theory. A monoidal category in the traditional sense is, as observed in Section 1.5, the same thing as a bicategory with only one object. Similarly, any proposed definition of weak n-category gives rise to a notion of monoidal category, defined as a one-object weak 2-category. So if we want to be able to compare the (many) proposed definitions of weak n-category then we will certainly need a firm grip on the various notions of monoidal category and how they are related. (This is something like a physicist's toy model: a manageably low-dimensional version of a higher-dimensional system.) Of course, if two definitions of weak n-category happen to induce equivalent definitions of monoidal category then this does not imply their equivalence in the general case, but the surprising variety of different notions of monoidal category means that it is a surprisingly good test.

In the classical definition of monoidal category, any pair (X_1, X_2) of objects has a specified tensor product $X_1 \otimes X_2$, and there is also a specified unit object. In Section 3.1 we consider a notion of monoidal category in which any sequence $X_1, \ldots, X_n$ of objects has a specified product; these are called 'unbiased monoidal categories'. More generally, one might have a category equipped with various different tensor products of various (finite) arities, but as long there are enough isomorphisms between derived products, this should make essentially no difference. We formalize and prove this in Section 3.2.

The definitions described so far are 'algebraic' in that tensor is an operation. For instance, in the classical definition, two objects X_1 and X_2 give rise to an actual, specified, object, $X_1 \otimes X_2$, not just an isomorphism class of objects. This goes somewhat against intuition: when using products of sets, for instance, it is not of the slightest importance to remember the standard set-theoretic definition of ordered pair $((x_1, x_2) = \{\{x_1\}, \{x_1, x_2\}\}$, as it happens); all that matters is the universal property of the product. So perhaps it is better to use a notion of monoidal category in which the tensor product of objects is only defined up to canonical isomorphism. Three such 'non-algebraic' notions are considered in Section 3.3, one using multicategories and the other two (closely related) using simplicial objects.

Almost everything we do for monoidal categories could equally be done for bicategories, as discussed in Section 3.4. The extension is mostly routine – there are few new ideas – but the explanations are a little easier in the special case of monoidal categories.

Some generality is sacrificed in the theorems asserting equivalence of different notions of monoidal category: these are equivalences of the categories of monoidal categories, but we could have made them equivalences of 2-categories (in other words, included monoidal transformations). I have not aimed to exhaust the subject, or the reader.

3.1 Unbiased monoidal categories

A classical monoidal category is a category A equipped with a functor $A^n \longrightarrow A$ for each of $n = 2$ and $n = 0$ (the latter being the unit object I), together with coherence data (Definition 1.2.5). An unbiased monoidal category is the same, but with n allowed to be any natural number. So in an unbiased monoidal category any four objects a_1, a_2, a_3, a_4 have a specified tensor product

$$(a_1 \otimes a_2 \otimes a_3 \otimes a_4),$$

but in a classical monoidal category there are only derived products such as

$$(a_1 \otimes a_2) \otimes (a_3 \otimes a_4), ((a_1 \otimes a_2) \otimes a_3) \otimes a_4,$$
$$a_1 \otimes ((I \otimes (a_2 \otimes I)) \otimes (a_3 \otimes a_4)).$$

The classical definition is 'biased' towards arities 2 and 0: it gives them a special status. Here we eliminate the bias.

The coherence data for an unbiased monoidal category consists of isomorphisms such as

$$((a_1 \otimes a_2 \otimes a_3) \otimes (a_4 \otimes (a_5)) \otimes a_6)$$
$$\xrightarrow{\sim} (a_1 \otimes (a_2 \otimes a_3 \otimes a_4 \otimes a_5) \otimes a_6).$$

In fact, all such isomorphisms can be built up from two families of special cases (the γ and ι of 3.1.1 below). We require, of course, that the coherence isomorphisms satisfy all sensible axioms.

We will show in the next section that the unbiased and classical definitions are equivalent, in a strong sense. The unbiased definition seems much the more natural (having, for instance, no devious coherence axioms), and much more useful for the purposes of theory. When verifying that a particular example of a category has a monoidal structure, it is sometimes easier to use one definition, sometimes the other; but the equivalence result means that we can take our pick.

Here is the definition of unbiased monoidal category. It is no extra work to define at the same time 'lax monoidal categories', in which the coherence maps need not be invertible.

Definition 3.1.1 A **lax monoidal category** A (or properly, $(A, \otimes, \gamma, \iota)$) consists of

- a category A
- for each $n \in \mathbb{N}$, a functor $\otimes_n : A^n \longrightarrow A$, called n-**fold tensor** and written

$$(a_1, \ldots, a_n) \longmapsto (a_1 \otimes \cdots \otimes a_n)$$

- for each $n, k_1, \ldots, k_n \in \mathbb{N}$ and double sequence $((a_1^1, \ldots, a_1^{k_1}), \ldots, (a_n^1, \ldots, a_n^{k_n}))$ of objects of A, a map

$$\gamma_{((a_1^1, \ldots, a_1^{k_1}), \ldots, (a_n^1, \ldots, a_n^{k_n}))} : ((a_1^1 \otimes \cdots \otimes a_1^{k_1}) \otimes \cdots \otimes (a_n^1 \otimes \cdots \otimes a_n^{k_n}))$$
$$\longrightarrow (a_1^1 \otimes \cdots \otimes a_1^{k_1} \otimes \cdots \otimes a_n^1 \otimes \cdots \otimes a_n^{k_n})$$

- for each object a of A, a map

$$\iota_a : a \longrightarrow (a),$$

with the following properties:

- $\gamma_{((a_1^1, \ldots, a_1^{k_1}), \ldots, (a_n^1, \ldots, a_n^{k_n}))}$ is natural in each of the a_i^js, and ι_a is natural in a
- associativity: for any $n, m_p, k_p^q \in \mathbb{N}$ and triple sequence

$(((a_{p,q,r})_{r=1}^{k_p^q})_{q=1}^{m_p})_{p=1}^n$ of objects, the diagram

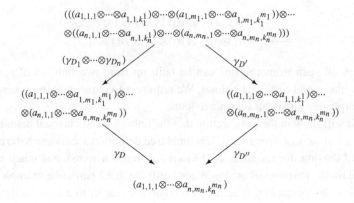

commutes, where the double sequences D_p, D, D', D'' are

$$D_p = ((a_{p,1,1}, \ldots, a_{p,1,k_p^1}), \ldots, (a_{p,m_p,1}, \ldots, a_{p,m_p,k_p^{m_p}})),$$
$$D = ((a_{1,1,1}, \ldots, a_{1,m_1,k_1^{m_1}}), \ldots, (a_{n,1,1}, \ldots, a_{n,m_n,k_n^{mn}})),$$
$$D' = (((a_{1,1,1} \otimes \cdots \otimes a_{1,1,k_1^1}), \ldots, (a_{1,m_1,1} \otimes \cdots \otimes a_{1,m_1,k_1^{m_1}})), \ldots,$$
$$((a_{n,1,1} \otimes \cdots \otimes a_{n,1,k_n^1}), \ldots, (a_{n,m_n,1} \otimes \cdots \otimes a_{n,m_n,k_n^{mn}}))),$$
$$D'' = ((a_{1,1,1}, \ldots, a_{1,1,k_1^1}), \ldots, (a_{n,m_n,1}, \ldots, a_{n,m_n,k_n^{mn}}))$$

* identity: for any $n \in \mathbb{N}$ and sequence $(a_1, \ldots, a_n)$ of objects, the diagrams

$$(a_1 \otimes \cdots \otimes a_n) \xrightarrow{(\iota_{a_1} \otimes \cdots \otimes \iota_{a_n})} ((a_1) \otimes \cdots \otimes (a_n))$$

with 1 and $\gamma_{((a_1), \ldots, (a_n))}$ to $(a_1 \otimes \cdots \otimes a_n)$

$$((a_1 \otimes \cdots \otimes a_n)) \xleftarrow{\iota_{(a_1 \otimes \cdots \otimes a_n)}} (a_1 \otimes \cdots \otimes a_n)$$

with $\gamma_{((a_1), \ldots, a_n))}$ and 1 to $(a_1 \otimes \cdots \otimes a_n)$

commute.

An **unbiased monoidal category** is a lax monoidal category $(A, \otimes, \gamma, \iota)$ in which each $\gamma_{((a_1^1, \ldots, a_1^{k_1}), \ldots, (a_n^1, \ldots, a_n^{k_n}))}$ and each ι_a is an isomorphism. An **unbiased strict monoidal category** is a lax monoidal category $(A, \otimes, \gamma, \iota)$ in which each of the γs and ιs is an identity map.

Remarks 3.1.2

a. The bark of the associativity axiom is far worse than its bite. All it says
   is that any two ways of removing brackets are equal: for instance, that the
   diagram

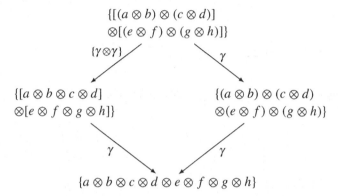

$$\{[(a \otimes b) \otimes (c \otimes d)]$$
$$\otimes [(e \otimes f) \otimes (g \otimes h)]\}$$

$$\{[a \otimes b \otimes c \otimes d]$$
$$\otimes [e \otimes f \otimes g \otimes h]\}$$

$$\{(a \otimes b) \otimes (c \otimes d)$$
$$\otimes (e \otimes f) \otimes (g \otimes h)\}$$

$$\{a \otimes b \otimes c \otimes d \otimes e \otimes f \otimes g \otimes h\}$$

   commutes. This is exactly the role of the associativity axiom for a monad
   such as 'free semigroup' on **Set**, as observed in Mac Lane (1971, VI.4, after
   Proposition 1).
b. The coherence axioms for an unbiased monoidal category are 'canonical'
   and rather obvious, in contrast to those for classical monoidal categories;
   they are the same shape as the diagrams expressing the associativity and
   unit axioms for a monoid or monad (1.1.3).
c. In an unbiased *strict* monoidal category, the coherence axioms (naturality, associativity and identity) hold automatically. Clearly, unbiased strict
   monoidal categories are in one-to-one correspondence with ordinary strict
   monoidal categories.

We have given a completely explicit definition of unbiased monoidal category, but a more abstract version is possible. First recall that if $\mathcal{C}$ is a strict
2-category then there is a notion of a **strict 2-monad** (T, μ, η) on $\mathcal{C}$, and there
are notions of **strict**, **weak** and **lax algebra** for such a 2-monad. (See Kelly
and Street, 1974, for instance; terminology varies between authors.) In particular, 'free strict monoidal category' is a strict 2-monad $((\ )^*, \mu, \eta)$ on **Cat**; the
functor $(\ )^*$ is given on objects of **Cat** by

$$A^* = \coprod_{n \in \mathbb{N}} A^n.$$

A (small) unbiased monoidal category is precisely a weak algebra for this
2-monad. Explicitly, this says that an unbiased monoidal category consists

of a category A together with a functor $\otimes : A^* \longrightarrow A$ and natural isomorphisms

$$
\begin{array}{ccc}
A^{**} & \xrightarrow{\mu_A} & A^* \\
\scriptstyle{\otimes^*}\downarrow & \nearrow\gamma & \downarrow\scriptstyle{\otimes} \\
A^* & \xrightarrow{\otimes} & A
\end{array}
\qquad
\begin{array}{ccc}
A & \xrightarrow{\eta_A} & A^* \\
& \nearrow\iota & \downarrow\scriptstyle{\otimes} \\
& \searrow^{1} & \\
& & A
\end{array}
\qquad (3.1)
$$

satisfying associativity and identity axioms: the diagrams

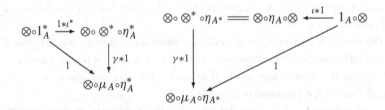

$$
\begin{array}{ccc}
\otimes\circ\otimes^*\circ\otimes^{**} & \xrightarrow{\gamma*1} & \otimes\circ\mu_A\circ\otimes^{**} == \otimes\circ\otimes^*\circ\mu_{A^*} \\
\scriptstyle{1*\gamma^*}\downarrow & & \downarrow\scriptstyle{\gamma*1} \\
\otimes\circ\otimes^*\circ\mu_A^* & \xrightarrow[\gamma*1]{} & \otimes\circ\mu_A\circ\mu_A^* == \otimes\circ\mu_A\circ\mu_{A^*}
\end{array}
$$

$$
\begin{array}{ccc}
\otimes\circ1_A^* & \xrightarrow{1*\iota^*} & \otimes\circ\otimes^*\circ\eta_A^* \\
& \searrow^{1} & \downarrow\scriptstyle{\gamma*1} \\
& & \otimes\circ\mu_A\circ\eta_A^*
\end{array}
\qquad
\begin{array}{ccc}
\otimes\circ\otimes^*\circ\eta_{A^*} == \otimes\circ\eta_A\circ\otimes \xleftarrow{\iota*1} 1_A\circ\otimes \\
\scriptstyle{\gamma*1}\downarrow \qquad\qquad \swarrow^{1} \\
\otimes\circ\mu_A\circ\eta_{A^*}
\end{array}
$$

commute. This may easily be verified. Similarly, a lax monoidal category is precisely a lax algebra for the 2-monad (for 'lax' means that the natural transformations γ and ι are no longer required to be isomorphisms) and an unbiased strict monoidal category is precisely a strict algebra (for 'strict' means that γ and ι are required to be identities, so that the diagrams (3.1) containing them commute).

A different abstract way of defining unbiased monoidal category will be explored in Section 3.2.

The next step is to define maps between (lax and) unbiased monoidal categories. Again, we could use the language of 2-monads to do this, but opt instead for an explicit definition.

Definition 3.1.3 Let A and A' be lax monoidal categories. Write $\otimes$ for the tensor and γ and ι for the coherence maps in both categories. A **lax monoidal functor** $(P, \pi) : A \longrightarrow A'$ consists of

- a functor $P : A \longrightarrow A'$
- for each $n \in \mathbb{N}$ and sequence $a_1, \ldots, a_n$ of objects of A, a map

$$\pi_{a_1,\ldots,a_n} : (Pa_1 \otimes \cdots \otimes Pa_n) \longrightarrow P(a_1 \otimes \cdots \otimes a_n),$$

such that

- $\pi_{a_1,\ldots,a_n}$ is natural in each a_i
- for each $n, k_i \in \mathbb{N}$ and double sequence $((a_1^1, \ldots, a_1^{k_1}), \ldots, (a_n^1, \ldots, a_n^{k_n}))$
  of objects of A, the diagram

commutes

- for each 1-cell a, the diagram

commutes.

A **weak monoidal functor** is a lax monoidal functor (P, π) for which each of the maps $\pi_{a_1,\ldots,a_n}$ is an isomorphism. A **strict monoidal functor** is a lax monoidal functor (P, π) for which each of the maps $\pi_{a_1,\ldots,a_n}$ is an identity map (in which case P preserves composites and identities strictly).

We remarked in 3.1.2(b) that the coherence axioms for an unbiased monoidal category were rather obvious, having the shape of the axioms for a monoid or monad. Perhaps the coherence axioms for an unbiased lax monoidal functor are a little less obvious; they are, however, the same shape as the axioms for a lax map of monads (= monad functor, p. 183), and in any case seem 'canonical'.

Lax monoidal functors can be composed in the evident way, and the weak and strict versions are closed under this composition. There are also the evident identities. So we obtain $3 \times 3 = 9$ possible categories, choosing one of 'strict', 'weak' or 'lax' for both the objects and the maps. In notation explained in a moment, the inclusions of subcategories are as follows:

$$\textbf{LaxMonCat}_{\text{str}} \subseteq \textbf{LaxMonCat}_{\text{wk}} \subseteq \textbf{LaxMonCat}_{\text{lax}}$$
$$\text{UI} \qquad\qquad\qquad \text{UI} \qquad\qquad\qquad \text{UI}$$
$$\textbf{UMonCat}_{\text{str}} \subseteq \textbf{UMonCat}_{\text{wk}} \subseteq \textbf{UMonCat}_{\text{lax}}$$
$$\text{UI} \qquad\qquad\qquad \text{UI} \qquad\qquad\qquad \text{UI}$$
$$\textbf{StrMonCat}_{\text{str}} \subseteq \textbf{StrMonCat}_{\text{wk}} \subseteq \textbf{StrMonCat}_{\text{lax}}.$$

For all three categories in the bottom (respectively, middle or top) row, the objects are small strict (respectively, unbiased or lax) monoidal categories. For all three categories in the left-hand (respectively, middle or right-hand) column, the maps are strict (respectively, weak or lax) monoidal functors. It is easy to check that the three categories in the bottom row are isomorphic to the corresponding three categories in the classical definition.

Of the nine categories, the three on the bottom-left to top-right diagonal are the most conceptually natural: a level of strictness has been chosen and stuck to. In this chapter our focus is on the middle entry, **UMonCat**$_{\text{wk}}$, where everything is weak.

Note that this 3×3 picture does not appear in the classical, 'biased', approach to monoidal categories. There the top row is obscured, as there is no very satisfactory way to laxify the classical definition of monoidal category. Admittedly it is possible to drop the condition that the associativity maps $(a \otimes b) \otimes c \longrightarrow a \otimes (b \otimes c)$ and unit maps $a \otimes 1 \longrightarrow a \longleftarrow 1 \otimes a$ are isomorphisms (as Borceux does in his (1994a), just after Definition 7.7.1), but somehow this does not seem quite right.

To complete the picture, and to make possible the definition of equivalence of unbiased monoidal categories, we define transformations.

Definition 3.1.4 Let $(P, \pi), (Q, \chi) : A \longrightarrow A'$ be lax monoidal functors between lax monoidal categories. A **monoidal transformation** $\sigma : (P, \pi) \longrightarrow (Q, \chi)$ is a natural transformation such that for all $a_1, \ldots, a_n \in A$, the diagram

$$
\begin{array}{ccc}
(Pa_1 \otimes \cdots \otimes Pa_n) & \xrightarrow{\pi_{a_1,\ldots,a_n}} & P(a_1 \otimes \cdots \otimes a_n) \\
{\scriptstyle (\sigma_{a_1} \otimes \cdots \otimes \sigma_{a_n})} \downarrow & & \downarrow {\scriptstyle \sigma_{(a_1 \otimes \cdots \otimes a_n)}} \\
(Qa_1 \otimes \cdots \otimes Qa_n) & \xrightarrow[\chi_{a_1,\ldots,a_n}]{} & Q(a_1 \otimes \cdots \otimes a_n)
\end{array}
$$

commutes.

(This time, there is only one possible level of strictness.)

Monoidal transformations can be composed in the expected ways, so that the nine categories above become strict 2-categories. In particular, **UMonCat**$_{wk}$ is a 2-category, so (1.5.7) there is a notion of equivalence of unbiased monoidal categories. Explicitly, A and A' are **equivalent** if there exist weak monoidal functors and invertible monoidal transformations

$$
A \overset{(P,\pi)}{\underset{(Q,\chi)}{\rightleftarrows}} A', \qquad A \underset{(Q,\chi)\circ(P,\pi)}{\overset{1}{\Downarrow\eta}} A, \qquad A' \underset{1}{\overset{(P,\pi)\circ(Q,\chi)}{\Downarrow\varepsilon}} A',
$$

and it makes no difference if we insist that $((P, \pi), (Q, \chi), \eta, \varepsilon)$ forms an adjunction in **UMonCat**$_{wk}$. As we might expect from the case of classical monoidal categories (1.2.14), there is the following alternative formulation.

Proposition 3.1.5 *Let A and A' be unbiased monoidal categories. Then A and A' are equivalent if and only if there exists a weak monoidal functor $(P, \pi) : A \longrightarrow A'$ whose underlying functor P is full, faithful and essentially surjective on objects.*

Proof *Mutatis mutandis*, this is the same as the proof of 1.2.14. □

We can now state and prove a coherence theorem for unbiased monoidal categories.

Theorem 3.1.6 *Every unbiased monoidal category is equivalent to a strict monoidal category.*

Proof Let A be an unbiased monoidal category. We construct an (unbiased) strict monoidal category **st**(A), the **strict cover** of A, and a weak monoidal functor $(P, \pi) : $ **st**$(A) \longrightarrow A$ whose underlying functor P is full, faithful and essentially surjective on objects. By the last proposition, this is enough.

An object of **st**(A) is a finite sequence $(a_1, \ldots, a_n)$ of objects of A (with $n \in \mathbb{N}$). A map $(a_1, \ldots, a_n) \longrightarrow (b_1, \ldots, b_m)$ in **st**(A) is a map $(a_1 \otimes \cdots$

$\otimes a_n) \longrightarrow (b_1 \otimes \cdots \otimes b_m)$ in A, and composition and identities in $\mathbf{st}(A)$ are as in A. The tensor in $\mathbf{st}(A)$ is given on objects by concatenation:

$$((a_1^1, \ldots, a_1^{k_1}) \otimes \cdots \otimes (a_n^1, \ldots, a_n^{k_n})) = (a_1^1, \ldots, a_1^{k_1}, \ldots, a_n^1, \ldots, a_n^{k_n}).$$

To define the tensor of maps in $\mathbf{st}(A)$, take maps

$$(a_1^1 \otimes \cdots \otimes a_1^{k_1}) \xrightarrow{\ f_1\ } (b_1^1 \otimes \cdots \otimes b_1^{l_1})$$
$$\vdots \qquad \vdots \qquad \vdots$$
$$(a_n^1 \otimes \cdots \otimes a_n^{k_n}) \xrightarrow{\ f_n\ } (b_n^1 \otimes \cdots \otimes b_n^{l_n})$$

in A; then their tensor product in $\mathbf{st}(A)$ is the composite map

$$(a_1^1 \otimes \cdots \otimes a_n^{k_n}) \xrightarrow{\ \gamma^{-1}\ } ((a_1^1 \otimes \cdots \otimes a_1^{k_1}) \otimes \cdots \otimes (a_n^1 \otimes \cdots \otimes a_n^{k_n}))$$
$$\xrightarrow{(f_1 \otimes \cdots \otimes f_n)} ((b_1^1 \otimes \cdots \otimes b_1^{l_1}) \otimes \cdots \otimes (b_n^1 \otimes \cdots \otimes b_n^{l_n}))$$
$$\xrightarrow{\ \gamma\ } (b_1^1 \otimes \cdots \otimes b_n^{l_n})$$

in A. It is absolutely straightforward and not too arduous to check that $\mathbf{st}(A)$ with this tensor forms a strict monoidal category.

The functor $P : \mathbf{st}(A) \longrightarrow A$ is defined by $(a_1, \ldots, a_n) \longmapsto (a_1 \otimes \cdots \otimes a_n)$ on objects, and 'is the identity' on maps – in other words, performs the identification

$$\mathbf{st}(A)((a_1, \ldots, a_n), (b_1, \ldots, b_m))$$
$$= A((a_1 \otimes \cdots \otimes a_n), (b_1 \otimes \cdots \otimes b_m)).$$

For each double sequence

$$D = ((a_1^1, \ldots, a_1^{k_1}), \ldots, (a_n^1, \ldots, a_n^{k_n}))$$

of objects of A, the isomorphism

$$\pi_D : (P(a_1^1, \ldots, a_1^{k_1}) \otimes \cdots \otimes P(a_n^1, \ldots, a_n^{k_n}))$$
$$\xrightarrow{\ \sim\ } P((a_1^1 \otimes \cdots \otimes a_1^{k_1}) \otimes \cdots \otimes (a_n^1 \otimes \cdots \otimes a_n^{k_n}))$$

is simply γ_D. It is also quick and straightforward to check that (P, π) is a weak monoidal functor.

P is certainly full and faithful. Moreover, for each $a \in A$ we have an isomorphism

$$\iota_a : a \xrightarrow{\ \sim\ } (a) = Pa,$$

and this proves that P is essentially surjective on objects. $\qquad\qquad\square$

As discussed on p. 36, coherence theorems take various forms, usually falling under one of two headings: 'all diagrams commute' or 'every weak thing is equivalent to a strict thing'. The one here is of the latter type. In Section 3.2 we will prove a coherence theorem for unbiased monoidal categories that is essentially of the former type.

The proof above was adapted from Joyal and Street (1993, p. 29). There the **st** construction was done for *classical* monoidal categories A, for which the situation is totally different. What happens is that the n-fold tensor product used in the definition of both $\mathbf{st}(A)$ and P must be replaced by some derived, non-canonical, n-fold tensor product, such as

$$(a_1, \ldots, a_n) \longmapsto a_1 \otimes (a_2 \otimes (a_3 \otimes \cdots \otimes (a_{n-1} \otimes a_n) \cdots));$$

and then in order to define both π and the tensor product of maps in $\mathbf{st}(A)$, it is necessary to use coherence isomorphisms such as

$$(a_1 \otimes (a_2 \otimes (a_3 \otimes a_4))) \otimes (a_5 \otimes (a_6 \otimes a_7))$$
$$\xrightarrow{\;\sim\;} a_1 \otimes (a_2 \otimes (a_3 \otimes (a_4 \otimes (a_5 \otimes (a_6 \otimes a_7))))).$$

It would be folly to attempt to define these coherence isomorphisms, and prove that $\mathbf{st}(A)$ and (P, π) have the requisite properties, without the aid of a coherence theorem for monoidal categories. That is, the work required to do this would be of about the same volume and kind as the work involved in the syntactic proof of the 'all diagrams commute' coherence theorem, so one might as well have proved that coherence theorem anyway. In contrast, the proof that every *unbiased* monoidal category is equivalent to a strict one is easy, short, and needs no supporting results.

This does not, however, provide a short cut to proving any kind of coherence result for classical monoidal categories. In the next section we will see that unbiased and classical monoidal categories are essentially the same, and it then follows from Theorem 3.1.6 that every classical monoidal category is equivalent to a strict one. However, as with any serious undertaking involving classical monoidal categories, the proof that they are the same as unbiased ones is close to impossible without the use of a coherence theorem.

3.2 Algebraic notions of monoidal category

We have already seen two notions of monoidal category: classical and unbiased. In one of them there was an n-fold tensor product just for $n \in \{0, 2\}$, and in the other there was an n-fold tensor product for all $n \in \mathbb{N}$. But what happens if we take a notion of monoidal category in which there is an n-fold tensor product for each n lying in some other subset of $\mathbb{N}$? More generally, what if we allow any number of different n-fold tensors (zero, one, or more) for each value of $n \in \mathbb{N}$?

For instance, we might choose to take a notion of monoidal category in which there are six unit objects, a single 3-fold tensor product, eight 11-fold tensor products, and $\aleph_4$ 38-fold tensor products. Just as long as we add in enough coherence isomorphisms to ensure that any two n-fold tensor products built up from the given ones are canonically isomorphic, this new notion of monoidal category ought to be essentially the same as the classical notion.

This turns out to be the case. Formally, we start with a 'signature' $\Sigma \in \mathbf{Set}^{\mathbb{N}}$. (In the example this was given by $\Sigma(0) = 6$, $\Sigma(1) = \Sigma(2) = 0$, $\Sigma(3) = 1$, and so on.) From this we define the category $\Sigma\text{-}\mathbf{MonCat}_{wk}$ of 'Σ-monoidal categories' and weak monoidal functors between them. We then show that, up to equivalence, $\Sigma\text{-}\mathbf{MonCat}_{wk}$ is independent of the choice of Σ, assuming only that Σ is large enough that we can build at least one n-fold tensor product for each $n \in \mathbb{N}$. We also show that the category $\mathbf{MonCat}_{wk}$ of classical monoidal categories is isomorphic to $\Sigma\text{-}\mathbf{MonCat}_{wk}$ for a certain value of Σ, and that the same goes for the unbiased version $\mathbf{UMonCat}_{wk}$ (for a different value of Σ); it follows that the classical and unbiased definitions are equivalent.

In the introduction to this chapter, I argued that comparing definitions of monoidal category is important for understanding higher-dimensional category theory. Here is a further reason why it is important, specific to this particular, 'algebraic', family of definitions of monoidal category.

Consider the definition of monoid. Usually a monoid is defined as a set equipped with a binary operation and a nullary operation, satisfying associativity and unit equations. 'Weakening' or 'categorifying' this definition, we obtain the classical definition of (weak) monoidal category. However, this process of categorification is dependent on presentation. That is, we could equally well have defined a monoid as a set equipped with one n-ary operation for each $n \in \mathbb{N}$, satisfying appropriate equations, and categorifying *this* gives a different notion of monoidal category – the unbiased one. So different presentations of the same 0-dimensional theory (monoids) give, under this process of categorification, different 1-dimensional theories (of monoidal category).

Thus, our purpose is to show that in this particular situation, the presentation-sensitivity of categorification disappears when we work up to equivalence.

More generally, a fully-developed theory of weak n-categories might include a formal process of weakening, which would take as input a theory of strict structures and give as output a theory of weak structures. If the weakening process depended on how the theory of strict structures was presented, then we would have to ask whether different presentations always gave equivalent theories of weak structures. We might hope so; but who knows?

Our first task is to define Σ-monoidal categories and maps between them, for an arbitrary $\Sigma \in \mathbf{Set}^{\mathbb{N}}$. Here it is in outline. A Σ-monoidal category should be a category A equipped with a tensor product $\otimes_\sigma : A^n \longrightarrow A$ for each $\sigma \in \Sigma(n)$. There should also be a coherence isomorphism between any pair of derived n-fold tensors: for instance, there should be a specified isomorphism

$$\otimes_{\sigma_1}(\otimes_{\sigma_2}(a_1, a_2, a_3), a_4, \otimes_{\sigma_3}(a_5)) \overset{\sim}{\longrightarrow} \otimes_{\sigma_4}(a_1, a_2, \otimes_{\sigma_5}(a_3, a_4), a_5)$$

for any

$$\sigma_1, \sigma_2 \in \Sigma(3), \sigma_3 \in \Sigma(1), \sigma_4 \in \Sigma(4), \sigma_5 \in \Sigma(2)$$

and any objects $a_1, \ldots, a_5$ of A. This isomorphism is naturally depicted as

$$(3.2)$$

Recall from Section 2.3 that labelled trees arise as operations in free operads: the two derived tensor products drawn above are elements of $(F\Sigma)(5)$, where $F\Sigma$ is the free operad on Σ. Hence $F\Sigma$ is the operad of (derived) tensor operations. To obtain the coherence isomorphisms, replace the set $(F\Sigma)(n)$ by the indiscrete category $I((F\Sigma)(n))$ whose objects are the elements of $(F\Sigma)(n)$; then the picture above shows a typical map in $I((F\Sigma)(5))$.

Also recall from p. 68 that there is a notion of '$\mathcal{V}$-operad' for any symmetric monoidal category $\mathcal{V}$, and an accompanying notion of an algebra (in $\mathcal{V}$) for any $\mathcal{V}$-operad. The categories $I((F\Sigma)(n))$ form a **Cat**-operad, an algebra for which consists of a category A, a functor $A^n \longrightarrow A$ for each derived n-fold tensor operation, and a natural isomorphism between any two functors $A^n \longrightarrow A$ so arising, all fitting together coherently. This is exactly what we want a Σ-monoidal category to be.

Precisely, define the functor

$$(\ )\text{-}\mathbf{MonCat}_{\mathrm{wk}} : \quad \mathbf{Set}^{\mathbb{N}} \longrightarrow \mathbf{CAT}^{\mathrm{op}}$$
$$\Sigma \longmapsto \Sigma\text{-}\mathbf{MonCat}_{\mathrm{wk}}$$

to be the composite of the functors

$$\mathbf{Set}^{\mathbb{N}} \overset{F}{\longrightarrow} \mathbf{Set\text{-}Operad} \overset{I_*}{\longrightarrow} \mathbf{Cat\text{-}Operad} \overset{\mathbf{Alg}_{\mathrm{wk}}}{\longrightarrow} \mathbf{CAT}^{\mathrm{op}},$$

where the terms involved are now defined in turn.

- **Set-Operad** is the category of ordinary (**Set-**)operads, usually just called **Operad** (2.2.1).
- **Cat-Operad** is the category of **Cat-**operads (p. 68).
- F is the free operad functor (p. 85).
- $I : \mathbf{Set} \longrightarrow \mathbf{Cat}$ is the functor assigning to each set the indiscrete category on it (p. 24), and since I preserves products, it induces a functor $I_* : \mathbf{Set\text{-}Operad} \longrightarrow \mathbf{Cat\text{-}Operad}$.
- For a **Cat**-operad R, the (large) category $\mathbf{Alg}_{\mathrm{wk}}(R)$ consists of R-algebras and **weak maps** between them. So by definition, an object of $\mathbf{Alg}_{\mathrm{wk}}(R)$ is a category A together with a sequence $(R(n) \times A^n \xrightarrow{\mathrm{act}_n} A)_{n \in \mathbb{N}}$ of functors, compatible with the composition and identities of the operad R, and a map $A \longrightarrow A'$ in $\mathbf{Alg}_{\mathrm{wk}}(R)$ is a functor $P : A \longrightarrow A'$ together with a natural isomorphism

$$
\begin{array}{ccc}
R(n) \times A^n & \xrightarrow{\ \mathrm{act}_n\ } & A \\
{\scriptstyle 1 \times P^n}\big\downarrow & \quad \nearrow\!\!\nearrow \pi_n & \big\downarrow{\scriptstyle P} \\
R(n) \times A'^n & \xrightarrow[\ \mathrm{act}_n\ ]{} & A'
\end{array}
$$

for each $n \in \mathbb{N}$, satisfying the coherence axioms in Fig. 3-A.
- For a map $H : R \longrightarrow S$ of **Cat**-operads, the induced functor $H^* : \mathbf{Alg}_{\mathrm{wk}}(S) \longrightarrow \mathbf{Alg}_{\mathrm{wk}}(R)$ is defined by composition with H: if A is an S-algebra then the resulting R-algebra has the same underlying category and R-action given by

$$
R(n) \times A^n \xrightarrow{H_n \times 1} S(n) \times A^n \xrightarrow{\mathrm{act}_n} A.
$$

In the construction above we used *weak* maps between algebras for a **Cat**-operad, but we could just as well have used **lax maps** (by dropping the insistence that the natural transformations π_n are isomorphisms) or **strict maps** (by insisting that the π_ns are identities). The resulting functors **Cat-Operad** $\longrightarrow$ **CAT**^{op} are, of course, called $\mathbf{Alg}_{\mathrm{lax}}$ and $\mathbf{Alg}_{\mathrm{str}}$, and so we have three functors

$$
(\ )\text{-}\mathbf{MonCat}_{\mathrm{lax}}, \ (\ )\text{-}\mathbf{MonCat}_{\mathrm{wk}}, \ (\ )\text{-}\mathbf{MonCat}_{\mathrm{str}} : \mathbf{Set}^{\mathbb{N}} \longrightarrow \mathbf{CAT}^{\mathrm{op}}.
$$

Definition 3.2.1 Let $\Sigma \in \mathbf{Set}^{\mathbb{N}}$. A Σ-**monoidal category** is an object of Σ-$\mathbf{MonCat}_{\mathrm{lax}}$ (or equivalently, of Σ-$\mathbf{MonCat}_{\mathrm{wk}}$ or Σ-$\mathbf{MonCat}_{\mathrm{str}}$). A **lax** (respectively, **weak** or **strict**) **monoidal functor** between Σ-monoidal categories is a map in Σ-$\mathbf{MonCat}_{\mathrm{lax}}$ (respectively, Σ-$\mathbf{MonCat}_{\mathrm{wk}}$ or Σ-$\mathbf{MonCat}_{\mathrm{str}}$).

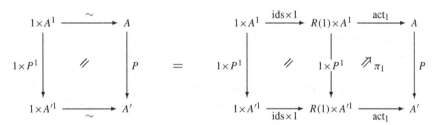

Fig. 3-A. Coherence axioms for a weak map of R-algebras

Now we can state the results. First, the notion of Σ-monoidal category really does generalize the notions of unbiased and classical monoidal category, as intended all along:

Theorem 3.2.2 (Coherence for unbiased monoidal categories and functors) *Writing* 1 *for the terminal object of* **Set**$^{\mathbb{N}}$, *there are isomorphisms of categories*

$$\mathbf{UMonCat}_{\mathrm{lax}} \cong 1\text{-}\mathbf{MonCat}_{\mathrm{lax}},$$
$$\mathbf{UMonCat}_{\mathrm{wk}} \cong 1\text{-}\mathbf{MonCat}_{\mathrm{wk}},$$
$$\mathbf{UMonCat}_{\mathrm{str}} \cong 1\text{-}\mathbf{MonCat}_{\mathrm{str}}.$$

Proof See Appendix B.1. It takes almost no calculation to see that there is a canonical functor 1-**MonCat**$_{\text{lax}}$ $\longrightarrow$ **UMonCat**$_{\text{lax}}$. To see that it is an isomorphism requires calculations using the coherence axioms for an unbiased monoidal category. Restricting to the weak and strict cases is simple.   □

Theorem 3.2.3 (Coherence for classical monoidal categories and functors)
Write Σ_c *for the object of* **Set**$^{\mathbb{N}}$ *given by* $\Sigma_c(n) = 1$ *for* $n \in \{0, 2\}$ *and* $\Sigma_c(n) = \emptyset$ *otherwise. Then there are isomorphisms of categories*

$$\textbf{MonCat}_{\text{lax}} \cong \Sigma_c\textbf{-MonCat}_{\text{lax}},$$

$$\textbf{MonCat}_{\text{wk}} \cong \Sigma_c\textbf{-MonCat}_{\text{wk}},$$

$$\textbf{MonCat}_{\text{str}} \cong \Sigma_c\textbf{-MonCat}_{\text{str}}.$$

Proof See Appendix B.2. The strategy is just the same as in 3.2.2, except that this time the calculations are in principle much more tricky (because of the irregularity of the data and axioms for a classical monoidal category) but in practice can be omitted (by relying on the coherence theorems of others).   □

It took a little work to define 'Σ-monoidal category', but I hope it will be agreed that it is a completely natural definition, free from *ad hoc* coherence axioms and correctly embodying the idea of a monoidal category with as many primitive tensor operations as are specified by Σ. So, for instance, the statement that the objects of **MonCat**$_{\text{wk}}$ correspond one-to-one with the objects of Σ_c-**MonCat**$_{\text{wk}}$ says that the coherence data and axioms in the classical definition of monoidal category are exactly right. Were there no such isomorphism, it would be the coherence data or axioms at fault, not the definition of Σ-monoidal category. This is why 3.2.2 and 3.2.3 are called coherence theorems.

The unbiased coherence theorem (3.2.2) also tells us that unbiased monoidal categories play a universal role: for any $\Sigma \in$ **Set**$^{\mathbb{N}}$, the unique map $\Sigma \longrightarrow 1$ induces a canonical map from unbiased monoidal categories to Σ-monoidal categories,

$$\textbf{UMonCat}_{\text{wk}} \cong 1\text{-}\textbf{MonCat}_{\text{wk}} \longrightarrow \Sigma\text{-}\textbf{MonCat}_{\text{wk}}.$$

Concretely, if we are given an unbiased monoidal category A then we can define a Σ-monoidal category by taking $\otimes_\sigma$ to be the n-fold tensor $\otimes_n$, for each $\sigma \in \Sigma(n)$.

Theorem 3.2.4 (Irrelevance of signature for monoidal categories) *For any plausible* $\Sigma, \Sigma' \in$ **Set**$^{\mathbb{N}}$, *there are equivalences of categories*

$$\Sigma\text{-}\textbf{MonCat}_{\text{lax}} \simeq \Sigma'\text{-}\textbf{MonCat}_{\text{lax}}, \qquad \Sigma\text{-}\textbf{MonCat}_{\text{wk}} \simeq \Sigma'\text{-}\textbf{MonCat}_{\text{wk}}.$$

Here $\Sigma \in \mathbf{Set}^{\mathbb{N}}$ is called **plausible** if $\Sigma(0) \neq \emptyset$ and for some $n \geq 2$, $\Sigma(n) \neq \emptyset$. This means that n-fold tensor products can be derived for all $n \in \mathbb{N}$. In contrast, if $\Sigma(0) = \emptyset$ then $(F\Sigma)(0) = \emptyset$, so $(IF\Sigma)(0) = \emptyset$, so in a Σ-monoidal category all the tensor operations are of arity $n \geq 1$ – there is no unit object. Dually, if $\Sigma(n) = \emptyset$ for all $n \geq 2$ then there is no derived binary tensor. In these cases we would not expect Σ-monoidal categories to be much like ordinary monoidal categories. So plausibility is an obvious minimal requirement.

Proof It is enough to prove the result in the case $\Sigma' = 1$. As can be seen from the explicit description of free operads on p. 85, plausibility of Σ says exactly that $(F\Sigma)(n) \neq \emptyset$ for each $n \in \mathbb{N}$, or equivalently that there exists a map $1 \longrightarrow UF\Sigma$ in $\mathbf{Set}^{\mathbb{N}}$, where $U : \mathbf{Set\text{-}Operad} \longrightarrow \mathbf{Set}^{\mathbb{N}}$ is the forgetful functor. By adjointness, this says that there is a map $F1 \longrightarrow F\Sigma$ of $\mathbf{Set}$-operads, giving a map $I_*F1 \longrightarrow I_*F\Sigma$ of $\mathbf{Cat}$-operads. On the other hand, 1 is the terminal object of $\mathbf{Set}^{\mathbb{N}}$, so we have maps $I_*F1 \rightleftarrows I_*F\Sigma$.

In brief, the rest of the proof runs as follows. **Cat-Operad** naturally has the structure of a 2-category, because **Cat** does; and if two objects of **Cat-Operad** are equivalent then so are their images under both $\mathbf{Alg}_{\mathrm{lax}}$ and $\mathbf{Alg}_{\mathrm{wk}}$. By the nature of indiscrete categories, the existence of maps $I_*F1 \rightleftarrows I_*F\Sigma$ implies that $I_*F1 \simeq I_*F\Sigma$. The result follows. However, the 2-categorical details are rather tiresome to check and the reader may prefer to avoid them. The main reason for including them below is that they reveal why the theorem holds at the lax and weak levels but not at the strict level.

So, for the conscientious, a **transformation**

of **Cat**-operads is a sequence

$$
\left(
\begin{array}{c}
 \\
R(n) \underset{H'_n}{\overset{H_n}{\rightleftarrows}} \Downarrow\!\alpha_n \; S(n)
\end{array}
\right)_{n \in \mathbb{N}}
$$

of natural transformations, such that

$$R(n) \times R(k_1) \times \cdots \times R(k_n) \xrightarrow{\ \text{comp}\ } R(k_1 + \cdots + k_n) \overset{H_{k_1 + \cdots + k_n}}{\underset{H'_{k_1 + \cdots + k_n}}{\Longrightarrow}} S(k_1 + \cdots + k_n)$$

$$= \ R(n) \times R(k_1) \times \cdots \times R(k_n) \overset{H_n \times H_{k_1} \times \cdots \times H_{k_n}}{\underset{H'_n \times H'_{k_1} \times \cdots \times H'_{k_n}}{\Longrightarrow}} S(n) \times S(k_1) \times \cdots \times S(k_n) \xrightarrow{\ \text{comp}\ } S(k_1 + \cdots + k_n),$$

where the unlabelled 2-cells are respectively $\alpha_{k_1 + \cdots + k_n}$ and $\alpha_n \times \alpha_{k_1} \times \cdots \times \alpha_{k_n}$, and

$$1 \xrightarrow{\ \text{ids}\ } R(1) \overset{H_1}{\underset{H'_1}{\Downarrow \alpha_1}} S(1) = 1 \overset{\text{ids}}{\underset{\text{ids}}{\Downarrow 1}} S(1).$$

With the evident compositions, **Cat-Operad** becomes a strict 2-category. The statement on indiscrete categories is easily proved; in fact, if R is any **Cat**-operad and Q any **Set**-operad then there is a unique transformation between any pair of maps $R \rightrightarrows I_* Q$.

The functor **Alg**$_{\text{lax}}$ becomes a strict map **Cat-Operad** $\longrightarrow$ **CAT**$^{\text{co op}}$ of 2-categories, where the codomain is **CAT** with both the 1-cells and the 2-cells reversed (Section 1.5). To see this, let α be a transformation of **Cat**-operads as above. We need to produce a natural transformation

$$\mathbf{Alg}_{\text{lax}}(R) \overset{H^*}{\underset{H'^*}{\Uparrow}} \mathbf{Alg}_{\text{lax}}(S),$$

which in turn means producing for each S-algebra A a lax map $H'^*(A) \longrightarrow H^*(A)$ of R-algebras. The composite natural transformation

$$R(n) \times A^n \overset{H_n}{\underset{H'_n}{\Downarrow \alpha_n}} S(n) \times A^n \xrightarrow{\ \text{act}^A_n\ } A$$

can be re-drawn as a natural transformation

$$
\begin{array}{ccc}
R(n) \times A^n & \xrightarrow{\ \operatorname{act}_n^{H'^*(A)}\ } & A \\
\downarrow{\scriptstyle 1} & \quad\nearrow & \downarrow{\scriptstyle 1} \\
R(n) \times A^n & \xrightarrow[\ \operatorname{act}_n^{H^*(A)}\ ]{} & A,
\end{array}
\tag{3.3}
$$

and this gives the desired lax map. (To make the necessary distinctions, superscripts have been added to each act_n naming the algebra concerned.) We thus obtain a strict map

$$\mathbf{Alg}_{\mathrm{lax}} : \mathbf{Cat\text{-}Operad} \longrightarrow \mathbf{CAT}^{\mathrm{co\,op}}$$

of 2-categories.

It follows immediately that if R and S are equivalent **Cat**-operads then $\mathbf{Alg}_{\mathrm{lax}}(R)$ and $\mathbf{Alg}_{\mathrm{lax}}(S)$ are equivalent categories. To see that $\mathbf{Alg}_{\mathrm{wk}}$ also preserves equivalence of objects, note that if α is an invertible transformation of **Cat**-operads then the natural transformation (3.3) is also invertible, and so defines a weak map of algebras. Put another way, $\mathbf{Alg}_{\mathrm{wk}}$ is a map from the 2-category (**Cat**-operads + maps + invertible transformations) into $\mathbf{CAT}^{\mathrm{co\,op}}$.

(However, (3.3) only induces a *strict* map $H'^*(A) \longrightarrow H^*(A)$ if it is the identity transformation, which means that if R and S are equivalent **Cat**-operads then $\mathbf{Alg}_{\mathrm{str}}(R)$ and $\mathbf{Alg}_{\mathrm{str}}(S)$ are not necessarily equivalent categories.) $\square$

Corollary 3.2.5 *There are equivalences of categories*

$$\mathbf{UMonCat}_{\mathrm{lax}} \simeq \mathbf{MonCat}_{\mathrm{lax}}, \qquad \mathbf{UMonCat}_{\mathrm{wk}} \simeq \mathbf{MonCat}_{\mathrm{wk}}.$$

Proof $3.2.2 + 3.2.3 + 3.2.4$. $\square$

We arrived at the equivalence of unbiased and classical monoidal categories by a roundabout route,

$$\mathbf{UMonCat}_{\mathrm{lax}} \cong 1\text{-}\mathbf{MonCat}_{\mathrm{lax}} \simeq \Sigma_{\mathrm{c}}\text{-}\mathbf{MonCat}_{\mathrm{lax}} \cong \mathbf{MonCat}_{\mathrm{lax}},$$

so it may be useful to consider a direct proof. Given an unbiased monoidal category $(A, \otimes, \gamma, \iota)$, we can canonically write down a classical monoidal category: the underlying category is A, the tensor is $\otimes_2$, the unit is $\otimes_0$, and the associativity and unit coherence isomorphisms are formed from certain components of γ and ι. The converse process is non-canonical: we have to choose

for each $n \in \mathbb{N}$ an n-fold tensor operation built up from binary tensor operations and the unit object. Put another way, we have to choose for each $n \in \mathbb{N}$ an
n-leafed classical tree (where 'classical' means that each vertex has either 2 or
0 outgoing edges, as in 2.3.4); and since $F\Sigma_c$ is exactly the operad of classical
trees, this corresponds to the step of the proof of 3.2.4 where we chose a map
$1 \longrightarrow UF\Sigma$ in $\mathbf{Set}^{\mathbb{N}}$.

All of the results above can be repeated with monoidal transformations
brought into the picture. Then Σ-monoidal categories form a strict 2-category,
and all the isomorphisms and equivalences of categories become isomorphisms
and equivalences of 2-categories. That we did not *need* to mention monoidal
transformations in order to prove the equivalence of the various notions of
monoidal category says something about the strength of our equivalence result. For suppose we start with a Σ-monoidal category, derive from it a Σ'-
monoidal category, and derive from that a second Σ-monoidal category. Then,
in our construction, the two Σ-monoidal categories are not just equivalent in
the 2-category Σ-$\mathbf{MonCat}_{wk}$, but isomorphic. So Theorem 3.2.4 is 'one level
better' than might be expected.

3.3 Non-algebraic notions of monoidal category

Coherence axioms have got a bad name for themselves: unmemorable, unen-
lightening, and unwieldy, they are often regarded as bureaucracy to be fought
through grimly before getting on to the real business. Some people would,
therefore, like to create a world where there are no coherence axioms at all –
or anyway, as few as possible.

Whatever the merits of this aspiration, it is a fact that it can be achieved
in some measure; that is, there exist approaches to various higher categorical
structures that involve almost no coherence axioms. In this chapter we look at
two different such approaches for monoidal categories. The first exploits the
relation between monoidal categories and multicategories. I will describe it in
some detail. The second is based on the idea of the nerve of a category, and has
its historical roots in the homotopy-algebraic structures known as Γ-spaces.
Since it has less to do with the main themes of this book, I will explain it more
sketchily.

So, we start by looking at monoidal categories vs. multicategories. One
might argue that multicategories are conceptually more primitive than monoidal categories: that an operation taking several inputs and producing one
output is a more basic idea than a set whose elements are ordered tuples. In
any case, the first example of a 'tensor product' that many of us learned was

implicitly introduced via multicategories – the tensor product $V \otimes W$ of two vector spaces being characterized as the codomain of a 'universal bilinear map' out of V, W. Now, the monoidal category of vector spaces contains no more or less information than the multicategory of vector spaces; given either one, the other can be derived in its entirety. So, with the thoughts in our head that multicategories are basic and coherence axioms are bad, we might hope to 'define' a monoidal category as a multicategory in which there are enough universal maps around. And this is what we do.

Formally, we have a functor V assigning to each monoidal category its underlying multicategory; we want to show that V gives an equivalence between (monoidal categories) and some subcategory $\mathcal{R}$ of **Multicat**; and we want, moreover, to describe $\mathcal{R}$. In the terms of the previous paragraph, $\mathcal{R}$ consists of those multicategories containing 'enough universal maps'.

Before starting we have to make precise something that has so far been left vague. In Example 2.1.3, we defined the underlying multicategory C of a monoidal category A to have the same objects as A and maps given by

$$C(a_1, \ldots, a_n; a) = A(a_1 \otimes \cdots \otimes a_n, a),$$

but we deferred the question of what exactly the expression $a_1 \otimes \cdots \otimes a_n$ meant. We answer it now, and so obtain a precise definition of the functor V.

If A is a strict monoidal category then the expression makes perfect sense. If A is not strict then it still makes perfect sense as long as we have chosen to use unbiased rather than classical monoidal categories: take $a_1 \otimes \cdots \otimes a_n$ to mean $\otimes_n(a_1, \ldots, a_n)$. Bringing into play the coherence maps of A, we can also define composition in C, and so obtain the entire multicategory structure of C without trouble; we arrive at a functor

$$V : \mathbf{UMonCat}_{\mathrm{wk}} \longrightarrow \mathbf{Multicat}.$$

What if we insist on starting from a classical monoidal category? We can certainly obtain a multicategory by passing first from classical to unbiased and then applying the functor V just mentioned. This passage amounts to a choice of an n-leafed classical tree $\tau_n \in \mathbf{ctr}(n)$ for each $n \in \mathbb{N}$ (see p. 109), and the resulting multicategory C has the same objects as A and maps given by

$$C(a_1, \ldots, a_n; a) = A(\otimes_{\tau_n}(a_1, \ldots, a_n), a)$$

where $\otimes_{\tau_n} : A^n \longrightarrow A$ is 'tensor according to the shape of τ_n' (defined formally in B.2). So, any such sequence $\tau_\bullet = (\tau_n)_{n \in \mathbb{N}}$ induces a functor

$$V_{\tau_\bullet} : \mathbf{MonCat}_{\mathrm{wk}} \longrightarrow \mathbf{Multicat}.$$

For instance, if we take one of the two most obvious choices of sequence $\tau_\bullet$ and write $C = V_{\tau_\bullet}(A)$ as usual, then

$$C(a_1, \ldots, a_n; a) = A(a_1 \otimes (a_2 \otimes (a_3 \otimes \cdots \otimes (a_{n-1} \otimes a_n) \cdots)), a).$$

But there is also a way of passing from classical monoidal categories to multi-categories without making any arbitrary choices. Let A be a classical monoidal category. For $n \in \mathbb{N}$ and $\tau, \tau' \in \mathbf{ctr}(n)$, let

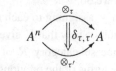

be the canonical isomorphism whose existence is asserted by the coherence theorem (Section 1.2). Now define a multicategory C by taking an object to be, as usual, just an object of A, and a map $a_1, \ldots, a_n \longrightarrow a$ to be a family $(f_\tau)_{\tau \in \mathbf{ctr}(n)}$ in which $f_\tau \in A(\otimes_\tau(a_1, \ldots, a_n), a)$ for each $\tau \in \mathbf{ctr}(n)$ and $f_{\tau'} \circ \delta_{\tau, \tau'} = f_\tau$ for all $\tau, \tau' \in \mathbf{ctr}(n)$. This yields another functor

$$V' : \mathbf{MonCat}_{\mathrm{wk}} \longrightarrow \mathbf{Multicat}.$$

However, a family $(f_\tau)_{\tau \in \mathbf{ctr}(n)}$ as above is entirely determined by any single f_τ, so the multicategory $C = V'(A)$ just constructed is isomorphic to the multi-category $V_{\tau_\bullet}(A)$ obtained by choosing a particular sequence $\tau_\bullet$ of trees. So $V' \cong V_{\tau_\bullet}$ for any $\tau_\bullet$, and henceforth we write V' or $V_{\tau_\bullet}$ as just V.

We now have ways of obtaining a multicategory from either an unbiased or a classical monoidal category, and it is a triviality to check that these are compatible with the equivalence between unbiased and classical: the diagram

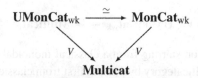

commutes up to canonical isomorphism. So we know unambiguously what it means for a multicategory to be 'the underlying multicategory of some monoi-dal category', and similarly for maps.

The results for which we hoped, exhibiting monoidal categories as special multicategories, can be phrased in various different ways.

Definition 3.3.1 A **representation** of a multicategory C consists of an object $\otimes(c_1, \ldots, c_n)$ and a map

$$u(c_1, \ldots, c_n) : c_1, \ldots, c_n \longrightarrow \otimes(c_1, \ldots, c_n)$$

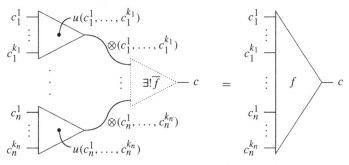

Fig. 3-B. Representation of a multicategory

for each $n \in \mathbb{N}$ and $c_1, \ldots, c_n \in C$, with the following factorization property (Fig. 3-B): for any objects $c_1^1, \ldots, c_1^{k_1}, \ldots, c_n^1, \ldots, c_n^{k_n}, c$ and any map $f : c_1^1, \ldots, c_n^{k_n} \longrightarrow c$, there is a unique map

$$\overline{f} : \otimes(c_1^1, \ldots, c_1^{k_1}), \ldots, \otimes(c_n^1, \ldots, c_n^{k_n}) \longrightarrow c$$

such that

$$\overline{f} \circ (u(c_1^1, \ldots, c_1^{k_1}), \ldots, u(c_n^1, \ldots, c_n^{k_n})) = f.$$

A multicategory is **representable** if it admits a representation.

Definition 3.3.2 A map $c_1, \ldots, c_n \xrightarrow{\ u\ } c'$ in a multicategory is **pre-universal** if (Fig. 3-C(a)) for any object c and map $c_1, \ldots, c_n \xrightarrow{\ f\ } c$, there is a unique map $c' \xrightarrow{\ \overline{f}\ } c$ such that $\overline{f} \circ u = f$.

Definition 3.3.3 A map $c_1, \ldots, c_n \xrightarrow{\ u\ } c'$ in a multicategory is **universal** if (Fig. 3-C(b)) for any objects $a_1, \ldots, a_p, b_1, \ldots, b_q, c$ and any map

$$a_1, \ldots, a_p, c_1, \ldots, c_n, b_1, \ldots, b_q \xrightarrow{\ f\ } c,$$

there is a unique map $a_1, \ldots, a_p, c', b_1, \ldots, b_q \xrightarrow{\ \overline{f}\ } c$ such that $\overline{f} \circ_{p+1} u = f$. (See Section 2.3 for the $\circ_{p+1}$ notation.)

Here is the main result.

Theorem 3.3.4

a. *The following conditions on a multicategory C are equivalent:*

  • *$C \cong V(A)$ for some monoidal category A*

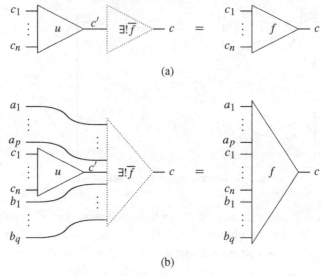

(a)

(b)

Fig. 3-C. (a) Pre-universal map, and (b) universal map

- *C is representable*
- *every sequence $c_1, \ldots, c_n$ of objects of C is the domain of some pre-universal map, and the composite of pre-universal maps is pre-universal*
- *every sequence $c_1, \ldots, c_n$ of objects of C is the domain of some universal map.*

Under these equivalent conditions, a map in C is universal if and only if it is pre-universal.

b. *Let A and A′ be monoidal categories. The following conditions on a map $H : V(A) \longrightarrow V(A')$ of multicategories are equivalent:*

- *$H = V(P, \pi)$ for some weak monoidal functor $(P, \pi) : A \longrightarrow A'$*
- *H preserves universal maps (that is, if u is a universal map in $V(A)$ then Hu is a universal map in $V(A')$).*

The functor V is faithful, and therefore provides an equivalence between **UMonCat**_{wk} or **MonCat**_{wk} (as you prefer) and the subcategory **RepMulti** of **Multicat** consisting of the representable multicategories and the universal-preserving maps.

The only subtle point here is that the existence of a pre-universal map for every given domain is *not* enough to ensure that the multicategory comes from a monoidal category. A specific example appears in Leinster (2003), but the point can be explained here in the familiar context of vector spaces. Suppose we are aware that for each pair (X, Y) of vector spaces, there is an object $X \otimes Y$

and a bilinear map $X, Y \xrightarrow{u_{X,Y}} X \otimes Y$ with the traditional universal property (which we are calling pre-universality). Then it does *not* follow for purely formal reasons that the tensor product is associative (up to isomorphism): one has to use some actual properties of vector spaces. Essentially, one either has to show that trilinear maps of the form $u_{X \otimes Y, Z} \circ (u_{X,Y}, 1_Z)$ and $u_{X, Y \otimes Z} \circ (1_X, u_{Y,Z})$ are pre-universal, or show that maps of the form $u_{X,Y}$ are in fact universal.

The energetic reader with plenty of time on her hands will have no difficulty in proving Theorem 3.3.4; the main ideas have been explained and it is just a matter of settling the details. So in a sense that is an end to the matter: monoidal categories can be recognized as multicategories with a certain property, and monoidal functors similarly, all as hoped for originally.

We can, however, take things further. With just a little more work than a direct proof would involve, Theorem 3.3.4 can be seen as a special case of a result in the theory of fibrations of multicategories. This theory is a fairly predictable extension of the theory of fibrations of ordinary categories, and the result of which 3.3.4 is a special case is the multicategorical analogue of a standard result on categorical fibrations.

The basic theory of fibrations of multicategories is laid out in Leinster (2003), which culminates in the deduction of 3.3.4 and some related facts on, for instance, *strict* monoidal categories as multicategories. Here is the short story. For any category D, a fibration (or really, opfibration) over D is essentially the same thing as a weak functor $D \longrightarrow \mathbf{Cat}$. (We looked at the case of *discrete* fibrations in Section 1.1.) With appropriate definitions, a similar statement can be made for multicategories. Taking D to be the terminal multicategory 1, we find that the unique map $C \longrightarrow 1$ is a fibration exactly when C is representable, and that weak functors $1 \longrightarrow \mathbf{Cat}$ are exactly unbiased monoidal categories. (Universal and pre-universal maps in C correspond to what are usually called cartesian and pre-cartesian maps.) So a representable multicategory is essentially the same thing as a monoidal category.

We now consider a different non-algebraic notion of monoidal category: 'homotopy monoidal categories'. The idea can be explained as follows.

Recall that every small category has a nerve, and that this allows categories to be described as simplicial sets satisfying certain conditions. Explicitly, if $n \in \mathbb{N}$ then let $[n] = \{0, 1, \ldots, n\}$, and let Δ be the category whose objects are $[0], [1], [2], \ldots$ and whose morphisms are all order-preserving functions; so Δ is equivalent to the category of non-empty finite totally ordered sets. A functor $\Delta^{\mathrm{op}} \longrightarrow \mathbf{Set}$ is called a **simplicial set**. (More generally, a functor $\Delta^{\mathrm{op}} \longrightarrow \mathcal{E}$ is called a **simplicial object in** $\mathcal{E}$.) Any ordered set $(I, \leq)$ can be regarded as a category with object-set I and with exactly one morphism $i \longrightarrow j$ if $i \leq j$, and none otherwise. This applies in particular to the ordered sets $[n]$, and so we may define the **nerve** NA of a small category A as the simplicial set

$$NA: \quad \Delta^{\mathrm{op}} \quad \longrightarrow \quad \mathbf{Set}$$
$$[n] \quad \longmapsto \quad \mathbf{Cat}([n], A).$$

This gives a functor $N : \mathbf{Cat} \longrightarrow [\Delta^{\mathrm{op}}, \mathbf{Set}]$, which turns out to be full and faithful. Hence **Cat** is equivalent to the full subcategory of $[\Delta^{\mathrm{op}}, \mathbf{Set}]$ whose objects are those simplicial sets X isomorphic to NA for some small category A. There are various intrinsic characterizations of such simplicial sets X. We do not need to think about the general case for now, only the special case of one-object categories, that is, monoids.

So: let $k, n_1, \ldots, n_k \in \mathbb{N}$, and for each $j \in \{1, \ldots, k\}$, define a map ι_j in Δ by

$$\iota_j: \quad [n_j] \quad \longrightarrow \quad [n_1 + \cdots + n_k]$$
$$p \quad \longmapsto \quad n_1 + \cdots + n_{j-1} + p.$$

Given also a simplicial set X, let

$$\xi_{n_1, \ldots, n_k} : X[n_1 + \cdots + n_k] \longrightarrow X[n_1] \times \cdots \times X[n_k]$$

be the map whose jth component is $X(\iota_j)$. Write

$$\xi^{(k)} = \xi_{1, \ldots, 1} : X[k] \longrightarrow X[1]^k.$$

Proposition 3.3.5 *The following conditions on a simplicial set X are equivalent:*

a. $\xi_{n_1, \ldots, n_k} : X[n_1 + \cdots + n_k] \longrightarrow X[n_1] \times \cdots \times X[n_k]$ *is an isomorphism for all $k, n_1, \ldots, n_k \in \mathbb{N}$*

b. $\xi_{m,n} : X[m + n] \longrightarrow X[m] \times X[n]$ *is an isomorphism for all $m, n \in \mathbb{N}$, and the unique map $X[0] \longrightarrow 1$ is an isomorphism*

c. $\xi^{(k)} : X[k] \longrightarrow X[1]^k$ *is an isomorphism for all $k \in \mathbb{N}$*

d. $X \cong NA$ *for some monoid A.*

Proof Straightforward. Note that (b) is just (a) restricted to $k \in \{0, 2\}$, and similarly that (c) is just (a) in the case $n_1 = \cdots = n_k = 1$.   □

Monoids are, therefore, the same thing as simplicial sets satisfying any of the conditions (a)–(c). The proposition can be generalized: replace **Set** by any category $\mathcal{E}$ possessing finite products to give a description of monoids in $\mathcal{E}$ as certain simplicial objects in $\mathcal{E}$. In particular, if we take $\mathcal{E} = \mathbf{Cat}$ then we obtain a description of strict monoidal categories as certain simplicial objects in **Cat**. This suggests that a 'righteous' (weak) notion of monoidal category could be obtained by changing the isomorphisms in Proposition 3.3.5 to equivalences.

Proposition 3.3.6 *The following conditions on a functor $X : \Delta^{\text{op}} \longrightarrow$* **Cat** *are equivalent:*

a. $\xi_{n_1,\ldots,n_k} : X[n_1 + \cdots + n_k] \longrightarrow X[n_1] \times \cdots \times X[n_k]$ *is an equivalence for all $k, n_1, \ldots, n_k \in \mathbb{N}$*

b. $\xi_{m,n} : X[m+n] \longrightarrow X[m] \times X[n]$ *is an equivalence for all $m, n \in \mathbb{N}$, and the unique map $X[0] \longrightarrow 1$ is an equivalence*

c. $\xi^{(k)} : X[k] \longrightarrow X[1]^k$ *is an equivalence for all $k \in \mathbb{N}$.*

Proof As noted above, both the implications (a) $\Rightarrow$ (b) and (a) $\Rightarrow$ (c) are trivial. Their converses are straightforward inductions. ☐

Definition 3.3.7 A **homotopy monoidal category** is a functor $X :$ $\Delta^{\text{op}} \longrightarrow$ **Cat** satisfying the equivalent conditions 3.3.6(a)–(c). **HMonCat** is the category of homotopy monoidal categories and natural transformations between them.

A similar definition can be made with **Top** replacing **Cat** and homotopy equivalences replacing categorical equivalences, to give a notion of 'topological monoid up to homotopy' (Leinster 1999b, §4). It can be shown that any loop space provides an example. In fact, loop space theory was where the idea first arose: there, topological monoids up to homotopy were called 'special Δ-spaces' or 'special simplicial spaces' (Segal 1974, Anderson 1971, p. 63 of Adams 1978).

Before I say anything about the comparison with ordinary monoidal categories, let me explain another route to the notion of homotopy monoidal category.

Let $\mathbb{D}$ be the augmented simplex category (1.2.2), whose objects are the (possibly empty) finite totally ordered sets $\mathbf{n} = \{1, \ldots, n\}$. The fact that Δ is $\mathbb{D}$ with the object $\mathbf{0}$ removed is a red herring: we will not use *this* connection between Δ and $\mathbb{D}$.

Now, $\mathbb{D}$ is the free monoidal category containing a monoid, in the sense that for any monoidal category $(\mathcal{E}, \otimes, I)$ (classical, say), there is an equivalence

$$\mathbf{MonCat}_{\text{wk}}((\mathbb{D}, +, 0), (\mathcal{E}, \otimes, I)) \simeq \mathbf{Mon}(\mathcal{E}, \otimes, I)$$

between the category of weak monoidal functors $\mathbb{D} \longrightarrow \mathcal{E}$ and the category of monoids in $\mathcal{E}$. For given a weak monoidal functor $\mathbb{D} \longrightarrow \mathcal{E}$, the image of any monoid in $\mathbb{D}$ is a monoid in $\mathcal{E}$, and in particular, the object $\mathbf{1}$ of $\mathbb{D}$ has a unique monoid structure, giving a monoid in $\mathcal{E}$. Conversely, given a monoid A in $\mathcal{E}$, there arises a weak monoidal functor $\mathbb{D} \longrightarrow \mathcal{E}$ sending $\mathbf{n}$ to A^n.

Taking $\mathcal{E} = $ **Cat** describes strict monoidal categories as weak monoidal functors $(\mathbb{D}, +, 0) \longrightarrow (\mathbf{Cat}, \times, 1)$. Such a weak monoidal functor is an ordinary functor $W : \mathbb{D} \longrightarrow$ **Cat** together with isomorphisms

$$\omega_{m,n} : W(\mathbf{m} + \mathbf{n}) \longrightarrow W\mathbf{m} \times W\mathbf{n}, \qquad \omega_{\cdot} : W\mathbf{0} \longrightarrow 1 \qquad (3.4)$$

$(m, n \in \mathbb{N})$ satisfying coherence axioms. By changing 'isomorphisms' to 'equivalences' in the previous sentence, we obtain another notion of weak monoidal category. Formally, given monoidal categories $\mathcal{D}$ and $\mathcal{E}$, write **MonCat**$_{\mathrm{colax}}(\mathcal{D}, \mathcal{E})$ for the category of colax monoidal functors $\mathcal{D} \longrightarrow \mathcal{E}$ (as defined in 1.2.10) and monoidal transformations between them. The new 'weak monoidal categories' are the objects of the category **HMonCat$'$** defined as follows.

Definition 3.3.8 **HMonCat$'$** is the full subcategory of

$$\mathbf{MonCat}_{\mathrm{colax}}((\mathbb{D}, +, 0), (\mathbf{Cat}, \times, 1))$$

consisting of the colax monoidal functors (W, ω) for which each of the functors (3.4) $(m, n \in \mathbb{N})$ is an equivalence of categories.

You might object that this is useless as a definition of monoidal category, depending as it does on pre-existing concepts of monoidal category and colax monoidal functor. There are several responses. One is that we could give an explicit description of what a colax monoidal functor $\mathbb{D} \longrightarrow$ **Cat** is, along the lines of the traditional description of cosimplicial objects by face and degeneracy maps, and this would eliminate the dependence. Another is that we will soon show that **HMonCat$'$** $\cong$ **HMonCat**, and **HMonCat** is defined without mention of monoidal categories. A third is that while 3.3.8 might not be good as a *definition* of monoidal category, it is a useful reformulation: for instance, if we change the monoidal category (**Cat**, $\times$, 1) to the monoidal category of chain complexes and change categorical equivalence to chain homotopy equivalence, then we obtain a reasonable notion of homotopy differential graded algebra. This exhibits an advantage of the $\mathbb{D}$ approach over the Δ approach: we can use it to discuss homotopy monoids in monoidal categories where the tensor is not cartesian product. Much more on this can be found in my (1999b) and (2000a).

To show that **HMonCat$'$** $\cong$ **HMonCat**, we first establish a connection between Δ and $\mathbb{D}$.

Proposition 3.3.9 *Let $\mathcal{E}$ be a category with finite products. Then there is an isomorphism of categories*

$$\mathbf{MonCat}_{\mathrm{colax}}((\mathbb{D}, +, 0), (\mathcal{E}, \times, 1)) \cong [\Delta^{\mathrm{op}}, \mathcal{E}].$$

Proof This is a special case of a general result on Kleisli categories (Leinster, 2000a, 3.1.6 ; beware the different notation). It can also be proved directly in the following way. A functor $W : \mathbb{D} \longrightarrow \mathcal{E}$ is conventionally depicted as a diagram

of objects and arrows in $\mathcal{E}$. A colax monoidal structure ω on W amounts to a pair of maps

$$W\mathbf{m} \longleftarrow W(\mathbf{m} + \mathbf{n}) \longrightarrow W\mathbf{n}$$

for each $m, n \in \mathbb{N}$, satisfying axioms implying, among other things, that all of these maps can be built up from the special cases

$$W\mathbf{m} \longleftarrow W(\mathbf{m} + 1), \qquad W(1 + \mathbf{n}) \longrightarrow W\mathbf{n}.$$

So a colax monoidal functor $(W, \omega) : \mathbb{D} \longrightarrow \mathcal{E}$ looks like

and this is the conventional picture of a simplicial object in $\mathcal{E}$. $\square$

So if $\mathcal{E}$ is a category with ordinary, cartesian, products, simplicial objects in $\mathcal{E}$ are the same as colax monoidal functors from $\mathbb{D}$ to $\mathcal{E}$. This fails when $\mathcal{E}$ is a monoidal category whose tensor product is not the cartesian product. It could be argued that in this situation, it would be better to define a simplicial object in $\mathcal{E}$ not as a functor $\Delta^{\mathrm{op}} \longrightarrow \mathcal{E}$, but rather as a colax monoidal functor $\mathbb{D} \longrightarrow \mathcal{E}$. For example, it was the colax monoidal version that made possible the definition of homotopy differential graded algebra referred to above.

In Proposition 3.3.9, if the colax monoidal functor (W, ω) corresponds to the simplicial object X then we have

$$W\mathbf{n} = X[n], \qquad \omega_{m,n} = \xi_{m,n}, \qquad \omega. = \xi.$$

(where $\xi. : X[0] \longrightarrow 1$ means $\xi_{n_1,\ldots,n_k}$ in the case $k = 0$). So by using condition (b) of Proposition 3.3.6 we obtain:

Corollary 3.3.10 *The isomorphism of Proposition 3.3.9 restricts to an isomorphism* **HMonCat′** $\cong$ **HMonCat**. $\square$

Our two ways of approaching homotopy monoidal categories, via either nerves in a finite product category or monoids in a monoidal category, are therefore equivalent in a strong sense. The next question is: are they also equivalent to the standard notion of monoidal category? I will stop short of a precise equivalence result, and instead just indicate how to pass back and forward between homotopy and 'ordinary' monoidal categories.

So, let us start from an unbiased monoidal category $(A, \otimes, \gamma, \iota)$ and define from it a homotopy monoidal category $X \in$ **HMonCat**. To define X we will need to use unbiased bicategories (defined in the next section). For each $n \in \mathbb{N}$ the ordered set $[n]$ may be regarded as a category, and so as an unbiased bicategory in which all 2-cells are identities. Also, the unbiased monoidal category A may be regarded as an unbiased bicategory with only one object. The homotopy monoidal category X is defined by taking $X[n]$ to be the category whose objects are all weak functors $[n] \longrightarrow A$ of unbiased bicategories and whose maps are transformations of a suitably-chosen kind. This is just a categorification of the usual nerve construction.

Conversely, let us start with $X \in$ **HMonCat** and derive from X an unbiased monoidal category $(A, \otimes, \gamma, \iota)$. The category A is $X[1]$. To obtain the rest of the data we first choose for each $n \in \mathbb{N}$ a functor $\psi^{(n)} : X[1]^n \longrightarrow X[n]$ and natural isomorphisms

$$\eta^{(n)} : 1 \overset{\sim}{\longrightarrow} \xi^{(n)} \circ \psi^{(n)}, \qquad \varepsilon^{(n)} : \psi^{(n)} \circ \xi^{(n)} \overset{\sim}{\longrightarrow} 1$$

such that $(\psi^{(n)}, \xi^{(n)}, \eta^{(n)}, \varepsilon^{(n)})$ forms an adjoint equivalence, which is possible since the functor $\xi^{(n)}$ is an equivalence (1.1.2). Define $\delta^{(n)} : [1] \longrightarrow [n]$ by $\delta^{(n)}(0) = 0$ and $\delta^{(n)}(1) = n$. Then define $\otimes_n : A^n \longrightarrow A$ as the composite

$$X[1]^n \overset{\psi^{(n)}}{\longrightarrow} X[n] \overset{X\delta^{(n)}}{\longrightarrow} X[1].$$

The coherence isomorphisms γ and ι are defined from the $\eta^{(n)}$s and $\varepsilon^{(n)}$s in a natural way. The coherence axioms follow from the fact that we chose *adjoint* equivalences – that is, they follow from the triangle identities. So we arrive at an unbiased monoidal category $(A, \otimes, \gamma, \iota)$.

The processes above determine functors **UMonCat**$_{wk}$ $\rightleftarrows$ **HMonCat**. It is fairly easy to see that composing one functor with the other does not yield a functor isomorphic to the identity, either way round. I believe, however, that if **HMonCat** is made into a 2-category in a suitable way then each composite functor is *equivalent* to the identity. This would mean that the notion of homotopy monoidal category is essentially the same as the other notions of monoidal category that we have discussed.

To summarize the chapter so far: we have formalized the idea of non-strict monoidal category in various ways, and, with the exception of homotopy monoidal categories, shown that all the formalizations are equivalent. Precisely, the following categories are equivalent:

MonCat$_{wk}$ (classical monoidal categories)

UMonCat$_{wk}$ (unbiased monoidal categories)

Σ**-MonCat**$_{wk}$ (Σ-monoidal categories, for any plausible Σ)

RepMulti (representable multicategories)

and these equivalences can presumably be extended to equivalences of 2-categories. The categories

HMonCat (homotopy monoidal categories, via simplicial objects)

HMonCat$'$ (homotopy monoidal categories, via monoidal functors)

are isomorphic, and there is reasonable hope that if they are made into 2-categories then they are equivalent to **MonCat**$_{wk}$. There are still more notions of monoidal category that we might contemplate – for instance, the anamonoidal categories of Makkai (1996) – but we leave it at that.

3.4 Notions of bicategory

Everything that we have done for monoidal categories can also be done for bicategories. This is usually at the expense of setting up some slightly more sophisticated language, which is why things so far have been done for monoidal categories only. Here we run through what we have done for monoidal categories and generalize it to bicategories, noting any wrinkles.

Unbiased bicategories

Definition 3.4.1 A **lax bicategory** B (or properly, $(B, \circ, \gamma, \iota)$) consists of

- a class B_0 (often assumed to be a set), whose elements are called **objects** or **0-cells** of B
- for each $a, b \in B_0$, a category $B(a, b)$, whose objects are called **1-cells** and whose morphisms are called **2-cells**
- for each $n \in \mathbb{N}$ and $a_0, \ldots, a_n \in B_0$, a functor

$$\circ_n : B(a_{n-1}, a_n) \times \cdots \times B(a_0, a_1) \longrightarrow B(a_0, a_n),$$

called n**-fold composition** and written

$$
\begin{aligned}
(f_n, \ldots, f_1) &\longmapsto (f_n \circ \cdots \circ f_1) \\
(\alpha_n, \ldots, \alpha_1) &\longmapsto (\alpha_n * \cdots * \alpha_1)
\end{aligned}
$$

where the f_is are 1-cells and the α_is are 2-cells

- a 2-cell

$$\gamma_{((f_n^{k_n},\ldots,f_n^1),\ldots,(f_1^{k_1},\ldots,f_1^1))} : \quad ((f_n^{k_n} \circ \cdots \circ f_n^1) \circ \cdots \circ (f_1^{k_1} \circ \cdots \circ f_1^1))$$
$$\longrightarrow (f_n^{k_n} \circ \cdots \circ f_n^1 \circ \cdots \circ f_1^{k_1} \circ \cdots \circ f_1^1)$$

for each $n, k_1, \ldots, k_n \in \mathbb{N}$ and double sequence $((f_n^{k_n}, \ldots, f_n^1), \ldots, (f_1^{k_1}, \ldots, f_1^1))$ of 1-cells for which the composites above make sense

- a 2-cell

$$\iota_f : f \longrightarrow (f)$$

for each 1-cell f,

satisfying naturality and coherence axioms analogous to those for unbiased monoidal categories (Definition 3.1.1).

A lax bicategory $(B, \circ, \gamma, \iota)$ is called an **unbiased bicategory** (respectively, an **unbiased strict 2-category**) if all the components of γ and ι are invertible 2-cells (respectively, identity 2-cells).

Remarks 3.4.2

a. A lax bicategory with exactly one object is, of course, just a lax monoidal category, and similarly for the weak and strict versions.
b. Unbiased strict 2-categories are in one-to-one correspondence with ordinary strict 2-categories, easily.

As in the case of monoidal categories, there is an abstract version of the definition of unbiased bicategory phrased in the language of 2-monads. Previously we took the 2-monad 'free strict monoidal category' on **Cat**; now we take the 2-monad 'free strict 2-category' on **Cat-Gph**, the strict 2-category of **Cat**-graphs. We met the ordinary category **Cat-Gph** earlier (1.3.1). The 2-category structure on **Cat** induces a 2-category structure on **Cat-Gph** as follows: given maps $P, Q : B \longrightarrow B'$ of **Cat**-graphs, there are only any 2-cells of the form

when $P_0 = Q_0 : B_0 \longrightarrow B_0'$, and in that case a 2-cell ζ is a family of natural transformations

$$\left(P_{a,b} \xrightarrow{\zeta_{a,b}} Q_{a,b} \right)_{a,b \in B_0}.$$

Now, there is a forgetful map **Str-2-Cat** $\longrightarrow$ **Cat-Gph** of 2-categories, and this has a left adjoint, so there is an induced 2-monad $((\ )^*, \mu, \eta)$ on **Cat-Gph**. An unbiased bicategory is exactly a weak algebra for this 2-monad, and the same applies in the lax and strict cases.

Definition 3.4.3 Let B and B' be lax bicategories. A **lax functor** (P, π) : $B \longrightarrow B'$ consists of

- a function $P_0 : B_0 \longrightarrow B'_0$ (usually just written as P)
- for each $a, b \in B_0$, a functor $P_{a,b} : B(a, b) \longrightarrow B'(Pa, Pb)$
- for each $n \in \mathbb{N}$ and composable sequence $f_1, \ldots, f_n$ of 1-cells of B, a 2-cell

$$\pi_{f_n,\ldots,f_1} : (Pf_n \circ \cdots \circ Pf_1) \longrightarrow P(f_n \circ \cdots \circ f_1),$$

satisfying axioms analogous to those in the definition of lax monoidal functor (3.1.3). A **weak functor** (respectively, **strict functor**) from B to B' is a lax functor (P, π) for which each component of π is an invertible 2-cell (respectively, an identity 2-cell).

As an example of the benefits of the unbiased approach, consider 'Hom-functors'. For any category A there is a functor

$$\text{Hom} : A^{\text{op}} \times A \longrightarrow \mathbf{Set}$$

defined on objects by $(a, b) \mapsto A(a, b)$ and on morphisms by composition – that is, morphisms $a' \xrightarrow{f} a$ and $b \xrightarrow{g} b'$ in A induce the function

$$\begin{aligned} \text{Hom}(f, g) : \quad \text{Hom}(a, b) \quad &\longrightarrow \quad \text{Hom}(a', b'), \\ p \quad &\longmapsto \quad g \circ p \circ f. \end{aligned}$$

Suppose we want to imitate this construction for bicategories, changing the category A to a bicategory B and looking for a weak functor $\text{Hom} : B^{\text{op}} \times B \longrightarrow \mathbf{Cat}$. If we use classical bicategories then we have a problem: there is no such composite as $g \circ p \circ f$, and the best we can do is to choose some substitute such as $(g \circ p) \circ f$ or $g \circ (p \circ f)$. Although we could, say, consistently choose the first option and so arrive at a weak functor Hom, this is an arbitrary choice. So there is no canonical Hom-functor in the classical world. In the unbiased world, however, we can simply take the ternary composite $(g \circ p \circ f)$, and everything runs smoothly.

By choosing different strengths of bicategory and of maps between them, we again obtain nine different categories:

$$\textbf{LaxBicat}_{str} \subseteq \textbf{LaxBicat}_{wk} \subseteq \textbf{LaxBicat}_{lax}$$
$$\cup| \qquad\qquad \cup| \qquad\qquad \cup|$$
$$\textbf{UBicat}_{str} \subseteq \textbf{UBicat}_{wk} \subseteq \textbf{UBicat}_{lax}$$
$$\cup| \qquad\qquad \cup| \qquad\qquad \cup|$$
$$\textbf{Str-2-Cat}_{str} \subseteq \textbf{Str-2-Cat}_{wk} \subseteq \textbf{Str-2-Cat}_{lax}.$$

For instance, **UBicat**$_{wk}$ is the category of unbiased bicategories and weak functors.

Differences from the theory of unbiased monoidal categories emerge when we try to define transformations between functors between unbiased bicategories. This should not come as a surprise given what we already know in the classical case about transformations and modifications of bicategories vs. transformations of monoidal categories (1.5.11). More mysteriously, there seems to be no satisfactorily unbiased way to formulate a definition of transformation or modification for unbiased bicategories; it seems that we are forced to grit our teeth and write down biased-looking definitions. This done, we obtain a notion of biequivalence of unbiased bicategories. Just as in the classical case (1.5.13), biequivalence amounts to the existence of a weak functor that is essentially surjective on objects and locally an equivalence. The **st** construction for monoidal categories (3.1.6) generalizes without trouble to give

Theorem 3.4.4 *Every unbiased bicategory is biequivalent to a strict 2-category.*   □

Example 3.4.5 Every topological space X has a fundamental 2-groupoid $\Pi_2 X$. We saw how to define $\Pi_2 X$ as a classical bicategory in 1.5.4. Here we consider the unbiased version of $\Pi_2 X$, in which n-fold composition is defined by choosing for each $n \in \mathbb{N}$ a reparametrization map $[0, 1] \longrightarrow [0, n]$ (the most obvious choice being multiplication by n). A 0-cell of the strict cover $\textbf{st}(\Pi_2 X)$ is a point of X, and a 1-cell is a pair (n, γ) where $n \in \mathbb{N}$ and $\gamma : [0, n] \longrightarrow X$ with $\gamma(0) = x$ and $\gamma(n) = y$. This is essentially the technique of Moore loops (Adams, 1978, p. 31), used to show that every loop space is homotopy equivalent to a strict topological monoid.

Algebraic notions of bicategory

Here we take the very general family of algebraic notions of monoidal category considered in Section 3.2 and imitate it for bicategories. In other words, we set up a theory of Σ-bicategories, where the 'signature' Σ is a sequence of sets.

Recall that we defined Σ-monoidal categories in three steps, the functor $(\ )\text{-}\mathbf{MonCat}_{wk}$ being the composite

$$\mathbf{Set}^{\mathbb{N}} \xrightarrow{\ F\ } \mathbf{Set\text{-}Operad} \xrightarrow{\ I_*\ } \mathbf{Cat\text{-}Operad} \xrightarrow{\ \mathbf{Alg}_{wk}\ } \mathbf{CAT}^{op}.$$

The first two steps create an operad consisting of all the derived tensor products arising from Σ and all the coherence isomorphisms between them. The third takes algebras for this operad (or, if you prefer, models for this theory), which in this case means forming the category of 'monoidal categories' with the kind of products described by the operad. It is therefore only the third step that we need to change here.

So, given a $\mathbf{Cat}$-operad R, define a category $\mathbf{CatAlg}_{wk}(R)$ as follows. An object is a **categorical R-algebra**, that is, a $\mathbf{Cat}$-graph B together with a functor

$$\mathrm{act}_n : R(n) \times B(a_{n-1}, a_n) \times \cdots \times B(a_0, a_1) \longrightarrow B(a_0, a_n)$$

for each $n \in \mathbb{N}$ and $a_0, \ldots, a_n \in B_0$, satisfying axioms very similar to the usual axioms for an algebra for an operad. (So if B_0 has only one element then B is just an R-algebra in the usual sense.) A **weak map** $B \longrightarrow B'$ of categorical R-algebras is a map $P : B \longrightarrow B'$ of $\mathbf{Cat}$-graphs together with a natural isomorphism

$$\begin{array}{ccc}
R(n) \times B(a_{n-1}, a_n) \times \cdots \times B(a_0, a_1) & \xrightarrow{\ \mathrm{act}_n\ } & B(a_0, a_n) \\
{\scriptstyle 1 \times P_{a_{n-1}, a_n} \times \cdots \times P_{a_0, a_1}}\Big\downarrow & \quad\nearrow\!\!\!\nearrow\ \pi_n & \Big\downarrow{\scriptstyle P_{a_0, a_n}} \\
R(n) \times B'(a_{n-1}, a_n) \times \cdots \times B'(a_0, a_1) & \xrightarrow[\mathrm{act}_n]{} & B'(Pa_0, Pa_n)
\end{array}$$

for each $n \in \mathbb{N}$ and $a_0, \ldots, a_n \in B_0$, satisfying axioms like the ones in the monoidal case (Fig. 3-A, p. 105). This defines a category $\mathbf{CatAlg}_{wk}(R)$. Defining $\mathbf{CatAlg}_{wk}$ on maps in the only sensible way, we obtain a functor

$$\mathbf{CatAlg}_{wk} : \mathbf{Cat\text{-}Operad} \longrightarrow \mathbf{CAT}^{op},$$

and we then define the functor $(\ )\text{-}\mathbf{Bicat}_{wk}$ as the composite

$$\mathbf{Set}^{\mathbb{N}} \xrightarrow{\ F\ } \mathbf{Set\text{-}Operad} \xrightarrow{\ I_*\ } \mathbf{Cat\text{-}Operad} \xrightarrow{\ \mathbf{CatAlg}_{wk}\ } \mathbf{CAT}^{op}.$$

The lax and strict cases are, of course, done similarly.

Definition 3.4.6 Let $\Sigma \in \mathbf{Set}^{\mathbb{N}}$. A **$\Sigma$-bicategory** is an object of $\Sigma\text{-}\mathbf{Bicat}_{lax}$ (or equivalently, of $\Sigma\text{-}\mathbf{Bicat}_{wk}$ or $\Sigma\text{-}\mathbf{Bicat}_{str}$). A **lax** (respectively, **weak** or **strict**) **functor** between Σ-bicategories is a map in $\Sigma\text{-}\mathbf{Bicat}_{lax}$ (respectively, $\Sigma\text{-}\mathbf{Bicat}_{wk}$ or $\Sigma\text{-}\mathbf{Bicat}_{str}$).

All of the equivalence results in Section 3.2 go through. Hence there are isomorphisms of categories

$$\textbf{UBicat}_- \cong 1\textbf{-Bicat}_-, \qquad \textbf{Bicat}_- \cong \Sigma_c\textbf{-Bicat}_-,$$

where '$-$' represents any of 'lax', 'wk' or 'str'. The proofs are as in the monoidal case (3.2.2, 3.2.3) with only cosmetic changes. Then, there is the irrelevance of signature theorem, analogous to 3.2.4:

$$\Sigma\textbf{-Bicat}_- \simeq \Sigma'\textbf{-Bicat}_-$$

for all plausible Σ and Σ', where '$-$' is either 'lax' or 'wk'. Again, the proof is essentially unchanged. As a corollary, unbiased bicategories are the same as classical bicategories:

$$\textbf{UBicat}_- \simeq \textbf{Bicat}_-$$

where '$-$' is either 'lax' or 'wk'.

We saw on p. 110 that the equivalence between unbiased and classical monoidal categories is 'one level better' than might be expected, because it does not refer upwards to transformations. In the case of bicategories it is *two* levels better, because modifications are not needed either.

Non-algebraic notions of bicategory

I will say much less about these.

Take representable multicategories first. To formulate a notion of a bicategory as a multicategory satisfying a representability condition, we need to use a new kind of multicategory: instead of the arrows looking like

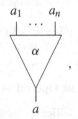

they should look like

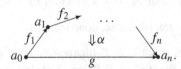

We will consider such multicategories later: they are a special kind of 'fc-multicategory' (Example 5.1.4), and they also belong to the world of opetopic structures (Chapter 7). All the results on monoidal categories as representable multicategories can be extended unproblematically to bicategories, and the same goes for the theory of fibrations of multicategories (Leinster, 2003).

Consider, finally, homotopy monoidal categories. A homotopy bicategory can be defined as a functor $\Delta^{\mathrm{op}} \longrightarrow \mathbf{Cat}$ satisfying conditions similar to (but, of course, looser than) those in Proposition 3.3.6. The new conditions involve pullbacks rather than products, and can be found by considering nerves of categories in general instead of just nerves of monoids. See Section 10.2 for further remarks.

Notes

The notion of unbiased monoidal category has been part of the collective consciousness for a long while (Kelly, 1974, and 9.1 of Hermida, 2000). Around 30 years ago Kelly and his collaborators began investigating 2-monads and 2-dimensional algebraic theories (see Blackwell, Kelly and Power, 1989, for instance), and they surely knew that unbiased monoidal categories were equivalent to classical monoidal categories in the way described above, although I have not been able to find anywhere this is made explicit before my own (2000b). If I am interpreting the (somewhat daunting) literature correctly, one can also deduce from it some of the more general results in Section 3.2 on algebraic notions of monoidal category, but there are considerable technical issues to be understood before one is in a position to do so. Here, in contrast, we have a quick and natural route to the results. Essentially, operads take the place of 2-monads as the means of describing an algebraic theory on **Cat**. In order even to *state* the problem at hand, in any approach, the explicit or implicit use of operads seems inevitable; an advantage of our approach is that we use nothing more.

The idea that a monoidal category is a multicategory with enough universal arrows goes back to Lambek's paper introducing multicategories (1969) and is implicit in the definition of weak n-category proposed by Baez and Dolan (1997), but as far as I know did not appear in print until Hermida (2000).

Homotopy monoidal categories were studied in my own (2000a) paper and its introductory companion (1999b), in the following wider context: given an operad P and a monoidal category $\mathcal{A}$ with a distinguished class of maps called 'equivalences', there is a notion of 'homotopy P-algebra in $\mathcal{A}$'. The case $P = 1$, $\mathcal{A} = \mathbf{Cat}$ gives homotopy monoidal categories; the case of homotopy differential graded algebras was mentioned on p. 118. Homotopy monoidal categories are related to the work of Simpson, Tamsamani, Toën, and Vezzosi mentioned in the Notes to Chapter 10.

PART II

Operads

Chapter 4

Generalized operads and multicategories: basics

> Three minutes' thought would suffice to find this out; but thought is irksome and three minutes is a long time.
>
> *A. E. Housman*

In a category, an arrow has a single object as its domain and a single object as its codomain. In a multicategory, an arrow has a finite sequence of objects as its domain and a single object as its codomain. What other things could we have for the domain of an arrow, keeping a single object as the codomain? Could we, for instance, have a tree or a many-dimensional array of objects? In different terms (logic or computer science): what can the input type of an operation be?

In this chapter – the central chapter of the book – we answer these questions. We formalize the idea of an input type, and for each input type we define a corresponding theory of operads and multicategories. For instance, the input type might be 'finite sequences', and this yields the theory of ordinary operads and multicategories.

From now on, operads and multicategories as defined in Chapter 2 will be called **plain operads** and **plain multicategories**. Some mathematicians are used to their operads coming equipped with symmetric group actions; they should take 'plain' as a pun on 'planar', to remind them that these operads do not.

The formal strategy is as follows. A small category C can be described as consisting of sets and functions

$$C_0 \xleftarrow{\text{dom}} C_1 \xrightarrow{\text{cod}} C_0 \qquad C_1 \times_{C_0} C_1 \xrightarrow{\text{comp}} C_1, \qquad C_0 \xrightarrow{\text{ids}} C_1,$$

satisfying associativity and identity axioms, which can be interpreted as commutative diagrams in **Set**. Here $C_1 \times_{C_0} C_1$ is a certain pullback, as explained on p. 30. Similarly, let $T : \textbf{Set} \longrightarrow \textbf{Set}$ be the functor sending a set A to the underlying set $\coprod_{n \in \mathbb{N}} A^n$ of the free monoid on A: then a multicategory can be described as consisting of sets and functions

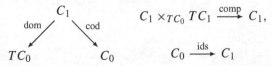

$$C_1 \times_{TC_0} TC_1 \xrightarrow{\text{comp}} C_1,$$

$$C_0 \xrightarrow{\text{ids}} C_1$$

satisfying associativity and identity axioms (expressed using the monad structure on T). Here C_0 is the set of all objects, C_1 is the set of all arrows, dom assigns to an arrow the sequence of objects that is its domain, and cod assigns to an arrow the single object that is its codomain. The crucial point is that this formalism works for any monad T on any category $\mathcal{E}$, as long as $\mathcal{E}$ and T satisfy some simple conditions concerning pullbacks. This gives a definition of T-multicategory for any such $\mathcal{E}$ and T; in the terms above, the pair $(\mathcal{E}, T)$ is the 'input type'. So when T is the identity monad on **Set**, a T-multicategory is an ordinary category, and when T is the free-monoid monad on **Set**, a T-multicategory is a plain multicategory.

We also define T-operads. As in the plain case, these are simply T-multicategories C with only one object – or formally, those in which C_0 is a terminal object of $\mathcal{E}$.

There is a canonical notion of an algebra for a T-multicategory. Like their plain counterparts, generalized operads and multicategories can be regarded as algebraic theories (single- and multi-sorted, respectively); algebras are the accompanying notion of model.

We start (Section 4.1) by describing the simple conditions on $\mathcal{E}$ and T needed to make the definitions work. Next (Section 4.2) are the definitions of T-multicategory and T-operad, and then (Section 4.3) the definition of algebra for a T-multicategory (or T-operad). There are many examples throughout, but some of the most important ones are done only very briefly; we do them in detail in later chapters.

4.1 Cartesian monads

In this section we introduce the conditions required of a monad (T, μ, η) on a category $\mathcal{E}$ in order that we may define the notions of T-multicategory and T-operad.

Definition 4.1.1

a. A category $\mathcal{E}$ is **cartesian** if it has all pullbacks.

b. A functor $\mathcal{E} \xrightarrow{T} \mathcal{F}$ is **cartesian** if it preserves pullbacks.

c. A natural transformation $\mathcal{F}$ is **cartesian** if for each map

$A \xrightarrow{f} B$ in $\mathcal{E}$, the naturality square

$$
\begin{array}{ccc}
SA & \xrightarrow{Sf} & SB \\
\alpha_A \downarrow & & \downarrow \alpha_B \\
TA & \xrightarrow[Tf]{} & TB
\end{array}
$$

is a pullback.

d. A monad $(T, T^2 \xrightarrow{\mu} T, 1 \xrightarrow{\eta} T)$ on a category $\mathcal{E}$ is **cartesian** if the category $\mathcal{E}$, the functor T, and the natural transformations μ and η are all cartesian.

Remarks 4.1.2

a. All of our examples of cartesian categories will have a terminal object, hence all finite limits.

b. When the category $\mathcal{E}$ has a terminal object, a (necessary and) sufficient condition for the natural transformation α of (c) to be cartesian is that for each object A of $\mathcal{E}$, the naturality square for the unique map $A \longrightarrow 1$ is a pullback.

c. Cartesian categories, cartesian functors and cartesian natural transformations form a sub-2-category **CartCat** of **Cat**, and a cartesian monad is exactly a monad in **CartCat**. (See p. 51 for the definition of monad in a 2-category.)

d. As is customary, we often write T to mean the whole monad (T, μ, η).

The rest of the section is examples.

Example 4.1.3 The identity monad on a cartesian category is cartesian.

Example 4.1.4 Let $\mathcal{E} = \textbf{Set}$ and let T be the free-monoid monad on $\mathcal{E}$. Certainly $\mathcal{E}$ is cartesian. An easy calculation shows that the monad T is cartesian too: Leinster (1998a, 1.4(ii)).

Example 4.1.5 A non-example. Let $\mathcal{E} = \textbf{Set}$ and let (T, μ, η) be the free commutative monoid monad. This fails to be cartesian on two counts: μ is not cartesian (for instance, its naturality square at the unique map $2 \longrightarrow 1$ is not a pullback), and the functor T does not preserve pullbacks. Let us show the latter in detail, using an argument of Weber (2001, 2.7.2). First note that for any set A,

$$TA = \coprod_{n \in \mathbb{N}} A^n / S_n$$

where S_n is the nth symmetric group acting on A^n in the natural way. Write $[a_1, \ldots, a_n]$ for the equivalence class of $(a_1, \ldots, a_n) \in A^n$ under this action, so that $[a_1, \ldots, a_n] = [b_1, \ldots, b_n]$ if and only if there exists $\sigma \in S_n$ such that $b_i = a_{\sigma i}$ for all i. Now let x, x', y, y', z be distinct formal symbols and consider applying T to the pullback square

$$
\begin{array}{ccc}
\{(x, y), (x, y'), (x', y), (x', y')\} & \longrightarrow & \{y, y'\} \\
\downarrow & & \downarrow \\
\{x, x'\} & \longrightarrow & \{z\}.
\end{array}
$$

There are distinct elements

$$[(x, y), (x', y')], \qquad [(x, y'), (x', y)]$$

of $T\{(x, y), (x, y'), (x', y), (x', y')\}$, yet both map to $[x, x'] \in T\{x, x'\}$ and to $[y, y'] \in T\{y, y'\}$. Hence the image of the square under T is not a pullback.

Example 4.1.6 Any algebraic theory gives rise to a monad on **Set** (its free algebra monad), and if the theory is strongly regular in the sense of 2.2.5 then the monad is cartesian. Carboni and Johnstone proved this first (1995); an alternative proof is in Appendix C. The result implies that the free monoid monad (4.1.4) is cartesian, and, as suspected in 2.2.5, that the theory of commutative monoids (4.1.5) is not strongly regular. Further examples appear below.

Example 4.1.7 Let $\mathcal{E} = $ **Set** and let T be the monad corresponding to the theory of monoids with involution. By definition, a **monoid with involution** is a monoid equipped with an endomorphism whose composite with itself is the identity; so this theory is defined by the usual operations and equations for the theory of monoids together with a unary operation $( )^\circ$ satisfying

$$x^{\circ\circ} = x, \qquad (x \cdot y)^\circ = x^\circ \cdot y^\circ, \qquad 1^\circ = 1.$$

These equations are strongly regular, so the monad T is cartesian. Any abelian group has an underlying monoid with involution, given by $x^\circ = x^{-1}$.

Example 4.1.8 More interestingly, let $\mathcal{E} = \mathbf{Set}$ and let T be the monad for **monoids with anti-involution**, where an anti-involution on a monoid is a unary operation $(\ )^\circ$ satisfying

$$x^{\circ\circ} = x, \qquad (x \cdot y)^\circ = y^\circ \cdot x^\circ, \qquad 1^\circ = 1.$$

Any group whatsoever has an underlying monoid with anti-involution, again given by $x^\circ = x^{-1}$. Now, this is *not* a strongly regular presentation of the theory, and in fact there is no strongly regular presentation, but T *is* a cartesian monad. So not all cartesian monads on **Set** arise from strongly regular theories. These assertions are proved in Example C.2.8; that T is cartesian was pointed out to me by Peter Johnstone.

Example 4.1.9 Let $\mathcal{E} = \mathbf{Set}$ and fix a set S. The endofunctor $S + -$ on $\mathcal{E}$ has a natural monad structure, and the monad is cartesian, corresponding to the algebraic theory consisting only of one constant for each member of S. In particular, if $S = 1$ then this is the theory of pointed sets.

Example 4.1.10 Algebraic theories generated by just unary operations correspond to monads on **Set** of the form $M \times -$, where M is a monoid and the unit and multiplication of the monad are given by those of the monoid. Since any equation formed from unary operations and a single variable is strongly regular, the monad $M \times -$ on **Set** is always cartesian.

Example 4.1.11 Let $\mathcal{E} = \mathbf{Set}^{\mathbb{N}}$ and let T be the monad 'free plain operad' on $\mathcal{E}$. As we saw in Section 2.3, the functor T forms trees with labelled vertices: for instance, if $A \in \mathcal{E}$, $a_1 \in A(3)$, $a_2 \in A(1)$, and $a_3 \in A(2)$, then

is an element of $(TA)(4)$. That T is cartesian follows from theory we develop later (6.5.5).

Example 4.1.12 Consider the finitary algebraic theory on **Set** generated by one n-ary operation for each $n \in \mathbb{N}$ and no equations. This theory is strongly regular, so the induced monad (T, μ, η) on **Set** is cartesian.

Given a set A, the set TA can be described inductively by

- if $a \in A$ then $a \in TA$
- if $t_1, \ldots, t_n \in TA$ then $(t_1, \ldots, t_n) \in TA$.

We have already looked at the case $A = 1$, where TA is the set of unlabelled trees (2.3.3). Similarly, an element of TA can be drawn as a tree whose leaves are labelled by elements of A: for instance,

$$((a_1, a_2, ()), a_3, (a_4, a_5)) \in TA$$

is drawn as

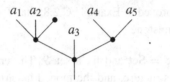

(Contrast this with the previous example, where the *vertices* were labelled.)
The unit $A \longrightarrow TA$ is

$$a \longmapsto \quad \overset{a}{\big|} \quad ,$$

and multiplication $T^2A \longrightarrow TA$ takes a tree whose leaves are labelled by elements of TA (for instance,

where

$$t_1 = \qquad , \qquad t_2 = \qquad)$$

and expands the labels to produce a tree whose leaves are labelled by elements of A (here,

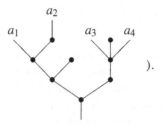

).

Example 4.1.13 Similar statements can be made for any free theory. That is, given any sequence $(\Sigma(n))_{n \in \mathbb{N}}$ of sets, the finitary algebraic theory on **Set** generated by one n-ary operation for each $\sigma \in \Sigma(n)$, and no equations, is strongly regular, so the induced monad (T, μ, η) on **Set** is cartesian. (The previous example was the case $\Sigma(n) = 1$ for all n.) Given a set A, an element of TA can be drawn as a tree in which the leaves are labelled by elements of A and the vertices with n branches coming up out of them are labelled by elements of $\Sigma(n)$: in other words, as a diagram like the domain or codomain of (3.2) (p. 103).

Example 4.1.14 The monad 'free topological monoid' on **Top**, whose functor part sends a space A to the disjoint union of cartesian powers $\coprod_{n \in \mathbb{N}} A^n$, is cartesian (by direct calculation).

Example 4.1.15 Similarly, the monad 'free strict monoidal category' (= free monoid) on **Cat** is cartesian. In fact, the free monoid monad on a category $\mathcal{E}$ is always cartesian provided that $\mathcal{E}$ satisfies the (co)limit conditions necessary to ensure that the usual free monoid construction $A \longmapsto \coprod_{n \in \mathbb{N}} A^n$ works in $\mathcal{E}$.

Example 4.1.16 The monad 'free symmetric strict monoidal category' on **Cat** is also cartesian, as a (lengthy) direct calculation shows. If we had taken strict symmetric monoidal categories instead – in other words, insisted that the symmetries were strict – then we would just be looking at commutative monoids in **Cat**, and the monad would fail to be cartesian by the argument of Example 4.1.5. So weakening the equality $x \otimes y = y \otimes x$ to an isomorphism makes a bad monad good.

Example 4.1.17 Let $\mathbb{H}$ be the category $(0 \underset{\tau}{\overset{\sigma}{\rightrightarrows}} 1)$, so that $[\mathbb{H}^{\mathrm{op}}, \mathbf{Set}]$ is the category of directed graphs. The forgetful functor $\mathbf{Cat} \longrightarrow [\mathbb{H}^{\mathrm{op}}, \mathbf{Set}]$ has a left adjoint and therefore induces a monad **fc** ('free category') on $[\mathbb{H}^{\mathrm{op}}, \mathbf{Set}]$. It follows from later theory (6.5.3) that **fc** is cartesian.

Example 4.1.18 In Part III we consider the free strict ω-category monad on the category of globular sets, and, for $n \in \mathbb{N}$, the free strict n-category monad on the category of n-globular sets. All of these monads are cartesian. The previous example is the case $n = 1$.

Example 4.1.19 In Section 1.4 we looked at strict double categories and, more generally, at strict n-tuple categories. In particular, we saw that a strict n-tuple category could be described as a functor $(\mathbb{H}^n)^{\mathrm{op}} \longrightarrow \mathbf{Set}$ with extra structure, where $\mathbb{H}$ is as in 4.1.17. The forgetful functor from the category of strict n-tuple categories and strict maps between them to the functor category $[(\mathbb{H}^n)^{\mathrm{op}}, \mathbf{Set}]$ has a left adjoint, the adjunction is monadic, and the induced monad on $[(\mathbb{H}^n)^{\mathrm{op}}, \mathbf{Set}]$ is cartesian. This can be shown by a similar method to that used for strict n-categories in Appendix F.

Alert readers may have noticed that nearly every one of the above examples of a cartesian monad on **Set** is, in fact, the free-algebra monad for a certain plain operad. This is no coincidence, as we discover in Section 6.2.

4.2 Operads and multicategories

We now define 'T-multicategory' and 'T-operad', for any cartesian monad T on a cartesian category $\mathcal{E}$. That is, for each such $\mathcal{E}$ and T we define a category T-**Multicat** of T-multicategories and a full subcategory T-**Operad** consisting of the T-operads. In the case $\mathcal{E} = \mathbf{Set}$ and $T = $ (free monoid) these are, respectively, **Multicat** and **Operad**; in the case $\mathcal{E} = \mathbf{Set}$ and $T = $ id they are **Cat** and **Monoid**.

The strategy for making these definitions is as described in the introduction to this chapter, dressed up a little: instead of handling the data and axioms for a T-multicategory directly, we introduce a bicategory $\mathcal{E}_{(T)}$ and define a T-multicategory as a monad in $\mathcal{E}_{(T)}$. This amounts to the same thing, as we shall see. (All we are doing is generalizing the description of a small category as a monad in the bicategory of spans: 5.4.3 of Bénabou, 1967)

Definition 4.2.1 For any cartesian monad (T, μ, η) on a cartesian category $\mathcal{E}$, the bicategory $\mathcal{E}_{(T)}$ is defined as follows:

0-cells are objects E of $\mathcal{E}$

1-cells $E \longrightarrow E'$ are diagrams

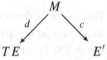

in $\mathcal{E}$

2-cells $(M, d, c) \longrightarrow (N, q, p)$ are maps $M \longrightarrow N$ in $\mathcal{E}$ such that

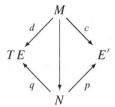

commutes

1-cell composition: the composite of 1-cells

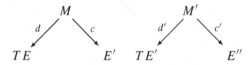

is given by composing along the upper slopes of the diagram

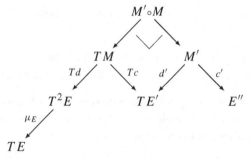

in $\mathcal{E}$, where the right-angle mark in the top square indicates that the square is a pullback and we assume from now on that a particular choice of pullbacks in $\mathcal{E}$ has been made

1-cell identities: the identity on E is

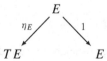

2-cell compositions and identities are defined in the evident way

coherence 2-cells: the associativity and unit 2-cells are defined using the universal property of pullback.

Since the choice of pullbacks in $\mathcal{E}$ was arbitrary, it is inevitable that composition of 1-cells in $\mathcal{E}_{(T)}$ does not obey strict associativity or unit laws. That it obeys them up to isomorphism is a consequence of T being cartesian. Changing the choice of pullbacks in $\mathcal{E}$ only changes the bicategory $\mathcal{E}_{(T)}$

up to isomorphism (in the category of bicategories and weak functors): see p. 150.

Here is the most important definition in this book. It is due to Burroni (1971), and in the form presented here uses the notion of monad in a bicategory (p. 51).

Definition 4.2.2 Let T be a cartesian monad on a cartesian category $\mathcal{E}$. A T-**multicategory** is a monad in the bicategory $\mathcal{E}_{(T)}$.

A T-multicategory C therefore consists of a diagram

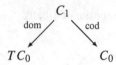

in $\mathcal{E}$ together with maps

$$C_1 \circ C_1 = C_1 \times_{TC_0} TC_1 \xrightarrow{\text{comp}} C_1, \qquad C_0 \xrightarrow{\text{ids}} C_1$$

satisfying associativity and identity axioms – exactly as promised at the start of the chapter.

Definition 4.2.3 Let T be a cartesian monad on a cartesian category $\mathcal{E}$. A T-**operad** is a T-multicategory C such that C_0 is a terminal object of $\mathcal{E}$.

Just as a T-multicategory is a generalized category, a T-operad is a generalized monoid. Explicitly, a monoid in a category $\mathcal{E}$ with finite limits consists of an object M of $\mathcal{E}$ together with maps

$$M \times M \xrightarrow{\text{mult}} M, \qquad 1 \xrightarrow{\text{unit}} M$$

in $\mathcal{E}$, satisfying associativity and identity axioms. Take a cartesian monad T on $\mathcal{E}$: then a T-operad consists of an object of $\mathcal{E}$ over $T1$, say $P \xrightarrow{d} T1$, together with maps

$$P \times_{T1} TP \xrightarrow{\text{comp}} P, \qquad 1 \xrightarrow{\text{ids}} P$$

over $T1$ in $\mathcal{E}$, again satisfying associativity and identity axioms.

A map of T-multicategories is a map of the underlying 'graphs' preserving composition and identities, as follows.

Definition 4.2.4 Let T be a cartesian monad on a cartesian category $\mathcal{E}$. A T-**graph** is a diagram

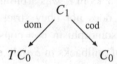

in $\mathcal{E}$ (that is, an endomorphism 1-cell in $\mathcal{E}_{(T)}$). A **map** $C \xrightarrow{f} C'$ **of** T**-graphs** is a pair $(C_0 \xrightarrow{f_0} C'_0,\ C_1 \xrightarrow{f_1} C'_1)$ of maps in $\mathcal{E}$ such that

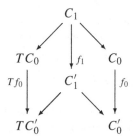

commutes. The category of T-graphs is written T-**Graph**.

This definition uses two different notions of a map between objects of $\mathcal{E}$: on the one hand, genuine maps in $\mathcal{E}$, and on the other, spans ($=$ 1-cells of $\mathcal{E}_{(T)}$). In Chapter 5 we will integrate the objects of $\mathcal{E}$ and these two different kinds of map into a single structure, an '**fc**-multicategory'.

Definition 4.2.5 A **map** $C \xrightarrow{f} C'$ **of** T**-multicategories** is a map f of their underlying graphs such that the diagrams

$$
\begin{array}{ccc}
C_0 & \xrightarrow{\ \mathrm{ids}\ } & C_1 \\
{\scriptstyle f_0}\downarrow & & \downarrow{\scriptstyle f_1} \\
C'_0 & \xrightarrow[\ \mathrm{ids}\ ]{} & C'_1
\end{array}
\qquad
\begin{array}{ccc}
C_1 \circ C_1 & \xrightarrow{\ \mathrm{comp}\ } & C_1 \\
{\scriptstyle f_1 * f_1}\downarrow & & \downarrow{\scriptstyle f_1} \\
C'_1 \circ C'_1 & \xrightarrow[\ \mathrm{comp}\ ]{} & C'_1
\end{array}
$$

commute, where $f_1 * f_1$ is the evident map induced by two copies of $C_1 \xrightarrow{f_1} C'_1$. The category of T-multicategories and maps between them is written T-**Multicat**. The full subcategory consisting of T-operads is written T-**Operad**.

So for any cartesian monad T, we have categories and functors

$$T\text{-}\mathbf{Operad} \hookrightarrow T\text{-}\mathbf{Multicat} \xrightarrow{\ \text{forgetful}\ } T\text{-}\mathbf{Graph}.$$

When the extra clarity is needed, we will refer to T-multicategories as $(\mathcal{E}, T)$-**multicategories** and to the category they form as $(\mathcal{E}, T)$-**Multicat**; similarly for operads and graphs.

We now look at some examples of generalized multicategories. In many of the most interesting ones the input of each operation/arrow forms quite a complicated shape, such as a diagram of pasted-together higher-dimensional cells. These examples are only described briefly here, with proper discussions postponed to later chapters.

We start with the two motivating cases.

Example 4.2.6 Let T be the identity monad on $\mathcal{E} = \textbf{Set}$. Then $\mathcal{E}_{(T)}$ is what is usually called the 'bicategory of spans' (Bénabou, 1967, 2.6), and a monad in $\mathcal{E}_{(T)}$ is just a small category. So

$$(\textbf{Set, id})\text{-}\textbf{Multicat} \simeq \textbf{Cat}, \qquad (\textbf{Set, id})\text{-}\textbf{Operad} \simeq \textbf{Monoid}.$$

More generally, if $\mathcal{E}$ is any cartesian category then $(\mathcal{E}, \text{id})$-multicategories are categories in $\mathcal{E}$ and $(\mathcal{E}, \text{id})$-operads are monoids in $\mathcal{E}$.

Example 4.2.7 Let T be the free monoid monad on the category $\mathcal{E}$ of sets. A T-graph

$$TC_0 = \coprod_{n \in \mathbb{N}} C_0^n \xleftarrow{\text{dom}} C_1 \xrightarrow{\text{cod}} C_0$$

amounts to a set C_0 'of objects' together with a set $C(a_1, \dots, a_n; a)$ for each $n \geq 0$ and $a_1, \dots, a_n, a \in C_0$. The composite 1-cell

$$
\begin{array}{ccc}
 & C_1 \circ C_1 & \\
\text{dom}' \swarrow & & \searrow \text{cod}' \\
TC_0 & & C_0
\end{array}
$$

in $\mathcal{E}_{(T)}$ is as follows: the set $C_1 \circ C_1$ at the apex is

$$C_1 \times_{TC_0} TC_1 = \coprod_{\substack{n, k_1, \dots, k_n \in \mathbb{N}, \\ a_i^j, a_i, a \in C_0}} C(a_1, \dots, a_n; a) \times C(a_1^1, \dots, a_1^{k_1}; a_1) \times \cdots \times C(a_n^1, \dots, a_n^{k_n}; a_n),$$

an element of which looks like the left-hand side of Fig. 2-B (p. 60); the function dom' sends this element to

$$(a_1^1, \dots, a_1^{k_1}, \dots, a_n^1, \dots, a_n^{k_n}) \in TC_0,$$

and cod' sends it to a. The identity 1-cell on C_0 in $\mathcal{E}_{(T)}$ is

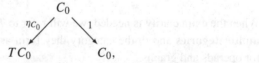

and η_{C_0} sends $a \in C_0$ to $(a) \in TC_0$. A T-multicategory structure on a T-graph C therefore consists of a function comp as in Fig. 2-B and a function ids assigning to each object $a \in C_0$ an 'identity' element $1_a \in C(a; a)$, obeying associativity and identity laws. So a T-multicategory is just a plain multicategory; indeed, there are equivalences of categories

$$T\text{-}\mathbf{Multicat} \simeq \mathbf{Multicat}, \qquad T\text{-}\mathbf{Operad} \simeq \mathbf{Operad}.$$

Example 4.2.8 Suppose we try to realize symmetric operads (2.2.21) as T-operads for some T. A first attempt might be to take the free commutative monoid monad T on **Set**. But this is both misguided and doomed to failure: misguided because if P is a symmetric operad then the maps $- \cdot \sigma :$ $P(n) \longrightarrow P(n)$ coming from permutations $\sigma \in S_n$ are only isomorphisms, not identities; and doomed to failure because T is not cartesian (4.1.5), which prevents us from making a definition of T-operad – in particular, from expressing associativity and identity laws. A better idea is to take the free symmetric strict monoidal category monad on **Cat**, thus replacing identities by isomorphisms: see 4.2.18 below.

Example 4.2.9 Let $\mathcal{E} = \mathbf{Set}$ and let T be the free monoid-with-involution monad on $\mathcal{E}$, as in 4.1.7. A T-multicategory looks like a plain multicategory except that the arrows are of the form

$$a_1^{\sigma_1}, \ldots, a_n^{\sigma_n} \longrightarrow a$$

where the a_is and a are objects and $\sigma_i \in \{-1, +1\}$. (It is sometimes convenient to write x^{-1} instead of x° and x^{+1} instead of x.) For instance:

a. There is a large T-multicategory **Cat** whose objects are all small categories and in which a map

$$A_1^{\sigma_1}, \ldots, A_n^{\sigma_n} \longrightarrow A$$

is a functor

$$A_1^{\sigma_1} \times \cdots \times A_n^{\sigma_n} \longrightarrow A$$

where

$$A_i^{\sigma_i} = \begin{cases} A_i^{\mathrm{op}} & \text{if } \sigma_i = -1 \\ A_i & \text{if } \sigma_i = +1. \end{cases}$$

Composition in **Cat** is usual composition of functors, taking opposites where necessary. That **Cat** does form a T-multicategory (and not just a plain multicategory) is a statement about the behaviour of contravariance with respect to products and functors. A 'calculus of substitution' of this kind was envisaged by Kelly in the introduction to his (1972) paper.

b. More generally, suppose that $(\mathcal{A}, \otimes, I)$ is a monoidal category and $(\ )^\circ$: $\mathcal{A} \longrightarrow \mathcal{A}$ a functor for which there are coherent natural isomorphisms

$$(A^\circ)^\circ \cong A, \qquad (A \otimes B)^\circ \cong A^\circ \otimes B^\circ, \qquad I^\circ \cong I$$

$(A, B \in \mathcal{A})$: then we obtain a T-multicategory in the same way as we did for **Cat** above. (Beware that although a typical duality operator $(\ )^\circ$, such as that for duals of finite-dimensional vector spaces, does satisfy the three displayed isomorphisms, it is a *contravariant* functor on $\mathcal{A}$, so does not give a T-multicategory.)

Example 4.2.10 Now consider the monad T for monoids with *anti*-involution, as in 4.1.8. T-multicategories are the same as in the previous example except that substitution reverses order. For instance:

a. Example 4.2.9(a) is also an example of a T-multicategory for the present T, since $A \times B \cong B \times A$ for categories A and B.
b. Example 4.2.9(b) can also be repeated, except that now we require $(\ )^\circ$ to reverse, rather than preserve, the order of the tensored factors:

$$(A \otimes B)^\circ \cong B^\circ \otimes A^\circ$$

naturally in $A, B \in \mathcal{A}$.
c. Loop spaces give an example. Fix a space X with a basepoint x. In 1.2.9 we met the monoidal category $\mathcal{A}$ whose objects are loops based at x and whose tensor $\otimes$ is concatenation of loops. There is a functor $(\ )^\circ : \mathcal{A} \longrightarrow \mathcal{A}$ sending a loop to the same loop run backwards, and this satisfies

$$(\gamma^\circ)^\circ = \gamma, \qquad (\gamma \otimes \delta)^\circ = \delta^\circ \otimes \gamma^\circ, \qquad \mathrm{const}_x^\circ = \mathrm{const}_x$$

for all loops γ and δ. So there is a resulting T-multicategory whose objects are loops and whose maps encode all the information about concatenation of loops, homotopy classes of homotopies between loops, and reversal of loops.

Example 4.2.11 Let $\mathcal{E} = \mathbf{Set}$ and let T be the monad $1 + -$ of 4.1.9. A T-graph is a diagram $1 + C_0 \xleftarrow{\text{dom}} C_1 \xrightarrow{\text{cod}} C_0$ of sets and functions. If we regard $1 + C_0$ as a subset of the free monoid $\coprod_{n \in \mathbb{N}} C_0^n$ on C_0 and recall Example 4.2.7 then it is clear that a T-multicategory is exactly a plain multicategory in which all arrows are either unary or nullary. So it is natural to draw an arrow θ of C as either

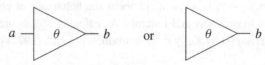

where in the first case $\mathrm{dom}(\theta) = a \in C_0$, in the second $\mathrm{dom}(\theta)$ is the unique element of 1, and in both $\mathrm{cod}(\theta) = b$. The unary arrows form a category D, and the nullary arrows define a functor $Y : D \longrightarrow \mathbf{Set}$ in which $Y(b)$ is the set of nullary arrows with codomain b. So a T-multicategory is the same thing as a small category D together with a functor $Y : D \longrightarrow \mathbf{Set}$, and in particular, a T-operad is a monoid acting on a set. Similarly, if S is any set then a $(\mathbf{Set}, S + -)$-multicategory is a small category D together with an S-indexed family of functors $(D \xrightarrow{Y_s} \mathbf{Set})_{s \in S}$.

Another way of putting this is that, when $T = 1 + -$, a T-multicategory is a discrete opfibration (p. 30). In fact, T-**Multicat** is equivalent to the category whose objects are discrete opfibrations between small categories and whose morphisms are commutative squares.

Example 4.2.12 As a kind of dual to the last example, let $\mathcal{E} = \mathbf{Set}$ and let T be the free semigroup monad, $TA = \coprod_{n \geq 1} A^n$. Then a T-multicategory is exactly a plain multicategory with no nullary arrows. In particular, a T-operad P is a family $(P(n))_{n \geq 1}$ of sets, indexed over positive numbers, equipped with composition and identities of the usual kind. Some authors prefer to exclude the possibility of nullary operations: see the Notes to Chapter 2.

Example 4.2.13 Fix a monoid M and let T be the monad $M \times -$ on **Set**, as in 4.1.10. Then a T-graph is a diagram $M \times C_0 \longleftarrow C_1 \longrightarrow C_0$, and by projecting onto the two factors of $M \times C_0$ we find that T-**Multicat** is isomorphic to the category $\mathbf{Cat}/M$ of categories over M. (Here we regard a monoid as a one-object category: p. 14.) In particular, T-**Operad** $\cong$ **Monoid**$/M$.

For a specific example, let M be the (large) monoid of all cardinals under multiplication. Let C be the (large) category of fields and homomorphisms between them (which are, of course, all injective). By taking degrees of extensions we obtain a functor $\pi : C \longrightarrow M$, making fields into an $(M \times -)$-multicategory.

Example 4.2.14 Let T be the free plain operad monad on $\mathcal{E} = \mathbf{Set}^{\mathbb{N}}$, as in 4.1.11. In a T-multicategory the objects form a graded set $(C_0(n))_{n \in \mathbb{N}}$ and the arrows look like

$(a_1 \in C_0(3), a_2 \in C_0(1), a_3 \in C_0(2), a \in C_0(4))$, where the tree in the codomain is always the corolla with the same number of leaves as the tree in the domain. A typical example of composition is that arrows

$$a_2^2 \quad a_2^3 \quad \xrightarrow{\theta_2} \quad a_2 \quad \xrightarrow{\theta} \quad a$$

$$a_2^1 \quad a_1 \quad$$

$$a_1^2 \quad \xrightarrow{\theta_1}$$

$$a_1^1$$

compose to give a single arrow

$$a_2^2 \quad a_2^3 \quad \xrightarrow{\theta \circ (\theta_1, \theta_2)} \quad a \quad .$$

$$a_2^1$$

$$a_1^2$$

$$a_1^1$$

In the case of T-operads the labels a_i vanish, so a T-operad consists of a family of sets $(P(\tau))_{\text{trees } \tau}$ equipped with composition and identities.

We have seen that when T is the identity monad on **Set**, a T-operad is exactly a monoid. We have seen that when T is the free monoid monad, a T-operad is exactly a plain operad. We have seen that when T is the free plain operad monad, a T-operad is as just described. This process can be iterated indefinitely, producing the shapes called 'opetopes'; that is the subject of Chapter 7. The T-operads of the present example are described in Section 7.1 under the name of 'T_2-operads'.

Example 4.2.15 A different example involving trees takes $\mathcal{E}$ to be **Set** and T to be the free algebraic theory of 4.1.12. In this context labels appear on leaves rather than vertices, and trees are amalgamated by grafting leaves to roots rather than by substituting trees into vertices. A T-multicategory consists of a set C_0 of objects and hom-sets like

$$C\left(\begin{array}{c} a_1 \\ \diagdown \\ a_2 \\ a \end{array}\right)$$

$(a_1, a_2, a \in C_0)$, together with an identity

$$1_a \in C\left(\begin{array}{c} a \\ | \\ a \end{array}\right)$$

for each $a \in C_0$ and composition functions like

$$C\left(\begin{array}{c} a_1 \\ a_2 \\ a \end{array}\right) \times C\left(\begin{array}{c} a_1^2 \\ a_1^1 \\ a_1 \end{array}\right) \times C\left(\begin{array}{c} a_2^1 \quad a_2^2 \\ a_2 \end{array}\right)$$

$$\longrightarrow C\left(\begin{array}{c} a_1^2 \\ a_1^1 \quad a_2^1 \quad a_2^2 \\ a \end{array}\right)$$

$(a_i^j, a_i, a \in C_0)$. Put another way, a T-multicategory C is a plain multicategory in which the hom-sets are graded by trees: to each $a_1, \ldots, a_n, a \in C_0$ and $\tau \in \mathbf{tr}(n)$ there is assigned a set $C_\tau(a_1, \ldots, a_n; a)$, composition consists of functions

$$C_\tau(a_1, \ldots, a_n; a) \times C_{\tau_1}(a_1^1, \ldots, a_1^{k_1}; a_1) \times \cdots \times C_{\tau_n}(a_n^1, \ldots, a_n^{k_n}; a_n)$$
$$\longrightarrow C_{\tau \circ (\tau_1, \ldots, \tau_n)}(a_1^1, \ldots, a_n^{k_n}; a),$$

and the identity on $a \in C_0$ is an element of $C_{\,|}(a; a)$. In particular, a T-operad is a family $(P(\tau))_{\text{trees } \tau}$ of sets together with compositions

$$P(\tau) \times P(\tau_1) \times \cdots \times P(\tau_n) \longrightarrow P(\tau_{\circ}(\tau_1, \ldots, \tau_n))$$

and an identity element of $P(\,|\,)$. T-multicategories are a simplified version of the 'relaxed multicategories' mentioned in 6.8.4.

Example 4.2.16 Recall that given any symmetric monoidal category $\mathcal{V}$, there is a notion of 'plain multicategory enriched in $\mathcal{V}$' (p. 64), the one-object version of which is 'plain operad in $\mathcal{V}$' (p. 68). If $\mathcal{E} = \mathbf{Top}$ and T is the free topological monoid monad (4.1.14) then a T-operad is precisely an operad in $\mathbf{Top}$. However, T-multicategories are *not* the same thing as multicategories enriched in $\mathbf{Top}$, as in a T-multicategory C there is a topology on the set of objects. A multicategory enriched in $\mathbf{Top}$ is a T-multicategory in which the set of objects has the discrete topology. This difference should not be found surprising or disappointing: it exhibits the tension between internal and enriched category theory, previously discussed in Section 1.3.

Example 4.2.17 Similarly, if $\mathcal{E} = \mathbf{Cat}$ and T is the free strict monoidal category monad (4.1.15) then a T-operad is exactly a $\mathbf{Cat}$-operad. We saw some examples of these in Section 3.2: the operads '$F\Sigma$' determining different theories of monoidal category. Another example is the $\mathbf{Cat}$-operad $(\mathbf{TR}(n))_{n \in \mathbb{N}}$, where $\mathbf{TR}(n)$ is the category of n-leafed trees and maps between them, defined and discussed in Section 7.3.

Example 4.2.18 Let T be the free symmetric strict monoidal category monad on $\mathbf{Cat}$, as in 4.1.16. Any symmetric multicategory A gives rise to a T-multicategory C as follows. The category C_0 is discrete, with the same objects as A. The objects of the category C_1 are the arrows of A; the arrows of C_1 are of the form

$$(a_{\sigma 1}, \ldots, a_{\sigma n} \xrightarrow{\ \theta \cdot \sigma\ } a) \qquad \xrightarrow{\ \sigma\ } \qquad (a_1, \ldots, a_n \xrightarrow{\ \theta\ } a)$$

– in other words, an arrow $\phi \longrightarrow \theta$ is a permutation σ such that $\phi = \theta \cdot \sigma$. The category TC_0 has as objects all finite sequences $(a_1, \ldots, a_n)$ of objects of A, and arrows of the form

$$(a_{\sigma 1}, \ldots, a_{\sigma n}) \qquad \xrightarrow{\ \sigma\ } \qquad (a_1, \ldots, a_n).$$

The rest of the structure of C is obvious.

This defines a full and faithful functor

$$\mathbf{SymMulticat} \longleftrightarrow T\text{-}\mathbf{Multicat},$$

so symmetric multicategories could equivalently be defined as T-multicategories with certain properties. In particular, symmetric operads are special T-operads. See also the comments on p. 255 on 'enhanced symmetric multicategories', where the category C_0 is not required to be discrete.

Example 4.2.19 The free category monad **fc** on the category of directed graphs (4.1.17) gives rise to a notion of **fc**-multicategory. This is the subject of Chapter 5.

Example 4.2.20 Let T be the free strict ω-category monad on the category $\mathcal{E}$ of globular sets (4.1.18). Then T-operads are exactly globular operads, which we study in depth in Part III and which can be used to specify different theories of ω-category. In Section 10.2 we consider briefly the more general T-multicategories.

Example 4.2.21 The cubical analogue of the previous example takes T to be the free strict n-tuple category monad of 4.1.19. A weak n-tuple category might be defined as an algebra for a certain T-operad. We look at weak double categories in Section 5.2, but otherwise do not pursue the cubical case.

Example 4.2.22 Let T be a cartesian monad on a cartesian category $\mathcal{E}$, let $X \in \mathcal{E}$, and let $h : TX \longrightarrow X$ be a map in $\mathcal{E}$. The T-graph $(TX \xleftarrow{\ 1\ } TX \xrightarrow{\ h\ } X)$ can be given the structure of a T-multicategory in at most one way, and such a structure exists if and only if $TX \xrightarrow{\ h\ } X$ is an algebra for the monad T. (If it does exist then comp $= \mu_X$ and ids $= \eta_X$.) This defines a full and faithful functor

$$(\ )^{+} : \mathcal{E}^{T} \longrightarrow T\text{-}\mathbf{Multicat}$$

turning algebras into multicategories. In Section 10.2 we will use the idea that a T-multicategory is a generalized T-algebra to formulate a notion of weak n-category.

In the situation of ordinary categories rather than generalized multicategories, not only is there the concept of a functor between categories, but also there are the concepts of a module between categories (p. 88) and a natural transformation between functors. The same goes for plain multicategories, as we saw in Section 2.3. In fact, both of these concepts make sense for T-multicategories in general. We meet the definitions and see the connection between them in Section 5.3.

$\mathcal{E}$ and T have so far been regarded as fixed. But we would expect some kind of functoriality: if T' is another cartesian monad on another category $\mathcal{E}'$ then a 'map' from $(\mathcal{E}, T)$ to $(\mathcal{E}', T')$ should induce a functor

$$(\mathcal{E}, T)\text{-}\mathbf{Multicat} \longrightarrow (\mathcal{E}', T')\text{-}\mathbf{Multicat}.$$

We prove this in Section 6.7. That the category $(\mathcal{E}, T)$-**Multicat** is independent (up to isomorphism) of the choice of pullbacks in $\mathcal{E}$ follows from this functoriality by considering the identity map from ($\mathcal{E}$ with one choice of pullbacks) to ($\mathcal{E}$ with another choice of pullbacks).

4.3 Algebras

Theories have models, groups have representations, categories have set-valued functors, and multicategories have algebras. Here we meet algebras for generalized (operads and) multicategories. There are several ways of framing the definition. I have chosen the one that seems most useful in practice; two alternatives are discussed in Sections 6.3 and 6.4.

Let us begin by considering algebras for a plain multicategory C. These are maps from C into the multicategory of sets, but this is not much use for generalization as there is not necessarily a sensible T-multicategory of sets for arbitrary cartesian T. However, as we saw in Chapter 2, algebras for plain multicategories can be described without explicit reference to the multicategory of sets. In the special case of plain operads P, an algebra is a set X together with a map

$$\coprod_{n \in \mathbb{N}} P(n) \times X^n \longrightarrow X$$

(usually written $(\theta, x_1, \ldots, x_n) \longmapsto \bar{\theta}(x_1, \ldots, x_n)$), satisfying axioms expressing compatibility with composition and identities in P. Let T_P be the endofunctor $X \longmapsto \coprod_{n \in \mathbb{N}} P(n) \times X^n$ on **Set**. Then the composition and identity of the operad P induce a monad structure on T_P: the multiplication has components

$$T_P^2 X \quad \cong \quad \coprod_{n,k_1,\ldots,k_n \in \mathbb{N}} P(n) \times P(k_1) \times \cdots \times P(k_n) \times X^{k_1 + \cdots + k_n}$$

$$\xrightarrow{\coprod \text{comp} \times 1} \coprod_{m \in \mathbb{N}} P(m) \times X^m = T_P X$$

and the unit has components

$$X \cong 1 \times X^1 \xrightarrow{\text{ids} \times 1} P(1) \times X^1 \hookrightarrow \coprod_{m \in \mathbb{N}} P(m) \times X^m = T_P X.$$

An algebra for the operad P is precisely an algebra for the monad T_P. More generally, an algebra for a plain multicategory C is a family $(X(a))_{a \in C_0}$ of sets together with a map

$$\coprod_{n \in \mathbb{N}, a_1, \ldots, a_n \in C_0} C(a_1, \ldots, a_n; a) \times X(a_1) \times \cdots \times X(a_n) \longrightarrow X(a)$$

for each $a \in C_0$, satisfying axioms, and again the endofunctor

$$T_C : (X(a))_{a \in C_0} \longmapsto$$

$$\left(\coprod_{a_1, \ldots, a_n \in C_0} C(a_1, \ldots, a_n; a) \times X(a_1) \times \cdots \times X(a_n) \right)_{a \in C_0}$$

on $\mathbf{Set}^{C_0}$ naturally has the structure of a monad, an algebra for which is exactly a C-algebra. Since there is an equivalence of categories $\mathbf{Set}^{C_0} \simeq \mathbf{Set}/C_0$ (p. 29), we have succeeded in expressing the definition of an algebra for a plain operad or multicategory in a completely internal way – that is, completely in terms of the objects and arrows of the category $\mathbf{Set}$ and the free-monoid monad on $\mathbf{Set}$. The strategy for arbitrary cartesian $\mathcal{E}$ and T is now clear: an algebra for a T-multicategory C should be defined as an algebra for a certain monad T_C on $\mathcal{E}/C_0$.

So, let (T, μ, η) be a cartesian monad on a cartesian category $\mathcal{E}$, and let C be a T-multicategory. If $X = (X \xrightarrow{p} C_0)$ is an object of $\mathcal{E}/C_0$ then let $T_C X = (T_C X \xrightarrow{p'} C_0)$ be the boxed composite in the diagram

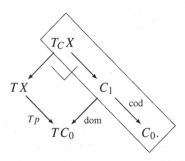

(So the object $T_C X$ of $\mathcal{E}$ is defined as a pullback.) This gives a functor T_C :

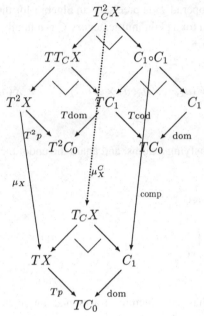

Fig. 4-A. Definition of μ_X^C

$\mathcal{E}/C_0 \longrightarrow \mathcal{E}/C_0$. The unit $\eta_X^C : X \longrightarrow T_C X$ is the unique map making

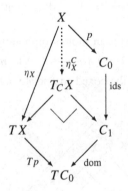

commute. Similarly, the multiplication $\mu_X^C : T_C^2 X \longrightarrow T_C X$ is the unique map making the diagram in Fig. 4-A commute. (That we do have pullbacks as in the top half of the diagram is an easy consequence of the definition of T_C, using the Pasting Lemma, 1.1.1.) This defines a monad (T_C, μ^C, η^C) on the category $\mathcal{E}/C_0$.

Definition 4.3.1 Let T be a cartesian monad on a cartesian category $\mathcal{E}$, and let C be a T-multicategory. Then the category **Alg**(C) of **algebras for** C is the category $(\mathcal{E}/C_0)^{T_C}$ of algebras for the monad T_C on $\mathcal{E}/C_0$.

For a more abstract derivation of the induced monad T_C, note that there is a weak functor $\mathcal{E}_{(T)} \longrightarrow$ **CAT** sending a 0-cell E to the category $\mathcal{E}/E$ and defined on 1- and 2-cells by pullback. Under this weak functor, any monad in $\mathcal{E}_{(T)}$ gives rise to a monad in **CAT**; thus, a T-multicategory C gives rise to a monad (T_C, μ^C, η^C) on $\mathcal{E}/C_0$.

Example 4.3.2 Let T be the identity monad on $\mathcal{E} = $ **Set** and let C be a T-multicategory, that is, a small category. Given a set $X \xrightarrow{\;p\;} C_0$ over C_0, write $X(a)$ for the fibre $p^{-1}\{a\}$ over $a \in C_0$. Then

$$(T_C X)(a) = \coprod_{b \in C_0} C(b, a) \times X(b),$$
$$(T_C^2 X)(a) = \coprod_{c, b \in C_0} C(c, b) \times C(b, a) \times X(c).$$

The multiplication map $\mu_X^C : (T_C^2 X)(a) \longrightarrow (T_C X)(a)$ is given by composition in C, and similarly $\eta_X^C : X(a) \longrightarrow (T_C X)(a)$ is given by the identity 1_a in C. A C-algebra is, therefore, a family $(X(a))_{a \in C_0}$ of sets together with a function

$$\coprod_{b \in C_0} C(b, a) \times X(b) \longrightarrow X(a) \tag{4.1}$$

for each $a \in C_0$, compatible with composition and identities in C. So **Alg**$(C) \simeq [C, \textbf{Set}]$.

Another way to put this is that for any category C, the forgetful functor $[C, \textbf{Set}] \longrightarrow [C_0, \textbf{Set}]$ is monadic and the induced monad on $[C_0, \textbf{Set}] \simeq \textbf{Set}/C_0$ is T_C. When C is a one-object category, regarded as a monoid M, the resulting monad T_C on **Set** is $M \times -$, and the category of algebras is the category of left M-sets.

More generally, let T be the identity monad on any cartesian category $\mathcal{E}$, so that an $(\mathcal{E}, T)$-multicategory C is a category in $\mathcal{E}$: then **Alg**(C) is the category of 'left C-objects' or 'diagrams on C' as defined in, for instance, Mac Lane and Moerdijk (1992, V.7).

Example 4.3.3 Let T be the free monoid monad on the category $\mathcal{E} = $ **Set** and let C be a T-multicategory, that is, a plain multicategory. A C-algebra consists of a family $(X(a))_{a \in C_0}$ of sets together with a function $(T_C X)(a) \longrightarrow X(a)$

for each $a \in C_0$, satisfying compatibility axioms. By definition of $T_C X$,

$$(T_C X)(a) = \coprod_{a_1,\ldots,a_n \in C_0} C(a_1, \ldots, a_n; a) \times X(a_1) \times \cdots \times X(a_n),$$

exactly as on p. 151, and we find that the category of algebras for the T-multicategory C is indeed equivalent to the category of algebras for the plain multicategory C.

Example 4.3.4 Any strongly regular algebraic theory (4.1.6) can be described by a plain operad. That is, if a monad on **Set** arises from a strongly regular theory then it is isomorphic to the monad T_P arising from some plain operad P; in fact, the converse holds too. The proofs are in Section C.1.

Example 4.3.5 When $\mathcal{E} = \mathbf{Set}$ and $T = 1 + -$, as in 4.2.11, a T-multicategory is an ordinary category D together with a functor $D \xrightarrow{Y} \mathbf{Set}$. A (D, Y)-algebra consists of a functor $D \xrightarrow{X} \mathbf{Set}$ together with a natural transformation

Example 4.3.6 Let T be the free semigroup monad on **Set**, so that a T-multicategory is a plain multicategory with no nullary arrows (4.2.12). Then an algebra for a T-multicategory C is exactly an algebra for the underlying plain multicategory. In fact, plain multicategories can be identified with pairs (C, X) where C is a T-multicategory and X is a C-algebra, by making elements of $X(a)$ correspond to nullary arrows into a.

Example 4.3.7 Let M be a monoid and let $(\mathcal{E}, T) = (\mathbf{Set}, M \times -)$, so that a T-multicategory is a category C together with a functor $C \xrightarrow{\phi} M$ (4.2.13). Then the category of algebras for (C, ϕ) is simply $[C, \mathbf{Set}]$, regardless of what ϕ is. This can be seen by working out T_C explicitly, but we will be able to understand the situation better after we have seen an alternative way of defining generalized multicategories and their algebras (Section 6.2, especially the remarks after Corollary 6.2.5.)

Example 4.3.8 Let T be any cartesian monad on any category $\mathcal{E}$ with finite limits. Then $(T1 \xleftarrow{1} T1 \xrightarrow{!} 1)$ is the terminal T-graph, and carries a unique multicategory structure, so is also the terminal T-multicategory. The induced monad on $\mathcal{E}/1 \cong \mathcal{E}$ is, inevitably, just (T, μ, η), and so an algebra for the terminal T-multicategory is just an algebra for T.

Example 4.3.9 Let T be the free plain operad monad on the category $\mathcal{E} = \mathbf{Set}^{\mathbb{N}}$ of sequences of sets (4.1.11, 4.2.14). Then by the previous example, an algebra for the terminal T-multicategory is precisely a plain operad. Compare Example 2.2.23, where we defined a *symmetric* multicategory whose algebras were plain operads. This is a minor theme of this book: objects equipped with symmetries are replaced by objects equipped with a more refined geometrical structure.

Example 4.3.10 Let T be the free algebraic theory on one operation of each arity (4.2.15). Let P be a T-operad. Then a P-algebra structure on a set X consists of a function $X^n \longrightarrow X$ for each n-leafed tree τ and each element of $P(\tau)$, satisfying axioms expressing compatibility with the composition and identity of P.

Example 4.3.11 If T is the free monoid monad on **Top** or **Cat**, as in 4.2.16 or 4.2.17, then an algebra for a T-operad P is an algebra in the usual sense, that is, a space or category X with continuous or functorial actions by the $P(n)$s.

Example 4.3.12 Historically, one of the most important plain operads has been the little intervals operad, that is, the little 1-discs operad of 2.2.16, whose algebras are roughly speaking the same thing as loop spaces. A plain operad is an operad for the free monoid monad, and a monoid is a category in which all arrows begin and end at the same point. If we are interested in paths rather than loops on a basepoint then it makes sense to replace monoids by arbitrary categories. Indeed, we show in 5.1.10 that if **fc** is the free category monad on the category $\mathcal{E}$ of directed graphs then there is a certain **fc**-operad P such that the paths in any fixed space naturally form a P-algebra. This solves the language problem posed in 2.2.18.

Example 4.3.13 Let T be the free strict ω-category monad on the category $\mathcal{E}$ of globular sets, so that a T-operad is a 'globular operad' (4.2.20). In Chapter 9 we construct a certain operad L, the initial 'operad-with-contraction', and define a weak ω-category to be an L-algebra. In Section 10.1 we consider some other possible definitions of weak ω-category, some of which are also of the form 'a weak ω-category is a P-algebra' for different choices of globular operad P.

Example 4.3.14 Let T be any monad on any category $\mathcal{E}$ and let $h = (TX \xrightarrow{h} X)$ be any T-algebra. Then there is a monad T/h on $\mathcal{E}/X$ whose functor part acts on objects by

$$
\begin{pmatrix} Y \\ \downarrow p \\ X \end{pmatrix}
\quad \longmapsto \quad
\begin{pmatrix} TY \\ \downarrow Tp \\ TX \\ \downarrow h \\ X \end{pmatrix}.
$$

Algebras for T/h are just T-algebras over h; precisely, $\mathcal{E}^{T/h} \cong \mathcal{E}^T/h$.

Now recall from 4.2.22 that when T is a cartesian monad on a cartesian category $\mathcal{E}$, any T-algebra $(TX \xrightarrow{h} X)$ defines a T-multicategory

$$
h^+ = (TX \xleftarrow{1} TX \xrightarrow{h} X).
$$

This induces a monad T_{h^+} on $\mathcal{E}/X$. So starting with $\mathcal{E}$, T, and $(TX \xrightarrow{h} X)$, we obtain the two monads T/h and T_{h^+} on $\mathcal{E}/X$; they are, inevitably, isomorphic. So $\mathbf{Alg}(h^+) \cong \mathcal{E}^T/h$. Example 4.3.8 is the special case where h is the terminal algebra.

Example 4.3.15 For any T-multicategory C, the object $(C_1 \xrightarrow{\mathrm{cod}} C_0)$ of $\mathcal{E}/C_0$ naturally has the structure of a T-algebra. When T is the identity monad on **Set**, so that C is a small category, this algebra is the functor

$$
\begin{aligned}
C &\longrightarrow \mathbf{Set}, \\
a &\longmapsto \coprod_{a' \in C_0} C(a', a)
\end{aligned}
$$

sometimes called the Cayley representation of C.

In this section we have seen how to assign to each T-multicategory C a category $\mathbf{Alg}(C)$. We would expect some kind of functoriality in C. When T is the identity monad on **Set**, a functor $C \longrightarrow C'$ between (T-multi)categories induces a functor in the opposite direction,

$$
\mathbf{Alg}(C') \simeq [C', \mathbf{Set}] \longrightarrow [C, \mathbf{Set}] \simeq \mathbf{Alg}(C).
$$

The same holds when T is the free monoid monad on **Set**, viewing C-algebras as multicategory maps $C \longrightarrow \mathbf{Set}$ as in the original definition, 2.1.12.

In fact, the construction works for an arbitrary cartesian monad T: any map $f : C \longrightarrow C'$ of T-multicategories induces a functor $\mathbf{Alg}(C') \longrightarrow \mathbf{Alg}(C)$. First, we have the functor $f_0^* : \mathcal{E}/C_0' \longrightarrow \mathcal{E}/C_0$ defined by pullback along

$f_0 : C_0 \longrightarrow C_0'$. Then there is a naturally-arising natural transformation

$$
\begin{array}{ccc}
\mathcal{E}/C_0' & \xrightarrow{\;T_{C'}\;} & \mathcal{E}/C_0' \\[1.5em]
{\scriptstyle f_0^*}\Big\downarrow & \nearrow \phi & \Big\downarrow{\scriptstyle f_0^*} \\[1.5em]
\mathcal{E}/C_0 & \xrightarrow[\;T_C\;]{} & \mathcal{E}/C_0,
\end{array}
$$

which the reader will easily be able to determine. This ϕ is compatible with the monad structures on $T_{C'}$ and T_C: in the terminology defined in Section 6.1, (f_0^*, ϕ) is a 'lax map of monads' from $T_{C'}$ to T_C. It follows that there is an induced functor on the categories of algebras for these monads, that is, from **Alg**(C') to **Alg**(C).

This construction defines a map

$$\textbf{Alg} : (T\text{-}\textbf{Multicat})^{\mathrm{op}} \longrightarrow \textbf{CAT}.$$

Since the induced functors are defined by pullback, it is inevitable that this map does not preserve composites and identities strictly, but only up to coherent isomorphism. Precisely, it is a weak functor from the 2-category $(T\text{-}\textbf{Multicat})^{\mathrm{op}}$ whose only 2-cells are identities to the 2-category **CAT**. If we also bring into play transformations between T-multicategories (Section 5.3), then T-**Multicat** becomes a strict 2-category and **Alg** a weak functor between 2-categories.

Notes

T-multicategories were introduced by Burroni in his (1971) paper, where they went by the name of T-categories. He showed how to define them for *any* monad T, though he concentrated on cartesian T (and it is not clear that T-categories are useful outside this case). The basic idea has been independently rediscovered on at least two occasions: by Hermida (2000) and by Leinster (1998a). The notion of an algebra for a T-multicategory seems not to have appeared before the latter paper.

The shape of this chapter is typical of much of this text: while the formalism is quite simple (in this case, the definition of T-multicategory and of algebra), it can take a long time to see what it means concretely in particular cases of interest. Indeed, in this chapter we have restricted ourselves to the simpler instances of T, leaving some of the more advanced examples to chapters where they can be explored at greater leisure (Chapters 5, 7, 8).

Hermida called $\mathcal{E}_{(T)}$ the 'Kleisli bicategory of spans' in (2000); the formal similarity between the definition of $\mathcal{E}_{(T)}$ and the usual construction of a Kleisli category is evident.

Dmitry Roytenberg suggested to me that something like Example 4.3.12 ought to exist.

Chapter 5

Example: fc-multicategories

A lot of people are afraid of heights. Not me. I'm afraid of widths.

Steven Wright

The generalized multicategories that we are interested in typically have some geometry to them. They are often 'higher-dimensional' in some sense. In this chapter we study a 2-dimensional example, **fc**-multicategories, which are T-multicategories when T is the free category monad **fc** on the category of directed graphs.

This case is interesting for a variety of reasons. First, **fc**-multicategories turn out to encompass a wide range of familiar 2-dimensional structures, including bicategories, double categories, monoidal categories and plain multicategories. Second, there are two well-known ideas for which **fc**-multicategories provide a cleaner and more general context than is traditional: the 'bimodules construction' (usually done on bicategories) and the enrichment of categories (usually done in monoidal categories). Third, these 2-dimensional structures are the second rung on an infinite ladder of higher-dimensional structures (the first rung being ordinary categories), and give us clues about the behaviour of the more difficult, less easily visualized higher rungs.

We start (Section 5.1) by unwinding the definition of **fc**-multicategory to give a completely elementary description. As mentioned above, various familiar structures arise as special kinds of **fc**-multicategory; we show how this happens.

A nearly-familiar structure that arises as a special kind of **fc**-multicategory is the 'weak double category', in which horizontal composition only obeys associativity and unit laws up to coherent isomorphism. We define these in Section 5.2 and give examples, of which there are many natural and pleasing ones.

158

Section 5.3 is on the 'bimodules construction' or, as we prefer to call it, the 'monads construction'. This takes an **fc**-multicategory C as input and produces a new **fc**-multicategory $\mathbf{Mon}(C)$ as output. For example, if C is the **fc**-multicategory of sets (and functions, and spans) then $\mathbf{Mon}(C)$ is the **fc**-multicategory of categories (and functors, and modules); if C is abelian groups (and homomorphisms) then $\mathbf{Mon}(C)$ is rings (and homomorphisms, and modules). We wait until Section 6.8 to see what a 'category enriched in an **fc**-multicategory' is.

5.1 fc-multicategories

Let $\mathbb{H}$ be the category $(0 \overset{\sigma}{\underset{\tau}{\rightrightarrows}} 1)$ and $\mathcal{E} = [\mathbb{H}^{\mathrm{op}}, \mathbf{Set}]$. Then $\mathcal{E}$ is the category of directed graphs, and there is a forgetful functor $U : \mathbf{Cat} \longrightarrow \mathcal{E}$. This has a left adjoint F: if $E \in \mathcal{E}$ then objects of FE are vertices of E, arrows in FE are strings

$$x_0 \xrightarrow{p_1} x_1 \xrightarrow{p_2} \cdots \xrightarrow{p_n} x_n$$

of edges in E (with $n \geq 0$), and composition is concatenation. The adjunction induces a monad $T = U \circ F$ on $\mathcal{E}$, which can be shown to be cartesian either by direct calculation or by applying some general theory (6.5.3). We write $T = \mathbf{fc}$, for 'free category', and so we have the notion of an **fc**-multicategory.

What is an **fc**-multicategory, explicitly? An **fc**-graph C is a diagram

$$C_1 = (C_{11} \rightrightarrows C_{10})$$

$$\mathbf{fc}(C_0) = (C'_{01} \rightrightarrows C_{00}) \qquad C_0 = (C_{01} \rightrightarrows C_{00})$$

where C_1 and C_0 are directed graphs and the diagonal arrows are maps of graphs, the C_{ij}s are sets and the horizontal arrows are functions, and C'_{01} is the set of strings of edges in C_0. Think of elements of C_{00} as **objects** or **0-cells**, elements of C_{01} as **horizontal 1-cells**, elements of C_{10} as **vertical 1-cells**, and elements of C_{11} as **2-cells**, as in the picture

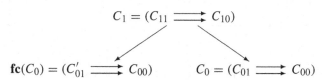

$$(5.1)$$

$(n \geq 0, \quad a_i, a, a' \in C_{00}, \quad m_i, m \in C_{01}, \quad f, f' \in C_{10}, \quad \theta \in C_{11})$. An **fc**-multicategory structure on the **fc**-graph C amounts to composition and identities of two types. First, the directed graph

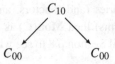

has the structure of a category; in other words, vertical 1-cells can be composed and there is an identity vertical 1-cell on each 0-cell. Second, there is a composition function for 2-cells,

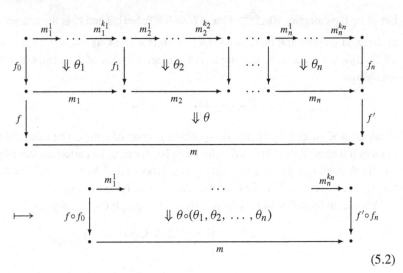

$$(5.2)$$

$(n \geq 0, k_i \geq 0$, with •s representing objects), and a function assigning an identity 2-cell to each horizontal 1-cell,

$$a \xrightarrow{\;m\;} a' \quad\longmapsto\quad
\begin{array}{c}
a \xrightarrow{\;m\;} a' \\
1_a \downarrow \quad \Downarrow 1_m \quad \downarrow 1_{a'} \\
a \xrightarrow[\;m\;]{} a'.
\end{array}$$

The composition and identities obey associativity and identity laws, which ensure that any diagram of pasted-together 2-cells with a rectangular boundary has a well-defined composite.

The pictures in the nullary case are worth a short comment. When $n = 0$, the 2-cell of diagram (5.1) is drawn as

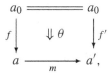

and the diagram of pasted-together 2-cells in the domain of (5.2) is drawn as

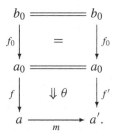

The composite of this last diagram will be written as $\theta \circ f_0$.

So in completely elementary terms, an **fc**-multicategory consists of

- a set of objects
- for each pair (a, a') of objects, a set of vertical 1-cells $\begin{array}{c} a \\ \big\downarrow \\ a' \end{array}$
- for each pair (a, a') of objects, a set of horizontal 1-cells $a \longrightarrow a'$
- for each $a_0, \ldots, a_n, a, a', m_1, \ldots, m_n, m, f, f'$ as in (5.1), a set of 2-cells θ
- composition and identity functions for vertical 1-cells, as described above
- composition and identity functions for 2-cells, as described above,

satisfying associativity and identity axioms. Having given this elementary description I will feel free to refer to large **fc**-multicategories, in which the collection of cells is a proper class.

Example 5.1.1 There is an **fc**-multicategory **Ring** in which

- 0-cells are (not necessarily commutative) rings
- vertical 1-cells are ring homomorphisms
- a horizontal 1-cell $A \longrightarrow A'$ is an (A', A)-bimodule (1.5.5)

- a 2-cell

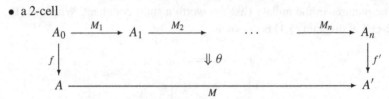

is an abelian group homomorphism

$$\theta : M_n \otimes_{A_{n-1}} M_{n-1} \otimes_{A_{n-2}} \cdots \otimes_{A_1} M_1 \longrightarrow M$$

satisfying

$$\theta(a_n \cdot m_n \otimes m_{n-1} \otimes \cdots \otimes m_2 \otimes m_1 \cdot a_1)$$
$$= f'(a_n) \cdot \theta(m_n \otimes \cdots \otimes m_1) \cdot f(a_1)$$

(that is, θ is a homomorphism of (A_n, A_0)-bimodules if M is given an (A_n, A_0)-bimodule structure via f and f')
- composition and identities are defined in the evident way.

Thus, rings, homomorphisms of rings, modules over rings, homomorphisms of modules, and tensor products of modules are integrated into a single structure. The category formed by the objects and vertical 1-cells is the ordinary category of rings and homomorphisms. When the distinction needs making, I will write **Ring₁** for this '1-dimensional' category and **Ring₂** for the '2-dimensional' **fc**-multicategory. Similar notation is extended to similar examples.

Many familiar 2-dimensional structures are degenerate **fc**-multicategories. The following examples demonstrate this; Fig. 5-A is a summary.

	No horizontal composition	*Weak horizontal composition*	*Strict horizontal composition*
No degeneracy	**fc**-multicategory	Weak double category	Strict double category
All vertical 1-cells are identities	Vertically discrete **fc**-multicategory	Bicategory	Strict 2-category
Only one object and one vertical 1-cell	Plain multicategory	Monoidal category	Strict monoidal category

Fig. 5-A. Some of the possible degeneracies of an **fc**-multicategory. The left-hand column refers to degeneracies in the category formed by the objects and vertical 1-cells. The top row says whether the **fc**-multicategory structure arises from a composition rule for horizontal 1-cells. See Examples 5.1.2–5.1.8

Example 5.1.2 Any strict double category gives rise to an **fc**-multicategory, in which a 2-cell as in (5.1) is a 2-cell

$$
\begin{array}{ccc}
a_0 & \xrightarrow{\ m_n \circ \cdots \circ m_1\ } & a_n \\
{\scriptstyle f}\big\downarrow & \Downarrow \theta & \big\downarrow{\scriptstyle f'} \\
a & \xrightarrow{\ m\ } & a'
\end{array}
$$

in the double category.

Example 5.1.3 The last example works just as well if we start with a double category in which the composition of horizontal 1-cells only obeys weak laws. We call these 'weak double categories' and discuss them in detail in the next section; a typical example is **Ring** (5.1.1). A similar example, and really the archetypal weak double category, is $\mathbf{Set}_2$, defined as follows:

- objects are sets
- vertical 1-cells are functions (and the category formed by objects and vertical 1-cells is the ordinary category $\mathbf{Set}_1$ of sets and functions; see 5.1.1 for the notation)
- horizontal 1-cells are spans: that is, a horizontal 1-cell $A \longrightarrow A'$ is a diagram

of sets and functions
- a 2-cell inside

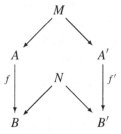

is a function $\theta : M \longrightarrow N$ making the diagram commute
- horizontal composition is by pullback.

That every weak double category has an underlying **fc**-multicategory is exactly analogous to every weak monoidal category having an underlying plain multicategory.

Example 5.1.4 Consider **fc**-multicategories C in which all vertical 1-cells are identities. This means that the category formed by the objects and vertical 1-cells is discrete, so we call C **vertically discrete**. A vertically discrete **fc**-multicategory consists of some objects $a, a', \ldots$, some 1-cells $m, m', \ldots$, and some 2-cells looking like

$$(5.3)$$

together with a composition function

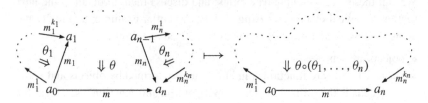

and an identity function

$$
a \xrightarrow{\ m\ } a' \quad \longmapsto \quad a \underset{m}{\overset{m}{\Longrightarrow 1_m}} a'
$$

obeying the inevitable associativity and identity laws. This leads us into the world of opetopes (Chapter 7).

Example 5.1.5 A bicategory is a weak double category in which the only vertical 1-cells are identities, and so gives rise to an **fc**-multicategory. Explicitly, if B is a bicategory then there is a vertically discrete **fc**-multicategory whose objects are those of B, whose horizontal 1-cells are the 1-cells of B, and whose 2-cells (5.3) are 2-cells

in B.

Example 5.1.6 Plain multicategories are the same as **fc**-multicategories with only one object and one vertical 1-cell. If the plain multicategory is called M

then we call the corresponding **fc**-multicategory ΣM, the **suspension** of M. Horizontal 1-cells in ΣM are objects of M, and 2-cells

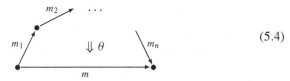

$$(5.4)$$

in ΣM are maps

$$m_1, \ldots, m_n \xrightarrow{\;\theta\;} m$$

in M.

Example 5.1.7 In particular, plain operads are **fc**-multicategories in which there is only one object, one vertical 1-cell and one horizontal 1-cell. An **fc**-operad is an **fc**-multicategory in which there is only one object and one horizontal 1-cell, so a plain operad is a special kind of **fc**-operad.

If we are going to take the suspension idea seriously then we should write

$$\Sigma : \mathbf{Operad} \hookrightarrow \mathbf{Multicat}$$

for the inclusion of operads as one-object multicategories. We also have (5.1.6) the suspension map

$$\Sigma : \mathbf{Multicat} \hookrightarrow \mathbf{fc\text{-}Multicat},$$

so the inclusion of operads into **fc**-multicategories should be written Σ^2: double suspension. Of course, we usually leave the first inclusion nameless.

Example 5.1.8 As a special case of both 5.1.5 and 5.1.6, any monoidal category M gives rise to an **fc**-multicategory ΣM in which there is one object and one vertical 1-cell. Horizontal 1-cells are objects of M, and 2-cells (5.4) are maps

$$\theta : (m_n \otimes \cdots \otimes m_1) \longrightarrow m$$

in M.

Example 5.1.9 Here is a family of **fc**-multicategories that are not usually degenerate in any of the ways listed above. Let V be a plain multicategory and define the **fc**-multicategory **Set**[V] as follows:

- objects are sets
- vertical 1-cells are functions
- a horizontal 1-cell $A \longrightarrow A'$ is a family $(m_{a,a'})_{a \in A, a' \in A'}$ of objects of V

- a 2-cell

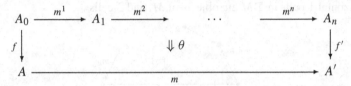

is a family

$$\left(m^1_{a_0,a_1}, \ldots, m^n_{a_{n-1},a_n} \xrightarrow{\;\theta_{a_0,\ldots,a_n}\;} m_{f(a_0),f'(a_n)} \right)_{a_0 \in A_0, \ldots, a_n \in A_n}$$

of maps in V

- composition and identities are obvious.

If $V = \mathbf{Set}$ then $\mathbf{Set}[V]$ is the **fc**-multicategory $\mathbf{Set}_2$ of 5.1.3. But $\mathbf{Set}[V]$ is not (the underlying **fc**-multicategory of) a weak double category unless V happens to be (the underlying plain multicategory of) a monoidal category, since $\mathbf{Set}[V]$ does not usually have 'enough universal 2-cells'. Compare the results on representable multicategories in Section 3.3.

We will not be much concerned with algebras for **fc**-multicategories, but let us look at them briefly. Given an object E of $\mathcal{E}$, that is, a directed graph, an object $(X \longrightarrow E)$ of $\mathcal{E}/E$ amounts to a set $X(a)$ for each vertex a of E and a span

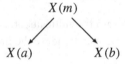

for each edge $a \xrightarrow{\;m\;} b$ of E. An algebra for an **fc**-multicategory C is an object $(X \longrightarrow C_0)$ of $\mathcal{E}/C_0$ together with an action of C on X, and so a C-algebra is just an **fc**-multicategory map from C into the (large) **fc**-multicategory $\mathbf{Set}_2$ defined in 5.1.3.

There are some interesting examples of algebras when C is the **fc**-multicategory coming from a plain operad P: in the notation of 5.1.7, $C = \Sigma P$. A (ΣP)-algebra is called a **categorical P-algebra** or a P-**category**, and consists of a directed graph X together with a function

$$P(n) \times X(x_{n-1}, x_n) \times \cdots \times X(x_0, x_1) \longrightarrow X(x_0, x_n) \qquad (5.5)$$

for each $n \in \mathbb{N}$ and sequence $x_0, \ldots, x_n$ of vertices of X, satisfying the evident axioms. (Here $X(x, x')$ denotes the set of edges from x to x'.) If $P = 1$

then a P-category is exactly a category; there are some less trivial examples too.

Example 5.1.10 Let E be the operad in which

$$E(n) = \{\text{endpoint-preserving continuous maps } [0, 1] \longrightarrow [0, n]\},$$

as in 2.2.18, so that any loop space is naturally an E-algebra. Then any path-space is naturally an E-category. In other words, take a topological space Y and let X be the graph whose vertices are points of Y and whose edges are continuous maps from $[0, 1]$ into Y: then X is naturally a E-category.

Example 5.1.11 Similarly, taking A_∞ to be the operad of chain complexes whose algebras are A_∞-algebras (p. 234) gives us the standard notion of an A_∞-**category**. For this to make sense, we must work in a world where everything is enriched in chain complexes: so X consists of a set X_0 of objects (or vertices) together with a chain complex $X(x, x')$ for each pair (x, x') of objects, and in (5.5) the $\times$s become $\otimes$s. We do not meet enriched generalized multicategories properly until Section 6.8, but it is clear how things should work in this particular situation.

5.2 Weak double categories

Generalizing both bicategories and strict double categories are 'weak double categories', introduced informally above. Recall that these are only weak in the horizontal direction: vertical composition still obeys strict laws. We do not consider double categories weak in both directions.

The definition of weak double category is a cross between that of strict double category and that of unbiased bicategory.

Definition 5.2.1 A **weak double category** D consists of some data subject to some axioms. The data is:

- A diagram

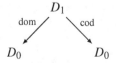

of categories and functors. The objects of D_0 are called the **0-cells** or **objects** of D, the maps in D_0 are the **vertical 1-cells** of D, the objects of D_1 are the **horizontal 1-cells** of D, and the maps in D_1 are the **2-cells** of D, as in the picture

$$
\begin{array}{ccc}
a & \xrightarrow{\;m\;} & a' \\
f\downarrow & \Downarrow\theta & \downarrow f' \\
b & \xrightarrow{\;p\;} & b'
\end{array}
\tag{5.6}
$$

where $a \xrightarrow{f} b$, $a' \xrightarrow{f'} b'$ are maps in D_0 and $m \xrightarrow{\theta} p$ is a map in D_1, with $\mathrm{dom}(m) = a$, $\mathrm{dom}(\theta) = f$, and so on.

• For each $n \geq 0$, a functor $\mathrm{comp}^{(n)} : D_1^{(n)} \longrightarrow D_1$ such that the diagram

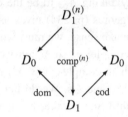

commutes, where $D_1^{(n)}$ is the limit of the diagram

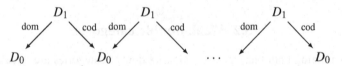

containing n copies of D_1. The functor $\mathrm{comp}^{(n)}$ is called n-**fold horizontal composition** and written

$$
\begin{array}{ccccccc}
a_0 & \xrightarrow{\;m_1\;} & a_1 & \xrightarrow{\;m_2\;} & \cdots & \xrightarrow{\;m_n\;} & a_n \\
f_0\downarrow & \Downarrow\theta_1 & f_1\downarrow & \Downarrow\theta_2 & & \Downarrow\theta_n & \downarrow f_n \\
b_0 & \xrightarrow{\;p_1\;} & b_1 & \xrightarrow{\;p_2\;} & \cdots & \xrightarrow{\;p_n\;} & b_n
\end{array}
$$

$$
\longmapsto \qquad
\begin{array}{ccc}
a_0 & \xrightarrow{(m_n\circ\cdots\circ m_1)} & a_n \\
f_0\downarrow & \Downarrow(\theta_n * \cdots * \theta_1) & \downarrow f_n \\
b_0 & \xrightarrow{(p_n\circ\cdots\circ p_1)} & b_n.
\end{array}
$$

• For each double sequence

$$
\mathbf{m} = ((m_1^1, \ldots, m_1^{k_1}), \ldots, (m_n^1, \ldots, m_n^{k_n}))
$$

of horizontal 1-cells such that the composites below make sense, an invertible 2-cell

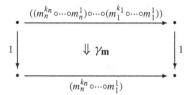

(where 'invertible' refers to vertical composition).

- For each horizontal 1-cell $a \xrightarrow{m} a'$, an invertible 2-cell

$$
\begin{array}{ccc}
a & \xrightarrow{m} & a' \\
1 \downarrow & \Downarrow \iota_m & \downarrow 1 \\
a & \xrightarrow[(m)]{} & a'.
\end{array}
$$

The axioms are:

- $\gamma_{\mathbf{m}}$ is natural in each of the m_i^js, and ι_m is natural in m. In the case of ι, this means that for each 2-cell θ as in (5.6) we have

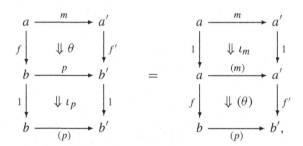

and similarly for γ.

- γ and ι satisfy associativity and identity coherence axioms analogous to those in the definition (3.1.1) of lax monoidal category.

It is clear how to define lax maps between weak double categories, again working by analogy with unbiased monoidal categories or unbiased bicategories, and there is a forgetful functor

$$
\text{(weak double categories and lax maps)} \longrightarrow \textbf{fc-Multicat}.
$$

Example 5.2.2 There are several degenerate cases. A weak double category in which the coherence cells γ and ι are all identities is exactly a strict double category. A weak double category whose only vertical 1-cells are identities is exactly an (unbiased) bicategory. A weak double category whose only horizontal 1-cells are identities is exactly a strict 2-category.

Example 5.2.3 The **fc**-multicategory $\mathbf{Ring}_2$ of 5.1.1 is a weak double category, horizontal composition being tensor of modules.

Example 5.2.4 The **fc**-multicategory $\mathbf{Set}_2$ of 5.1.3 is also a weak double category. It is formed from sets, functions, and spans, which is reminiscent of the definition of T-multicategories and maps between them in Section 4.2. Indeed, let T be a cartesian monad on a cartesian category $\mathcal{E}$, and define a weak double category as follows:

- objects are objects of $\mathcal{E}$
- vertical 1-cells are maps in $\mathcal{E}$
- horizontal 1-cells $E \longrightarrow E'$ are diagrams

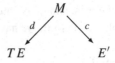

in $\mathcal{E}$
- 2-cells θ are commutative diagrams

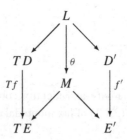

- vertical composition is as in $\mathcal{E}$
- horizontal composition is by pullback: the composite of horizontal 1-cells

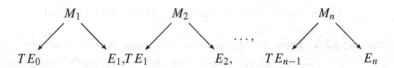

is the limit of the diagram

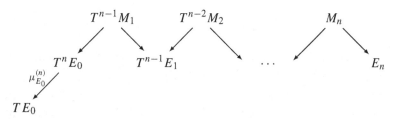

where $\mu^{(n)}$ is the n-fold multiplication of the monad T; horizontal composition of 2-cells works similarly.

Any weak double category yields a bicategory by discarding all the vertical 1-cells except for the identities, and applying this to the weak double category just defined yields the bicategory $\mathcal{E}_{(T)}$ (Definition 4.2.1). It is therefore reasonable to call our weak double category $\mathcal{E}_{(T)}$ too. In particular, $\mathbf{Set}_{(\mathrm{id})} = \mathbf{Set}_2$.

Example 5.2.5 Categories, functors and modules are integrated in the following weak double category $\mathbf{Cat}_2$:

- objects are small categories
- vertical 1-cells are functors
- horizontal 1-cells are modules (Section 2.3): that is, a horizontal 1-cell $A \longrightarrow A'$ is a functor $A^{\mathrm{op}} \times A' \longrightarrow \mathbf{Set}$
- a 2-cell

$$
\begin{array}{ccc}
A & \xrightarrow{\;M\;} & A' \\
{\scriptstyle F}\downarrow & \Downarrow & \downarrow{\scriptstyle F'} \\
B & \xrightarrow[\;P\;]{} & B'
\end{array}
$$

is a natural transformation

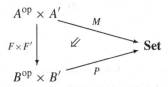

- vertical composition is ordinary composition of functors and natural transformations
- horizontal composition is tensor of modules: the composite $(M_n \otimes \cdots \otimes M_1)$ of

$$
A_0 \xrightarrow{\;M_1\;} A_1 \xrightarrow{\;M_2\;} \quad \cdots \quad \xrightarrow{\;M_n\;} A_n
$$

is given by the coend formula

$$(M_n \otimes \cdots \otimes M_1)(a_0, a_n)$$
$$= \int^{a_1, \dots, a_{n-1}} M_n(a_{n-1}, a_n) \times \cdots \times M_1(a_0, a_1)$$

($a_0 \in A_0, a_n \in A_n$), or when $n = 0$ by $I_A(a, a') = A(a, a')$; similarly 2-cells. (A coend is a colimit of sorts: Mac Lane, 1971, Ch. IX.)

The weak double category $\mathbf{Cat}_2$ incorporates not only categories, functors and modules, but also natural transformations: for by a Yoneda argument, a 2-cell

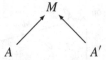

amounts to a natural transformation $F \longrightarrow F'$.

In the next section we will see that T-multicategories and maps, transformations and modules between them form an **fc**-multicategory, but not in general a weak double category.

Example 5.2.6 There is a weak double category $\mathbf{Monoid}_2$ made up of monoids, homomorphisms of monoids, bimodules over monoids, and maps between them. This is just like $\mathbf{Ring}_2$ but with all the additive structure removed. Alternatively, it is the substructure of $\mathbf{Cat}_2$ whose 0-cells are just the one-object categories, and with all 1- and 2-cells between them.

Example 5.2.7 We have considered using spans as horizontal 1-cells; we can also use cospans. Thus, if $\mathcal{E}$ is any category in which all pushouts exist then there is a weak double category whose objects and vertical 1-cells form the category $\mathcal{E}$, whose horizontal 1-cells $A \longrightarrow A'$ are diagrams

$$\begin{array}{ccc} & M & \\ & \nearrow \; \nwarrow & \\ A & & A' \end{array}$$

in $\mathcal{E}$, and with the rest of the structure defined in the obvious way.

For instance, let $\mathcal{E}$ be the category of topological spaces and embeddings: then M is a space containing copies of A and A' as subspaces, and horizontal composition in the weak double category is gluing along subspaces.

Example 5.2.8 Adapting the previous example slightly, we obtain a weak double category n-$\mathbf{Mfd}_2$ of n-manifolds and cobordisms, for each $n \geq 0$. An object is an n-dimensional manifold (topological, say, and without boundary).

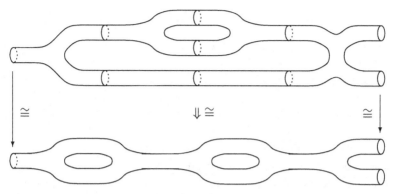

Fig. 5-B. 2-cell in the **fc**-multicategory 1-**Mfd**$_2$, with four horizontal 1-cells along the top row

A vertical 1-cell is a continuous map. A horizontal 1-cell $A \longrightarrow A'$ is an $(n + 1)$-manifold M together with a homeomorphism h between the boundary ∂M of M and the disjoint union $A \amalg A'$. A 2-cell

$$
\begin{array}{ccc}
A & \xrightarrow{\;(M,h)\;} & A' \\
{\scriptstyle f}\big\downarrow & \Downarrow & \big\downarrow{\scriptstyle f'} \\
B & \xrightarrow[\;(P,k)\;]{} & B'
\end{array}
$$

is a continuous map $\theta : M \longrightarrow P$ making the diagram

$$
\begin{array}{ccc}
\partial M & \xrightarrow[\sim]{h} & A \amalg A' \\
{\scriptstyle \theta}\big\downarrow & & \big\downarrow{\scriptstyle f \amalg f'} \\
\partial P & \xrightarrow[k]{\sim} & B \amalg B'
\end{array}
$$

commute. Vertical composition is composition of functions; horizontal composition is by gluing.

Fig. 5-B shows a 2-cell in the underlying **fc**-multicategory of 1-**Mfd**$_2$, in a case where all the continuous maps involved are homeomorphisms.

Similar constructions can be made for other types of manifold: oriented, smooth, holomorphic, ... There is a slight problem with identities for horizontal composition, as the horizontal identity on A 'ought' to be just A itself (compare 5.2.7) but this is not usually counted as an $(n + 1)$-manifold with boundary. Whether or not we can fix this, there is certainly an underlying **fc**-multicategory: the problem is purely one of representability.

5.3 Monads, monoids and modules

Modules have loomed large in our examples of **fc**-multicategories, appearing as horizontal 1-cells. Monoids and their cousins (such as rings and categories) have also been prominent, appearing as 0-cells. Here we show how any **fc**-multicategory C gives rise to a new **fc**-multicategory **Mon**(C), whose 0-cells are monoids/monads in C and whose horizontal 1-cells are modules between them.

As I shall explain, this construction has traditionally been carried out in a narrower context than **fc**-multicategories, which has meant working under certain technical restrictions. If we expand to the wider context of **fc**-multicategories then the technicalities vanish.

For an example of the traditional construction, start with the monoidal category (**Ab**, $\otimes$, $\mathbb{Z}$) of abelian groups. A monoid therein is just a ring. Given two monoids A, B in a monoidal category, a (B, A)-**module** is an object M equipped with compatible left and right actions

$$B \otimes M \xrightarrow{\ \mathrm{act_L}\ } M, \qquad M \otimes A \xrightarrow{\ \mathrm{act_R}\ } M,$$

and there is an obvious notion of map between (B, A)-modules. In this particular case these are (bi)modules and their homomorphisms, in the usual sense. So we have almost arrived at the bicategory of 1.5.5: 0-cells are rings, 1-cells are modules, and 2-cells are maps of modules. The only missing ingredient is the tensor product of modules. If M is a (B, A)-module and N a (C, B)-module then $N \otimes_B M$ is a quotient of the abelian group tensor $N \otimes M$; categorically, it is the (reflexive) coequalizer

$$N \otimes B \otimes M \mathrel{\substack{\xrightarrow{\mathrm{act_R} \otimes 1_M} \\[-0.3em] \xrightarrow[1_N \otimes \mathrm{act_L}]{}}} N \otimes M \longrightarrow N \otimes_B M.$$

The abelian group $N \otimes M$ acquires a (C, A)-module structure just as long as the endofunctors $C \otimes -$ and $- \otimes A$ of **Ab** preserve (reflexive) coequalizers, which they do. The rest of the bicategory structure comes easily. So from the monoidal category (**Ab**, $\otimes$, $\mathbb{Z}$) we have derived the bicategory of rings, modules and maps of modules.

In the same way, any monoidal category ($\mathcal{A}$, $\otimes$, I) gives rise to a bicategory **Mon**($\mathcal{A}$) of monoids and modules in $\mathcal{A}$, as long as $\mathcal{A}$ has reflexive coequalizers and these are preserved by the functors $A \otimes -$ and $- \otimes A$ for all $A \in \mathcal{A}$. Generalizing, any bicategory $\mathcal{B}$ gives rise to a bicategory **Mon**($\mathcal{B}$) of monads and modules in $\mathcal{B}$, as long as $\mathcal{B}$ locally has reflexive coequalizers and these are preserved by the functors $f \circ -$ and $- \circ f$ for all 1-cells f.

There are two unsatisfactory aspects to this construction. One is the necessity of the coequalizer conditions. The other is that homomorphisms of monoids are conspicuous by their absence: for instance, ring homomorphisms are not an explicit part of **Mon(Ab)** (although they can be recovered as those 1-cells of **Mon(Ab)** that have a right adjoint). The following definition for **fc**-multicategories solves both problems.

Definition 5.3.1 Let C be an **fc**-multicategory. The **fc**-multicategory **Mon(C)** is defined as follows.

- A 0-cell of **Mon(C)** is an **fc**-multicategory map $1 \longrightarrow C$. That is, it is a 0-cell a of C together with a horizontal 1-cell $a \xrightarrow{\; t \;} a$ and 2-cells

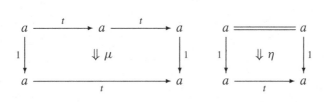

satisfying the usual axioms for a monad, $\mu \circ (\mu, 1_t) = \mu \circ (1_t, \mu)$ and $\mu \circ (\eta, 1_t) = 1_t = \mu \circ (1_t, \eta)$.

- A vertical 1-cell

$$(a, t, \mu, \eta)$$
$$\downarrow$$
$$(\hat{a}, \hat{t}, \hat{\mu}, \hat{\eta})$$

in **Mon(C)** is a vertical 1-cell

in C together with a 2-cell

$$\begin{array}{ccc} a & \xrightarrow{\; t \;} & a \\ f \downarrow & \Downarrow \omega & \downarrow f \\ \hat{a} & \xrightarrow[\; \hat{t} \;]{} & \hat{a} \end{array}$$

such that $\omega \circ \mu = \hat{\mu} \circ (\omega, \omega)$ and $\omega \circ \eta = \hat{\eta} \circ f$. (The notation $\hat{\eta} \circ f$ is explained on p. 161.)

- A horizontal 1-cell $(a, t, \mu, \eta) \longrightarrow (a', t', \mu', \eta')$ is a horizontal 1-cell

$a \xrightarrow{m} a'$ in C together with 2-cells

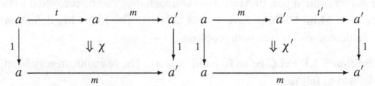

satisfying the usual module axioms, $\chi \circ (\mu, 1_m) = \chi \circ (1_t, \chi)$, $\chi \circ (\eta, 1_m) = 1_m$, and dually for χ', and the 'commuting actions' axiom, $\chi' \circ (\chi, 1_{t'}) = \chi \circ (1_t, \chi')$.

• A 2-cell

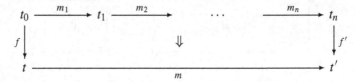

in **Mon**(C), where t stands for (a, t, μ, η), m for (m, χ, χ'), f for (f, ω), and so on, is a 2-cell

$$
\begin{array}{ccccccc}
a_0 & \xrightarrow{m_1} & a_1 & \xrightarrow{m_2} & & \xrightarrow{m_n} & a_n \\
\Big\downarrow f & & & \Downarrow \theta & & & \Big\downarrow f' \\
a & & & \xrightarrow{\quad m \quad} & & & a'
\end{array}
$$

in C satisfying the 'external equivariance' axioms

$$\theta \circ (\chi_1, 1_{m_2}, \ldots, 1_{m_n}) = \chi \circ (\omega, \theta)$$
$$\theta \circ (1_{m_1}, \ldots, 1_{m_{n-1}}, \chi'_n) = \chi' \circ (\theta, \omega')$$

and the 'internal equivariance' axioms

$$\theta \circ (1_{m_1}, \ldots, 1_{m_{i-2}}, \chi'_{i-1}, 1_{m_i}, 1_{m_{i+1}}, \ldots, 1_{m_n})$$
$$= \theta \circ (1_{m_1}, \ldots, 1_{m_{i-2}}, 1_{m_{i-1}}, \chi_i, 1_{m_{i+1}}, \ldots, 1_{m_n})$$

for $2 \leq i \leq n$.

• Composition and identities for both 2-cells and vertical 1-cells in **Mon**(C) are just composition and identities in C.

With the obvious definition on maps of **fc**-multicategories, this gives a functor

$$\textbf{Mon} : \textbf{fc-Multicat} \longrightarrow \textbf{fc-Multicat}.$$

Example 5.3.2 Our new **Mon** generalizes the traditional **Mon** in the following sense. Let B be a bicategory satisfying the conditions on local reflexive coequalizers mentioned above, so that it is possible to construct the bicategory **Mon**(B) in the traditional way. Let C be the **fc**-multicategory corresponding to B (Example 5.1.5), with only trivial vertical 1-cells. Then a 0-cell of **Mon**(C) is a monad in B, a horizontal 1-cell $t \longrightarrow t'$ is a (t', t)-module in B, and a 2-cell of the form

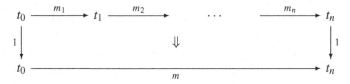

is a map

$$m_n \otimes_{t_{n-1}} \cdots \otimes_{t_2} m_2 \otimes_{t_1} m_1 \longrightarrow m$$

of (t_n, t_0)-modules – that is, a 2-cell in **Mon**(B). So if we discard the non-identity vertical 1-cells of **Mon**(C) then we obtain the **fc**-multicategory corresponding to **Mon**(B).

Example 5.3.3 Take the monoidal category (**Ab**, $\otimes$, $\mathbb{Z}$) and the corresponding **fc**-multicategory Σ**Ab** (Example 5.1.8). Then **Mon**(Σ**Ab**) is **Ring**$_2$, the **fc**-multicategory of Example 5.1.1 in which 0-cells are rings, vertical 1-cells are homomorphisms of rings, horizontal 1-cells are modules between rings, and 2-cells are module maps of a suitable kind.

Example 5.3.4 Similarly, if we take the monoidal category (**Set**, $\times$, 1) then **Mon**(Σ**Set**) is (the underlying **fc**-multicategory of) the weak double category **Monoid**$_2$ (Example 5.2.6).

Example 5.3.5 Take the **fc**-multicategory **Set**$_2$ of sets, functions and spans (5.1.3). Then a 0-cell of **Mon**(**Set**$_2$) is a monad in **Set**$_2$, that is, a small category. A vertical 1-cell is a functor; a horizontal 1-cell is a module. A 2-cell

$$
\begin{array}{ccccccc}
A_0 & \xrightarrow{M_1} & A_1 & \xrightarrow{M_2} & & \xrightarrow{M_n} & A_n \\
F \downarrow & & & \Downarrow \theta & & & \downarrow F' \\
A & & & \xrightarrow{M} & & & A'
\end{array}
$$

consists of a function

$$\theta_{a_0,\ldots,a_n} : M_n(a_{n-1}, a_n) \times \cdots \times M_1(a_0, a_1) \longrightarrow M(Fa_0, F'a_n)$$

for each $a_0 \in A_0, \ldots, a_n \in A_n$, such that this family is natural in the a_is. So **Mon(Set$_2$)** is **Cat$_2$**, the weak double category of 5.2.5.

This last example leads us to definitions of module and transformation for generalized multicategories. Let T be a cartesian monad on a cartesian category $\mathcal{E}$. Recall the **fc**-multicategory $\mathcal{E}_{(T)}$ of 5.2.4, and consider the **fc**-multicategory **Mon($\mathcal{E}_{(T)}$)**. A 0-cell of **Mon($\mathcal{E}_{(T)}$)** is a monad in $\mathcal{E}_{(T)}$, that is, a T-multicategory, so we write T-**Multicat** for **Mon($\mathcal{E}_{(T)}$)**. A vertical 1-cell is a map of T-multicategories. A horizontal 1-cell $A \longrightarrow B$ will be called a (B, A)-**module**. To see what modules are explicitly, note that the underlying graph

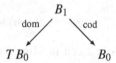

of B is a horizontal 1-cell in $\mathcal{E}_{(T)}$, composition with which defines an endofunctor $B \circ -$ of the category $\mathcal{E}_{(T)}(A_0, B_0)$ of diagrams of the form

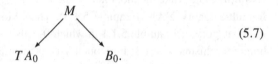

$$(5.7)$$

Moreover, the multicategory structure of B gives $B \circ -$ the structure of a monad on $\mathcal{E}_{(T)}(A_0, B_0)$. Dually, the multicategory A also induces a monad $- \circ A$ on $\mathcal{E}_{(T)}(A_0, B_0)$, and there is a canonical isomorphism $(B \circ -) \circ A \cong B \circ (- \circ A)$ compatible with the monad structures (formally, a distributive law: 6.1.2). So a (B, A)-module is a diagram (5.7) in $\mathcal{E}$ together with maps $\text{act}_L : B \circ M \longrightarrow M$ and $\text{act}_R : M \circ A \longrightarrow M$ such that (M, act_L) and (M, act_R) are algebras for the monads $B \circ -$ and $- \circ A$ respectively, and the following diagram commutes:

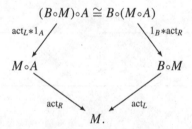

This definition generalizes the definition for plain multicategories (2.3). Just as there, it also makes sense to define left and right modules for a

T-multicategory: do not take a whole span M, but only the right-hand or left-hand half. A **left B-module** is nothing other than a B-algebra. A **right A-module** (or **A-coalgebra**) is an object M of $\mathcal{E}/T A_0$ (thought of as a span

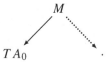

with a phantom right leg) together with a map $h : M \circ A \longrightarrow M$ making M into an algebra for the monad $- \circ A$ on $\mathcal{E}/T A_0$.

In general, T-**Multicat** is not a weak double category: there is no tensor product of modules. For to form the tensor product of modules we would need $\mathcal{E}$ to possess reflexive coequalizers and T to preserve them, and although the requirement on $\mathcal{E}$ is satisfied in almost all of the examples that we consider, the requirement on T is not – for instance, when $T = \mathbf{fc}$.

There is, however, an 'identity' horizontal 1-cell on each object. Given a T-multicategory A, this is the (A, A)-module I_A whose underlying span is

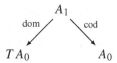

and whose actions are both composition in A. Recalling the remarks on natural transformations in Example 5.2.5, we make the following definition. Let $A \underset{f'}{\overset{f}{\rightrightarrows}} B$ be maps of T-multicategories. A **transformation**

$$(5.8)$$

of T-multicategories is a 2-cell

$$
\begin{array}{ccc}
A & \xrightarrow{\ I_A\ } & A \\
{\scriptstyle f}\downarrow & \Downarrow \alpha & \downarrow{\scriptstyle f'} \\
B & \xrightarrow[\ I_B\ ]{} & B
\end{array}
$$

in T-**Multicat**. Explicitly, a transformation $f \longrightarrow f'$ is a map $\alpha : A_1 \longrightarrow$ B_1 such that the diagrams

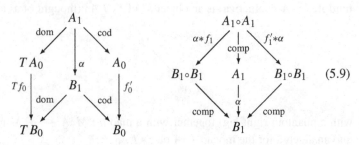

$$(5.9)$$

commute, where the map $\alpha * f_1 : A_1 \circ A_1 \longrightarrow B_1 \circ B_1$ is induced by the maps $\alpha : A_1 \longrightarrow B_1$ and $f_1 : A_1 \longrightarrow B_1$, and similarly $f'_1 * \alpha$.

This may come as a surprise: if A and B are ordinary categories and α a natural transformation as in (5.8) then we would usually think of α as a map $A_0 \longrightarrow B_1$, not $A_1 \longrightarrow B_1$. Observe, however, that α assigns to each map $a \xrightarrow{\theta} a'$ in A a map $fa \xrightarrow{\alpha_\theta} f'a'$ in B, the diagonal of the naturality square for α at θ. Since α_a can be recovered as α_{1_a} for each $a \in A$, it must be possible to define a natural transformation (5.8) as a map $A_1 \longrightarrow B_1$ satisfying axioms. The same goes for transformations of plain multicategories, where now α_θ is either side of the equation in Fig. 2-I (p. 88). For generalized multicategories we also have the choice between viewing transformations as maps $A_0 \longrightarrow B_1$ or as maps $A_1 \longrightarrow B_1$; the axioms for the $A_0 \longrightarrow B_1$ version can easily be worked out (or looked up: Leinster, 1999a, 1.1.1).

For general cartesian T, suppose that we discard from the **fc**-multicategory T-**Multicat** all of the horizontal 1-cells except for the 'identities' (those of the form I_A). Then we obtain a weak double category whose only horizontal 1-cells are identities – in other words (5.2.2), a strict 2-category. Concretely: transformations can be composed vertically and horizontally, and this makes T-**Multicat** into a strict 2-category.

Notes

Many **fc**-multicategories are weak double categories. Many **fc**-multicategories also have the property that every vertical 1-cell $f : a \longrightarrow b$ gives rise canonically to cells

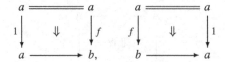

(as is familiar in **Ring**$_2$, for instance). These two properties combine to say that any diagram

has a 2-cell filling it in (canonically, up to isomorphism): the weak double category structure provides a fill-in when f and f' are both identities, the vertical-to-horizontal property provides a fill-in when $n = 0$ and either f or f' is an identity, and the general case can be built up from these special cases. The exact situation is still unclear, but there are obvious similarities between this, the representations of plain multicategories in Section 3.3, and the universal opetopic cells in Section 7.5.

The **fc**-multicategory **Set**[V] of 5.1.9 appears in a different guise in Carboni, Kasangian and Walters (1987), under the name of 'matrices'. Weak double categories have also been studied by Grandis and Paré (1999, 2003).

Chapter 6

Generalized operads and multicategories: further theory

The last paragraph plays on the postmodern fondness for 'multidimensionality' and 'nonlinearity' by inventing a nonexistent field: 'multidimensional (nonlinear) logic'.

Sokal and Bricmont (1998)

This chapter is an assortment of topics in the theory of T-multicategories. Some are included because they answer natural questions, some because they connect to established concepts for classical operads, and some because we will need them later. The reader who wants to get on to geometrically interesting structures should skip this chapter and come back later if necessary; there are no pictures here.

In Section 6.1 we 'recall' some categorical language: maps between monads, mates under adjunctions, and distributive laws. This language has nothing intrinsically to do with generalized multicategories, but it will be efficient to use it in some parts of this chapter (Sections 6.2 and 6.7) and later chapters.

The first three proper sections each recast one of the principal definitions. In Section 6.2 we find an alternative definition of generalized multicategory, which amounts to characterizing a T-multicategory C by its free-algebra monad T_C plus one small extra piece of data. Both Sections 6.3 and 6.4 are alternative ways of defining an algebra for a T-multicategory. The former generalizes the categorical fact that **Set**-valued functors can be described as discrete fibrations, and the latter the fact that a classical operad-algebra is a map into an endomorphism operad (often taken as the *definition* of algebra by 'working mathematicians' using operads).

The next two sections are also generalizations. The free T-multicategory on a T-graph is discussed in Section 6.5. Abstract as this may seem, it is crucial to the way in which geometry arises spontaneously from category theory: witness the planar trees of Section 2.3 and the opetopes of Chapter 7. Then we make a definition to complete the phrase: 'plain multicategories are to monoidal categories as T-multicategories are to ...'; we call these things 'T-structured categories' (Section 6.6).

So far all is generalization, but the final two sections are genuinely new. A choice of cartesian monad T specifies the type of input shape that the operations in a T-multicategory will have, so changing T amounts to a change of shape. In Section 6.7 we show how to translate between different shapes, in other words, how a relation between two monads T and T' induces a relation between the classes of T- and T'-multicategories. Finally, in Section 6.8 we take a short look at enrichment of generalized multicategories. As mentioned in the Introduction, there is much more to this than one might guess, and in particular there is more than we have room for; this is just a taste.

6.1 More on monads

To compare approaches to higher categorical structures using different shapes or of different dimensions we will need a notion of map $(\mathcal{E}, T) \longrightarrow (\mathcal{E}', T')$ between monads. Actually, there are various such notions: lax, colax, weak and strict. Such comparisons lead to functors between categories of structures, and to discuss adjunctions between such functors it will be convenient to use the language of mates (an Australian creation, of course). It will also make certain later proofs (9.4.1, F.1.1) easier if we know a little about distributive laws, which are recipes for gluing together two monads T, T' on the same category to give a monad structure on the composite functor $T' \circ T$.

Nothing here is new. I learned this material from Street (1972) and Kelly and Street (1974), although I have changed some terminology. Distributive laws were introduced by Beck (1969). Where Street discusses monads in an arbitrary 2-category $\mathcal{V}$, we stick to the case $\mathcal{V} = \mathbf{CAT}$, since that is all we need.

Let $T = (T, \mu, \eta)$ be a monad on a category $\mathcal{E}$ and $T' = (T', \mu', \eta')$ a monad on a category $\mathcal{E}'$. A **lax map of monads** $(\mathcal{E}, T) \longrightarrow (\mathcal{E}', T')$ is a functor $Q : \mathcal{E} \longrightarrow \mathcal{E}'$ together with a natural transformation

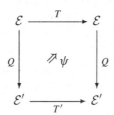

making the diagrams

$$T'^2Q \xrightarrow{\;T'\psi\;} T'QT \xrightarrow{\;\psi T\;} QT^2 \qquad\qquad Q =\!=\!=\!=\!= Q$$

$$\mu'Q \downarrow \qquad\qquad\qquad\qquad \downarrow Q\mu \qquad \eta'Q \downarrow \qquad\qquad \downarrow Q\eta$$

$$T'Q \xrightarrow[\quad\psi\quad]{} QT \qquad\qquad T'Q \xrightarrow[\;\psi\;]{} QT$$

commute. If $(\mathcal{E}, T) \xrightarrow{(\tilde{Q}, \tilde{\psi})} (\mathcal{E}', T')$ is another lax map of monads then a **transformation** $(Q, \psi) \longrightarrow (\tilde{Q}, \tilde{\psi})$ is a natural transformation $Q \xrightarrow{\;\alpha\;} \tilde{Q}$ such that

$$T'Q \xrightarrow{\;\psi\;} QT$$

$$T'\alpha \downarrow \qquad\qquad \downarrow \alpha T$$

$$T'\tilde{Q} \xrightarrow[\;\tilde{\psi}\;]{} \tilde{Q}T$$

commutes. There is a strict 2-category $\mathbf{Mnd}_{\text{lax}}$ whose 0-cells are pairs $(\mathcal{E}, T)$, whose 1-cells are lax maps of monad, and whose 2-cells are transformations.

Dually, if T and T' are monads on categories $\mathcal{E}$ and $\mathcal{E}'$ respectively then a **colax map of monads** $(\mathcal{E}, T) \longrightarrow (\mathcal{E}', T')$ consists of a functor $P : \mathcal{E} \longrightarrow \mathcal{E}'$ together with a natural transformation

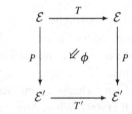

satisfying axioms dual to those for lax maps. With the accompanying notion of **transformation** between colax maps of monads, we obtain another strict 2-category $\mathbf{Mnd}_{\text{colax}}$.

A **weak map of monads** is a lax map (Q, ψ) of monads in which ψ is an isomorphism (or equivalently, a colax map (P, ϕ) in which ϕ is an isomorphism), and a **strict map of monads** is a lax map (Q, ψ) in which ψ is the identity (and so $T'Q = QT$).

Often $\mathcal{E} = \mathcal{E}'$ and the functor $\mathcal{E} \longrightarrow \mathcal{E}'$ is the identity. A natural transformation $\psi : T' \longrightarrow T$ **commutes with the monad structures** if (id, ψ) is a lax map of monads $(\mathcal{E}, T) \longrightarrow (\mathcal{E}, T')$ or, equivalently, a colax map $(\mathcal{E}, T') \longrightarrow (\mathcal{E}, T)$.

A crucial property of lax maps of monads is that they induce maps between categories of algebras: $(Q, \psi) : (\mathcal{E}, T) \longrightarrow (\mathcal{E}', T')$ induces the functor

$$Q_* = (Q, \psi)_* : \qquad \mathcal{E}^T \longrightarrow \mathcal{E}'^{T'},$$

$$\begin{pmatrix} TX \\ \downarrow h \\ X \end{pmatrix} \longmapsto \begin{pmatrix} T'QX \\ \downarrow \psi_X \\ QTX \\ \downarrow Qh \\ QX \end{pmatrix}.$$

(Dually, but less usefully for us, a colax map of monads (P, ϕ) : $(\mathcal{E}, T) \longrightarrow (\mathcal{E}', T')$ induces a functor $\mathcal{E}_T \longrightarrow \mathcal{E}'_{T'}$ between Kleisli categories.) In fact we have:

Lemma 6.1.1 *Let* $T = (T, \mu, \eta)$ *and* $T' = (T', \mu', \eta')$ *be monads on categories* $\mathcal{E}$ *and* $\mathcal{E}'$, *respectively. Then there is a one-to-one correspondence between lax maps of monads* $(\mathcal{E}, T) \longrightarrow (\mathcal{E}', T')$ *and pairs* (Q, R) *of functors such that the square*

$$\begin{array}{ccc} \mathcal{E}^T & \xrightarrow{\;R\;} & \mathcal{E}'^{T'} \\ \text{forgetful} \downarrow & & \downarrow \text{forgetful} \\ \mathcal{E} & \xrightarrow{\;Q\;} & \mathcal{E}' \end{array} \qquad (6.1)$$

commutes, with a lax map (Q, ψ) *corresponding to the pair* (Q, Q_*).

Proof Given (Q, ψ), the square (6.1) with $R = Q_*$ plainly commutes. Conversely, take a pair (Q, R) such that (6.1) commutes. For each $X \in \mathcal{E}$, write $(T'QTX \xrightarrow{\chi_X} QTX)$ for the image under R of the free algebra $(T^2X \xrightarrow{\mu_X} TX)$, then put

$$\psi_X = (T'QX \xrightarrow{T'Q\eta_X} T'QTX \xrightarrow{\chi_X} QTX).$$

This defines a lax map of monads (Q, ψ). It is easily checked that the two processes described are mutually inverse. $\qquad \square$

Lax and colax maps can be related as 'mates'. Suppose we have an adjunction

$$\begin{array}{c} \mathcal{D} \\ P \uparrow \dashv \downarrow Q \\ \mathcal{D}' \end{array}$$

and functors $\mathcal{D} \xrightarrow{T} \mathcal{D}$, $\mathcal{D}' \xrightarrow{T'} \mathcal{D}'$. Then there is a one-to-one correspondence between natural transformations ϕ and natural transformations ψ with domains and codomains as shown:

This is given by

$$\psi = \left(T'Q \xrightarrow{\gamma T'Q} QPT'Q \xrightarrow{Q\phi Q} QTPQ \xrightarrow{QT\delta} QT \right),$$
$$\phi = \left(PT' \xrightarrow{PT'\gamma} PT'QP \xrightarrow{P\psi P} PQTP \xrightarrow{\delta TP} TP \right)$$

where γ and δ are the unit and counit of the adjunction. We call ψ the **mate** of ϕ and write $\psi = \overline{\phi}$; dually, we call ϕ the **mate** of ψ and write $\phi = \overline{\psi}$. The world of mates is strictly monogamous: everybody has exactly one mate, and your mate's mate is you ($\overline{\overline{\phi}} = \phi$, $\overline{\overline{\psi}} = \psi$). All imaginable statements about mates are true. In particular, if T and T' have the structure of monads then (P, ϕ) is a colax map of monads if and only if $(Q, \overline{\phi})$ is a lax map of monads.

We finish by showing how to glue monads together. Given two monads (S, μ, η) and (S', μ', η') on the same category $\mathcal{C}$, how can we give the composite functor $S' \circ S$ the structure $(\widehat{\mu}, \widehat{\eta})$ of a monad on $\mathcal{C}$? The unit is easy –

$$\widehat{\eta} = \left(1 \xrightarrow{\eta' * \eta} S' \circ S \right) \tag{6.2}$$

– but for the multiplication we need some extra data –

$$\widehat{\mu} = \left(S' \circ S \circ S' \circ S \xrightarrow{S' \circ ? \circ S} S' \circ S' \circ S \circ S \xrightarrow{\mu' * \mu} S' \circ S \right) \tag{6.3}$$

– and that is provided by a distributive law.

Definition 6.1.2 Let S and S' be monads on the same category. A **distributive law** $\lambda : S \circ S' \longrightarrow S' \circ S$ is a natural transformation such that (S', λ) is a lax map of monads $S \longrightarrow S$ and (S, λ) is a colax map of monads $S' \longrightarrow S'$.

Lemma 6.1.3 Let $\lambda : S \circ S' \longrightarrow S' \circ S$ be a distributive law between monads (S, μ, η) and (S', μ', η') on a category $\mathcal{C}$. Then the formulas (6.2) and (6.3) (with $? = \lambda$) define a monad structure $(\widehat{\mu}, \widehat{\eta})$ on the functor $S' \circ S$. If the monads S and S' and the transformation λ are all cartesian then so too is the monad $S' \circ S$. $\qquad\square$

The distributive law λ determines a lax map of monads $(S', \lambda) : S \longrightarrow S$, hence a functor $\widetilde{S} : C^S \longrightarrow C^S$. More incisively, we have the following.

Lemma 6.1.4 *Let S and S' be monads on a category C. Then there is a one-to-one correspondence between distributive laws $\lambda : S \circ S' \longrightarrow S' \circ S$ and monads $\widetilde{S}$ on C^S such that the forgetful functor $U : C^S \longrightarrow C$ is a strict map of monads $\widetilde{S} \longrightarrow S'$:*

$$
\begin{array}{ccc}
C^S & \xrightarrow{\widetilde{S}} & C^S \\
{\scriptstyle U}\downarrow & & \downarrow{\scriptstyle U} \\
C & \xrightarrow[S']{} & C.
\end{array}
$$

If the monad S' is cartesian then so too is the monad $\widetilde{S}$.

Proof Straightforward, using Lemma 6.1.1. □

A distributive law $S \circ S' \longrightarrow S' \circ S$ therefore gives two new categories of algebras, $C^{S' \circ S}$ and $(C^S)^{\widetilde{S}}$; but they are isomorphic by general principles of coherence, or more rigorously by

Lemma 6.1.5 *Let S and S' be monads on a category C, let $\lambda : S \circ S' \longrightarrow S' \circ S$ be a distributive law, and let $\widetilde{S}$ be the corresponding monad on C^S. Then there is a canonical natural transformation*

$$
\begin{array}{ccc}
C^S & \xrightarrow{\widetilde{S}} & C^S \\
{\scriptstyle U}\downarrow & {\scriptstyle \nearrow \psi} & \downarrow{\scriptstyle U} \\
C & \xrightarrow[S' \circ S]{} & C.
\end{array}
$$

This makes (U, ψ) into a lax map of monads $\widetilde{S} \longrightarrow S' \circ S$, and the induced functor $(U, \psi)_* : (C^S)^{\widetilde{S}} \longrightarrow C^{S' \circ S}$ is an isomorphism of categories.

Proof The transformation ψ is $S' U \varepsilon$, where ε is the counit of the free-forgetful adjunction $F \dashv U$ for S-algebras. Algebras for both $\widetilde{S}$ and $S' \circ S$ can be described as triples (X, h, h') where $X \in C$, h and h' are respectively S-algebra and S'-algebra structures on X, and the following diagram commutes:

$$
\begin{array}{ccc}
SS'X & \xrightarrow{\ Sh'\ } & SX \\
{\scriptstyle \lambda_X}\swarrow & & \big\downarrow{\scriptstyle h} \\
S'SX & & \\
{\scriptstyle S'h}\big\downarrow & & \\
S'X & \xrightarrow[\ h'\ ]{} & X.
\end{array}
$$

The details of the proof are, again, straightforward.   □

6.2 Multicategories via monads

Operads are meant to be regarded as algebraic theories of a special kind. Monads are meant to be regarded as algebraic theories of a general kind. It is therefore natural to ask whether operads can be re-defined as monads satisfying certain conditions.

We show here that the answer is nearly 'yes': for any cartesian monad T on a cartesian category $\mathcal{E}$, a T-operad is the same thing as a cartesian monad S on $\mathcal{E}$ together with a cartesian natural transformation $S \longrightarrow T$ commuting with the monad structures. (The transformation really is necessary, by the results of Appendix C.) An algebra for a T-operad is then just an algebra for the corresponding monad S. More generally, a version holds for T-multicategories, to be regarded as *many-sorted* algebraic theories.

We will need to know a little about slice categories. If E is an object of a category $\mathcal{E}$ then the forgetful functor $U_E : \mathcal{E}/E \longrightarrow \mathcal{E}$ creates pullbacks in the following strict sense: if

$$
\begin{array}{ccc}
P & \longrightarrow & X \\
\big\downarrow & & \big\downarrow \\
Y & \longrightarrow & Z
\end{array}
$$

is a pullback square in $\mathcal{E}$ and $(Z \longrightarrow E)$ is a map in $\mathcal{E}$ then the evident square

$$
\begin{array}{ccc}
\left(\begin{array}{c} P \\ \downarrow \\ E \end{array}\right) & \longrightarrow & \left(\begin{array}{c} X \\ \downarrow \\ E \end{array}\right) \\
\big\downarrow & & \big\downarrow \\
\left(\begin{array}{c} Y \\ \downarrow \\ E \end{array}\right) & \longrightarrow & \left(\begin{array}{c} Z \\ \downarrow \\ E \end{array}\right)
\end{array}
$$

in $\mathcal{E}/E$ is also a pullback. Hence U_E reflects pullbacks, and if $\mathcal{E}$ is cartesian then the category $\mathcal{E}/E$ and the functor U_E are also cartesian.

Proposition 6.2.1 *Let T be a cartesian monad on a cartesian category $\mathcal{E}$, and let C be a T-multicategory. Then the induced monad T_C on $\mathcal{E}/C_0$ is cartesian, and there is a cartesian natural transformation*

$$
\begin{array}{ccc}
\mathcal{E}/C_0 & \xrightarrow{\ T_C\ } & \mathcal{E}/C_0 \\
\downarrow{\scriptstyle U_{C_0}} & \overset{\pi^C}{\Swarrow} & \downarrow{\scriptstyle U_{C_0}} \\
\mathcal{E} & \xrightarrow[\ T\ ]{} & \mathcal{E}
\end{array}
$$

such that (U_{C_0}, π^C) is a colax map of monads $(\mathcal{E}/C_0, T_C) \longrightarrow (\mathcal{E}, T)$.

Proof For $X = (X \xrightarrow{\ p\ } C_0) \in \mathcal{E}/C_0$, let π^C_X be the map in the diagram

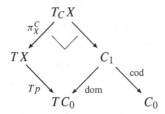

defining $T_C X$. It is easy to check that π^C is natural and that (U_{C_0}, π^C) forms a colax map of monads. Using the Pasting Lemma (1.1.1) it is also easy to check that π^C is a cartesian natural transformation, and from this that T_C is a cartesian monad. □

The proposition says that any $(\mathcal{E}, T)$-multicategory C gives rise to a triple $(E, S, \pi) = (C_0, T_C, \pi^C)$ where

a. E is an object of $\mathcal{E}$
b. S is a cartesian monad on $\mathcal{E}/E$
c. π is a cartesian natural transformation

$$
\begin{array}{ccc}
\mathcal{E}/E & \xrightarrow{\ S\ } & \mathcal{E}/E \\
\downarrow{\scriptstyle U_E} & \overset{\pi}{\Swarrow} & \downarrow{\scriptstyle U_E} \\
\mathcal{E} & \xrightarrow[\ T\ ]{} & \mathcal{E}
\end{array}
$$

such that (U_E, π) is a colax map of monads $S \longrightarrow T$.

It turns out that this captures exactly what a T-multicategory is: every triple (E, S, π) satisfying these three conditions arises from a T-multicategory, and the whole multicategory structure of C can be recovered from the associated triple (C_0, T_C, π^C). We will prove this in a moment. In the case of T-operads, assuming that $\mathcal{E}$ has a terminal object 1, this says that a T-operad is a pair (S, π) where S is a cartesian monad on $\mathcal{E}$ and $\pi : S \longrightarrow T$ is a cartesian natural transformation commuting with the monad structures – just as promised in the introduction to this section.

In particular, the monad T_P on **Set** induced by a plain operad P is always cartesian, which gives a large class of cartesian monads on **Set**. (Not all cartesian monads on **Set** are of this type, but all strongly regular theories are: see Appendix C.) The natural transformation π^P induces the obvious functor from **Monoid** to **Alg**(P).

We now give the alternative definition of T-multicategory. Notation: if $E \xrightarrow{e} \widetilde{E}$ is a map in a category $\mathcal{E}$ then $e_!$ is the functor $\mathcal{E}/E \longrightarrow \mathcal{E}/\widetilde{E}$ given by composition with e.

Definition 6.2.2 Let T be a cartesian monad on a cartesian category $\mathcal{E}$. Define a category T-**Multicat**$'$ as follows:

objects are triples (E, S, π) as in (a)–(c) above
maps $(E, S, \pi) \longrightarrow (\widetilde{E}, \widetilde{S}, \widetilde{\pi})$ are pairs (e, ϕ) where $e : E \longrightarrow \widetilde{E}$ is a map in $\mathcal{E}$ and

$$
\begin{array}{ccc}
\mathcal{E}/E & \xrightarrow{\ S\ } & \mathcal{E}/E \\
e_! \downarrow & \swarrow \phi & \downarrow e_! \\
\mathcal{E}/\widetilde{E} & \xrightarrow[\widetilde{S}]{} & \mathcal{E}/\widetilde{E}
\end{array}
$$

is a natural transformation such that $(e_!, \phi)$ is a colax map of monads $S \longrightarrow \widetilde{S}$ and

$$
\begin{array}{ccc}
\mathcal{E}/E \xrightarrow{\ S\ } \mathcal{E}/E & & \mathcal{E}/E \xrightarrow{\ S\ } \mathcal{E}/E \\
e_! \downarrow\ \swarrow\phi\ \downarrow e_! & & U_E \downarrow\ \swarrow\pi\ \downarrow U_E \\
\mathcal{E}/\widetilde{E} \xrightarrow{\widetilde{S}} \mathcal{E}/\widetilde{E} & = & \mathcal{E} \xrightarrow[T]{} \mathcal{E}. \qquad (6.4) \\
U_{\widetilde{E}} \downarrow\ \swarrow\widetilde{\pi}\ \downarrow U_{\widetilde{E}} & & \\
\mathcal{E} \xrightarrow[T]{} \mathcal{E} & &
\end{array}
$$

The natural transformation ϕ in the definition of map is automatically carte-sian: this follows from equation (6.4), the fact that π and $\tilde{\pi}$ are cartesian, the fact that $U_{\tilde{E}}$ reflects pullbacks, and the Pasting Lemma (1.1.1). The functor $e_!$ is also cartesian.

Proposition 6.2.3 *For any cartesian monad T on a cartesian category $\mathcal{E}$, there is an equivalence of categories*

$$T\text{-}\mathbf{Multicat} \simeq T\text{-}\mathbf{Multicat}'.$$

Proof A T-multicategory C gives rise to an object (C_0, T_C, π^C) of T-**Multicat**$'$, as in Proposition 6.2.1. Conversely, take an object (E, S, π) of T-**Multicat**$'$, and define

$$C_0 = E,$$
$$(C_1 \xrightarrow{\text{cod}} C_0) = S(C_0 \xrightarrow{1} C_0),$$
$$\text{dom} = \pi_{1_{C_0}} : C_1 \longrightarrow TC_0.$$

This specifies a T-graph C, and there is a T-multicategory structure on C given by

$$\text{comp} = \mu_{1_{C_0}}, \qquad \text{ids} = \eta_{1_{C_0}}.$$

(It takes a little work to see that these make sense.) The associativity and iden-tity axioms for the multicategory follow from the coherence axioms for the colax map of monads (U_E, π). It is straightforward to check that this extends to an equivalence of categories. □

Corollary 6.2.4 *Let T be a cartesian monad on a category $\mathcal{E}$ with finite limits. Then T-**Operad** is equivalent to the category T-**Operad**$'$ in which*

- *an object is a cartesian monad S on $\mathcal{E}$ together with a cartesian natural transformation $S \longrightarrow T$ commuting with the monad structures*
- *a map $(S, \pi) \longrightarrow (\tilde{S}, \tilde{\pi})$ is a cartesian natural transformation $\phi :$ $S \longrightarrow \tilde{S}$ commuting with the monad structures and satisfying $\tilde{\pi} \circ \phi = \pi$.*

Proof Restrict the proof of 6.2.3 to the case $C_0 = E = 1$. □

Since the monad T_C arising from a T-multicategory C is cartesian, it makes sense to ask what T_C-multicategories are. The answer is simple:

Corollary 6.2.5 *Let T be a cartesian monad on a cartesian category $\mathcal{E}$, and let C be a T-multicategory. Then there is an equivalence of categories*

$$T_C\text{-}\mathbf{Multicat} \simeq T\text{-}\mathbf{Multicat}/C,$$

and if a T_C-multicategory D corresponds to a T-multicategory $\overline{D}$ over C then $\mathbf{Alg}(D) \cong \mathbf{Alg}(\overline{D})$.

Proof For the first part, it is enough to show that

$$T_C\text{-}\mathbf{Multicat}' \simeq T\text{-}\mathbf{Multicat}'/(C_0, T_C, \pi^C).$$

An object of the right-hand side is an object (E, S, π) of T-**Multicat**$'$ together with a map $E \xrightarrow{\;e\;} C_0$ and a cartesian natural transformation ϕ such that $(e_!, \phi)$ is a colax map of monads and π is the pasting of ϕ and π^C. In other words, it is just an object $(E \xrightarrow{\;e\;} C_0, S, \phi)$ of T_C-**Multicat**$'$. The first part follows; and for the second part, $\mathbf{Alg}(D) \cong (\mathcal{E}/E)^S \cong \mathbf{Alg}(\overline{D})$. □

This explains many of the examples in Sections 4.2 and 4.3 (generalized multicategories and their algebras). Take, for instance, 4.2.13 and 4.3.7, where we fixed a monoid M and considered $(M \times -)$-multicategories and their algebras. Write T for the free monoid monad on **Set**, and let P be the T-operad (= plain operad) with $P(1) = M$ and $P(n) = \emptyset$ for $n \neq 1$. Then $T_P = (M \times -)$, so by Corollary 6.2.5 an $(M \times -)$-multicategory is a plain multicategory over P. Evidently a plain multicategory over P can only have unary arrows, so in fact we have

$$(M \times -)\text{-}\mathbf{Multicat} \simeq \mathbf{Cat}/M.$$

Moreover, the second part of 6.2.5 tells us that if an $(M \times -)$-multicategory corresponds to an object $(C \xrightarrow{\;\phi\;} M)$ of **Cat**$/M$ then its category of algebras is just $\mathbf{Alg}(C) \simeq [C, \mathbf{Set}]$, as claimed in 4.3.7.

This reformulation of T-multicategories in terms of monads and colax maps between them has a dual, using *lax* maps. As we saw in Section 6.1, colax and lax maps can be related using mates. The adjunctions involved are

$$
\begin{array}{cc}
\mathcal{E} & \mathcal{E}/\tilde{E} \\
U_E \uparrow \dashv \downarrow (-\times E) & e_! \uparrow \dashv \downarrow e^* \\
\mathcal{E}/E, & \mathcal{E}/E.
\end{array}
\qquad (6.5)
$$

In the first adjunction E is an object of a category $\mathcal{E}$ with finite limits, and the right adjoint to the forgetful functor U_E sends $X \in \mathcal{E}$ to $(X \times E \xrightarrow{\mathrm{pr}_2} E) \in \mathcal{E}/E$. In the second $e : E \longrightarrow \tilde{E}$ is a map in $\mathcal{E}$, and the right adjoint to $e_!$ is the functor e^* defined by pullback along e. Actually, the first is just the second in the case $\tilde{E} = 1$.

Definition 6.2.6 Let T be a cartesian monad on a category $\mathcal{E}$ with finite limits. The category T-**Multicat**$''$ is defined as follows:

objects are triples (E, S, ρ) where $E \in \mathcal{E}$, S is a cartesian monad on $\mathcal{E}/E$, and ρ is a cartesian natural transformation such that $(- \times E, \rho)$ is a lax map of monads $T \longrightarrow S$

maps $(E, S, \rho) \longrightarrow (\widetilde{E}, \widetilde{S}, \widetilde{\rho})$ are pairs (e, ψ) where $e : E \longrightarrow \widetilde{E}$ in $\mathcal{E}$ and ψ is a cartesian natural transformation such that (e^*, ψ) is a lax map of monads $\widetilde{S} \longrightarrow S$ and an equation dual to (6.4) in Definition 6.2.2 holds.

Proposition 6.2.7 *For any cartesian monad T on a category $\mathcal{E}$ with finite limits, there is an isomorphism of categories*

$$T\text{-}\mathbf{Multicat}'' \cong T\text{-}\mathbf{Multicat}'.$$

Proof Just take mates throughout. All the categories, functors and natural transformations involved in the adjunctions (6.5) are cartesian, so under these adjunctions, the mate of a cartesian natural transformation is also cartesian. □

It follows that T-**Multicat**$'' \simeq T$-**Multicat**. So given any T-multicategory C, there is a corresponding lax map of monads $(- \times C_0, \rho^C) : T \longrightarrow T_C$, and this induces a functor $\mathcal{E}^T \longrightarrow \mathbf{Alg}(C)$. For instance, any monoid M yields an algebra for any plain multicategory C; concretely, this algebra X is given by $X(a) = M$ for all objects a of C and by $X(\theta) = (n\text{-fold multiplication})$ for all n-ary maps θ in C. Similarly, any map $C \longrightarrow \widetilde{C}$ of T-multicategories corresponds to a lax map of monads $T_{\widetilde{C}} \longrightarrow T_C$, and this induces the functor $\mathbf{Alg}(\widetilde{C}) \longrightarrow \mathbf{Alg}(C)$ that we constructed directly at the end of Chapter 4.

6.3 Algebras via fibrations

For a small category C, the functor category $[C, \mathbf{Set}]$ is equivalent to the category of discrete opfibrations over C, as we saw in Section 1.1. Here we extend this result from categories to generalized multicategories.

By definition (p. 30), a functor $g : D \longrightarrow C$ between ordinary categories is a discrete opfibration if for each object b of D and arrow $g(b) \overset{\theta}{\longrightarrow} a$ in C, there is a unique arrow $b \overset{\chi}{\longrightarrow} b'$ in D such that $g(\chi) = \theta$. Another way of

saying this is that in the diagram

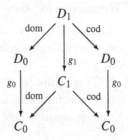

depicting g, the left-hand 'square' is a pullback.

Generalizing to any cartesian monad T on any cartesian category $\mathcal{E}$, let us call a map $D \xrightarrow{g} C$ of T-multicategories a **discrete opfibration** if the square

$$
\begin{array}{ccc}
TD_0 & \xleftarrow{\mathrm{dom}} & D_1 \\
{\scriptstyle Tg_0}\downarrow & & \downarrow{\scriptstyle g_1} \\
TC_0 & \xleftarrow{\mathrm{dom}} & C_1
\end{array}
$$

is a pullback. We obtain, for any T-multicategory C, the category **DOpfib**(C) of discrete opfibrations over C: an object is a discrete opfibration with codomain C, and a map from $(D \xrightarrow{g} C)$ to $(D' \xrightarrow{g'} C)$ is a map $D \xrightarrow{f} D'$ of T-multicategories such that $g' \circ f = g$. Such an f is automatically a discrete opfibration too, by the Pasting Lemma (1.1.1).

Theorem 6.3.1 *Let T be a cartesian monad on a cartesian category $\mathcal{E}$, and let C be a T-multicategory. Then there is an equivalence of categories*

$$\mathbf{DOpfib}(C) \simeq \mathbf{Alg}(C).$$

Proof A C-algebra is an algebra for the monad T_C on $\mathcal{E}/C_0$, which sends an object $X = (X \xrightarrow{p} C_0)$ of $\mathcal{E}/C_0$ to the boxed composite in the diagram

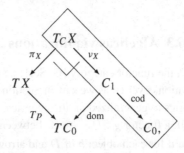

and therefore consists of an object $(X \xrightarrow{p} C_0)$ of $\mathcal{E}/C_0$ together with a map $h : T_C X \longrightarrow X$ over C_0, satisfying axioms.

So, given a C-algebra $(X \xrightarrow{p} C_0, h)$ we obtain a commutative diagram

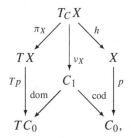

the left-hand half of which is a pullback square. The top part of the diagram
defines a T-graph D, and there is a map $g : D \longrightarrow C$ defined by $g_0 = p$ and
$g_1 = \nu_X$. With some calculation we see that D is naturally a T-multicategory
and g a map of T-multicategories. So we have constructed from the C-algebra
X a discrete opfibration over C.

This defines a functor from $\mathbf{Alg}(C)$ to $\mathbf{DOpfib}(C)$, which is easily checked
to be full, faithful and essentially surjective on objects. $\square$

Let us look more closely at the T-multicategory D corresponding to a
C-algebra $h = (X \xrightarrow{p} C_0, h)$. Generalizing the terminology for ordinary
categories (p. 29), we call D the **multicategory of elements** of h and write
$D = C/h$.

Proposition 6.3.2 *Let T be a cartesian monad on a cartesian category $\mathcal{E}$, let C
be a T-multicategory, and let h be a C-algebra. Then there is an isomorphism
of categories*

$$\mathbf{DOpfib}(C/h) \cong \mathbf{DOpfib}(C)/h.$$

Proof Follows from the definition of $\mathbf{DOpfib}$, using the observation above that
maps in $\mathbf{DOpfib}(C)$ are automatically discrete opfibrations. $\square$

Hence $\mathbf{Alg}(C/h) \simeq \mathbf{Alg}(C)/h$, generalizing Proposition 1.1.7. In fact, this
equivalence is an isomorphism. To see this, recall from 4.3.14 the process of
slicing a monad by an algebra: for any monad S on a category $\mathcal{F}$ and any
S-algebra k, there is a monad S/k on $\mathcal{D}$ with the property that $\mathcal{F}^{S/k} \cong \mathcal{F}^S/k$.

Proposition 6.3.3 *Let T be a cartesian monad on a cartesian category $\mathcal{E}$,
let C be a T-multicategory, and let h be a C-algebra. Then there is an
isomorphism of monads $T_{C/h} \cong T_C/h$ and an isomorphism of categories
$\mathbf{Alg}(C/h) \cong \mathbf{Alg}(C)/h$.*

Proof The first assertion is easily verified, and the second follows immediately. $\square$

As an example, let C be the terminal T-multicategory 1. We have $T_1 \cong T$ and so $\mathbf{Alg}(1) \cong \mathbf{Alg}(T)$ (4.3.8). Given a T-algebra $h = (TX \xrightarrow{h} X)$, we therefore obtain a T-multicategory $1/h$. Plausibly enough, this is the same as the T-multicategory h^+ of 4.2.22, with graph

$$TX \xleftarrow{\ 1\ } TX \xrightarrow{\ h\ } X.$$

So by the results above, $T_{h^+} = T_{1/h} \cong T/h$ and $\mathbf{Alg}(h^+) = \mathbf{Alg}(1/h) \cong \mathbf{Alg}(T)/h$; compare 4.3.14.

We could also define **opalgebras** for a T-multicategory C as discrete brations over C: that is, as maps $D \xrightarrow{\ g\ } C$ of T-multicategories such that the right-hand square

$$
\begin{array}{ccc}
D_1 & \xrightarrow{\ \mathrm{cod}\ } & D_0 \\
{\scriptstyle g_1}\downarrow & & \downarrow{\scriptstyle g_0} \\
C_1 & \xrightarrow{\ \mathrm{cod}\ } & C_0
\end{array}
$$

of the diagram depicting g is a pullback. In the case of ordinary categories C, an opalgebra for C is a functor $C^{\mathrm{op}} \longrightarrow \mathbf{Set}$. In the case of plain multicategories C, an opalgebra is a family $(X(a))_{a \in C_0}$ of sets together with a function

$$X(a) \longrightarrow X(a_1) \times \cdots \times X(a_n)$$

for each map $a_1, \ldots, a_n \longrightarrow a$ in C, satisfying the obvious axioms. Note that these are different from the 'coalgebras' or 'right modules' mentioned in Sections 2.3 and 5.3; we do not discuss them any further.

6.4 Algebras via endomorphisms

An action of a monoid on a set is a homomorphism from the monoid to the monoid of endomorphisms of the set. A representation of a Lie algebra is a homomorphism from it into the Lie algebra of endomorphisms of some vector space. An algebra for a plain operad is often defined as a map from it into the operad of endomorphisms of some set (2.2.8). Here we show that algebras for generalized multicategories can be described in the same way, assuming some mild properties of the base category $\mathcal{E}$.

First recall from 2.1.16 what happens for plain multicategories: given any family $(X(a))_{a \in E}$ of sets, there is an associated plain multicategory $\mathbf{End}(X)$ with object-set E and with

$$(\mathbf{End}(X))(a_1, \ldots, a_n; a) = \mathbf{Set}(X(a_1) \times \cdots \times X(a_n), X(a)), \qquad (6.6)$$

and if C is a plain multicategory with object-set C_0 then a C-algebra amounts to a family $(X(a))_{a \in C_0}$ of sets together with a map $C \longrightarrow \mathbf{End}(X)$ of multicategories leaving the objects unchanged.

To extend this to generalized multicategories we need to rephrase the definition of $\mathbf{End}(X)$. Let T be the free monoid monad on $\mathbf{Set}$. Recall that given a set E, a family $(X(a))_{a \in E}$ amounts to an object $X \xrightarrow{p} E$ of $\mathbf{Set}/E$. Then note that $X(a_1) \times \cdots \times X(a_n)$ is the fibre over $(a_1, \ldots, a_n)$ in the map $TX \xrightarrow{Tp} TE$, or equivalently that it is the fibre over $((a_1, \ldots, a_n), a)$ in the map $TX \times E \xrightarrow{Tp \times 1} TE \times E$. On the other hand, $X(a)$ is the fibre over $((a_1, \ldots, a_n), a)$ in the map $TE \times X \xrightarrow{1 \times p} TE \times E$. So if we define objects

$$G_1(X) = \begin{pmatrix} TX \times E \\ \downarrow Tp \times 1 \\ TE \times E \end{pmatrix}, \qquad G_2(X) = \begin{pmatrix} TE \times X \\ \downarrow 1 \times p \\ TE \times E \end{pmatrix} \qquad (6.7)$$

of the category $\mathbf{Set}/(TE \times E)$, then (6.6) says that the underlying T-graph of the multicategory $\mathbf{End}(X)$ is the exponential $G_2(X)^{G_1(X)}$.

It is now clear what the definition of endomorphism multicategory in the general case must be. Let T be a cartesian monad on a cartesian category $\mathcal{E}$. Assume further that $\mathcal{E}$ is **locally cartesian closed**: for each object D of $\mathcal{E}$, the slice category $\mathcal{E}/D$ is cartesian closed (has exponentials). This is true when $\mathcal{E}$ is a presheaf category, as in the majority of our examples. (For the definition of cartesian closed, see, for instance, Mac Lane (1971, IV.6). For a more full account, including a proof that presheaf categories are cartesian closed, see Mac Lane and Moerdijk (1992, I.6); our Proposition 1.1.7 then implies that each slice is cartesian closed.) Given $E \in \mathcal{E}$, define functors $G_1, G_2 : \mathcal{E}/E \longrightarrow \mathcal{E}/(TE \times E)$ by the formulas of (6.7) above.

The short story is that for any $X \in \mathcal{E}/E$, there is a natural T-multicategory structure on the T-graph $\mathbf{End}(X) = G_2(X)^{G_1(X)}$, and that if C is any T-multicategory then a C-algebra amounts to an object X of $\mathcal{E}/C_0$ together with a map $C \longrightarrow \mathbf{End}(X)$ of T-multicategories fixing the objects.

Here is the long story. Given T and $\mathcal{E}$ as above and $E \in \mathcal{E}$, define a functor

$$\mathbf{Hom} : \quad (\mathcal{E}/E)^{\mathrm{op}} \times \mathcal{E}/E \quad \longrightarrow \quad \mathcal{E}/(TE \times E)$$
$$(X, Y) \quad \longmapsto \quad G_2(Y)^{G_1(X)}.$$

Consider also the functor

$$\mathcal{E}/(TE \times E) \times \mathcal{E}/E \longrightarrow \mathcal{E}/E$$

sending a pair

$$
C = \left(\begin{array}{c} C_1 \\ \text{dom} \swarrow \quad \searrow \text{cod} \\ TE \qquad\qquad E \end{array} \right), \qquad X = \left(\begin{array}{c} X \\ \downarrow p \\ E \end{array} \right)
$$

to the boxed diagonal in the pullback diagram

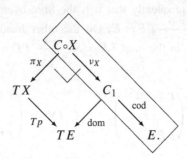

This functor is written more shortly as $(C, X) \mapsto C \circ X$; of course, when C has the structure of a T-multicategory, we usually write $C \circ -$ as T_C.

Proposition 6.4.1 *Let T be a cartesian monad on a cartesian, locally cartesian closed category $\mathcal{E}$, and let E be an object of $\mathcal{E}$. Then there is an isomorphism*

$$
\frac{\mathcal{E}}{TE \times E}(C, \mathbf{Hom}(X, Y)) \cong \frac{\mathcal{E}}{E}(C \circ X, Y) \qquad (6.8)
$$

natural in $C \in \mathcal{E}/(TE \times E)$ and $X, Y \in \mathcal{E}/E$.

Proof Write $X = (X \xrightarrow{p} E)$ and $Y = (Y \xrightarrow{q} E)$. Product in $\mathcal{E}/(TE \times E)$ is pullback over $TE \times E$ in $\mathcal{E}$, so the left-hand side of (6.8) is naturally isomorphic to

$$
\frac{\mathcal{E}}{TE \times E}(C \times_{TE \times E} G_1(X), G_2(Y)),
$$

where $C \times_{TE \times E} G_1(X) \longrightarrow TE \times E$ is the diagonal of the pullback square

$$
\begin{array}{ccc}
C \times_{TE \times E} G_1(X) & \longrightarrow & C_1 \\
\downarrow & & \downarrow {\scriptstyle (\text{dom,cod})} \\
TX \times E & \xrightarrow{\ Tp \times 1\ } & TE \times E.
\end{array}
$$

But by an easy calculation, we also have a pullback square

$$
\begin{array}{ccc}
C{\circ}X & \xrightarrow{\;\nu_X\;} & C_1 \\
{\scriptstyle(\pi_X,\mathrm{cod}{\circ}\nu_X)}\big\downarrow\;\;\lrcorner & & \big\downarrow{\scriptstyle(\mathrm{dom},\mathrm{cod})} \\
TX \times E & \xrightarrow[\;Tp\times 1\;]{} & TE \times E,
\end{array}
$$

so in fact an element of the left-hand side of (6.8) is a map $C{\circ}X \longrightarrow G_2(Y)$ in $\mathcal{E}/(TE \times E)$. This is a map $C{\circ}X \longrightarrow TE \times Y$ in $\mathcal{E}$ such that

$$
\begin{array}{ccc}
C{\circ}X & \longrightarrow & TE \times Y \\
{\scriptstyle(\mathrm{dom}{\circ}\nu_X,\mathrm{cod}{\circ}\nu_X)}\searrow & & \swarrow{\scriptstyle 1\times q} \\
& TE \times E &
\end{array}
$$

commutes, and this in turn is a map $C{\circ}X \longrightarrow Y$ in $\mathcal{E}/E$. $\qquad\square$

Next observe that $\mathcal{E}/(TE \times E)$ is naturally a monoidal category: it is the full sub-bicategory of $\mathcal{E}_{(T)}$ whose only object is E. Tensor product of objects of $\mathcal{E}/(TE \times E)$ is composition $\circ$ of 1-cells in $\mathcal{E}_{(T)}$, and a monoid in $\mathcal{E}/(TE \times E)$ is a T-multicategory C with $C_0 = E$. The functor

$$
\begin{array}{ccc}
\mathcal{E}/(TE \times E) \times \mathcal{E}/E & \longrightarrow & \mathcal{E}/E \\
(C, X) & \longmapsto & C{\circ}X
\end{array}
$$

then becomes an action of the monoidal category $\mathcal{E}/(TE \times E)$ on the category $\mathcal{E}/E$, in the sense of 1.2.12: there are coherent natural isomorphisms

$$
D{\circ}(C{\circ}X) \xrightarrow{\;\sim\;} (D{\circ}C){\circ}X, \qquad X \xrightarrow{\;\sim\;} 1_E{\circ}X,
$$

for $C, D \in \mathcal{E}/(TE \times E)$, $X \in \mathcal{E}/E$.

Proposition 6.4.2 *Let T, $\mathcal{E}$ and E be as in Proposition 6.4.1. For each $X \in \mathcal{E}/E$, the T-graph $\mathbf{End}(X) = \mathbf{Hom}(X, X)$ naturally has the structure of a T-multicategory.*

Proof We have to define a composition map $\mathbf{End}(X){\circ}\mathbf{End}(X) \longrightarrow \mathbf{End}(X)$. First let $\mathrm{ev}_X : \mathbf{End}(X){\circ}X \longrightarrow X$ be the map corresponding under Proposition 6.4.1 to the identity $\mathbf{End}(X) \longrightarrow \mathbf{Hom}(X, X)$. Then define composition to be the map corresponding under 6.4.1 to the composite

$$
\mathbf{End}(X){\circ}\mathbf{End}(X){\circ}X \xrightarrow{\;1*\mathrm{ev}_X\;} \mathbf{End}(X){\circ}X \xrightarrow{\;\mathrm{ev}_X\;} X.
$$

The definition of identities is similar but easier. The associativity and identity axioms follow from the axioms for an action of a monoidal category (1.2.12). $\qquad\square$

We can now express the alternative definition of algebra. Given $\mathcal{E}$ and T as above and a T-multicategory C, let **Alg$'$(C)** be the category in which

objects are pairs (X, h) where $X \in \mathcal{E}/C_0$ and $h : C \longrightarrow \mathbf{End}(X)$ is a homomorphism of monoids in $\mathcal{E}/(TC_0 \times C_0)$
maps $(X, h) \longrightarrow (Y, k)$ are maps $f : X \longrightarrow Y$ in $\mathcal{E}/C_0$ such that

$$
\begin{array}{ccc}
C & \xrightarrow{\ \ h\ \ } & \mathbf{Hom}(X, X) \\
{\scriptstyle k}\downarrow & & \downarrow{\scriptstyle \mathbf{Hom}(1, f)} \\
\mathbf{Hom}(Y, Y) & \xrightarrow[\mathbf{Hom}(f, 1)]{} & \mathbf{Hom}(X, Y)
\end{array}
$$

commutes.

(A homomorphism between monoids in $\mathcal{E}/(TC_0 \times C_0)$ is just a map f between the corresponding multicategories such that $f_0 : C_0 \longrightarrow C_0$ is the identity.)

Theorem 6.4.3 *Let T be a cartesian monad on a cartesian, locally cartesian closed category $\mathcal{E}$. Let C be a T-multicategory. Then there is an isomorphism of categories* **Alg$'$(C)** $\cong$ **Alg(C)**.

Proof Let $X \in \mathcal{E}/C_0$. Proposition 6.4.1 in the case $Y = X$ gives a bijection between T-graph maps $h : C \longrightarrow \mathbf{End}(X)$ and maps $\overline{h} : T_C(X) = C \circ X \longrightarrow X$ in $\mathcal{E}/C_0$. Under this correspondence, h is a homomorphism of monoids if and only if $\overline{h}$ is an algebra structure on X. So we have a bijection between the objects of **Alg$'$(C)** and those of **Alg(C)**. The remaining checks are straightforward. $\qquad\square$

6.5 Free multicategories

Any directed graph freely generates a category: objects are vertices and maps are chains of edges. More generally, any 'graph' in which each 'edge' has a finite sequence of inputs and a single output freely generates a plain multicategory, as explained in Section 2.3: objects are vertices and maps are trees of edges. In this short section we extend this to generalized multicategories.

The construction for plain multicategories involves an infinite recursive process, so we cannot hope to generalize to arbitrary cartesian $\mathcal{E}$ and T – after all, the category $\mathcal{E}$ being cartesian only means that it admits certain finite limits. There is, however, a class of cartesian categories $\mathcal{E}$ and a class of cartesian monads T for which the free $(\mathcal{E}, T)$-multicategory construction is possible, the so-called **suitable** categories and monads. The definition of suitability is quite complicated, but fortunately can be treated as a black box: all the properties of suitable monads that we need are stated in this section, and the details are confined to Appendix D.

First, we have a good stock of suitable categories and monads:

Theorem 6.5.1 *Any presheaf category is suitable. Any finitary cartesian monad on a cartesian category is suitable.*

A functor is said to be **finitary** if it preserves filtered colimits (themselves defined in Mac Lane, 1971, IX.1); a monad (T, μ, η) is said to be **finitary** if the functor T is finitary. In almost all of the examples in this book, T is a finitary monad on a presheaf category.

Second, suitability is a sufficient condition for the existence of free multicategories:

Theorem 6.5.2 *Let T be a suitable monad on a suitable category $\mathcal{E}$. Then the forgetful functor*

$$(\mathcal{E}, T)\text{-}\mathbf{Multicat} \longrightarrow \mathcal{E}^+ = (\mathcal{E}, T)\text{-}\mathbf{Graph}$$

has a left adjoint, the adjunction is monadic, and if T^+ is the induced monad on $\mathcal{E}^+$ then both T^+ and $\mathcal{E}^+$ are suitable.

Example 6.5.3 The category of sets and the identity monad are suitable (6.5.1). In this case Theorem 6.5.2 tells us that there is a free category monad **fc** on the category of directed graphs, and that it is suitable. In particular it is cartesian, so it makes sense to talk about **fc**-multicategories, as we did in Chapter 5.

Taking the free category on a directed graph leaves the set of objects (vertices) unchanged, and the corresponding fact for generalized multicategories is expressed in a variant of the theorem. Notation: if E is an object of $\mathcal{E}$ then $(\mathcal{E}, T)\text{-}\mathbf{Multicat}_E$ is the subcategory of $(\mathcal{E}, T)\text{-}\mathbf{Multicat}$ whose objects C satisfy $C_0 = E$ and whose morphisms f satisfy $f_0 = 1_E$. Observe that $\mathcal{E}/(TE \times E)$ is the category of T-graphs with fixed object-of-objects E.

Theorem 6.5.4 *Let T be a suitable monad on a suitable category $\mathcal{E}$, and let $E \in \mathcal{E}$. Then the forgetful functor*

$$(\mathcal{E}, T)\text{-}\mathbf{Multicat}_E \longrightarrow \mathcal{E}_E^+ = \mathcal{E}/(TE \times E)$$

has a left adjoint, the adjunction is monadic, and if T_E^+ is the induced monad on $\mathcal{E}_E^+$ then both T_E^+ and $\mathcal{E}_E^+$ are suitable.

Example 6.5.5 The free monoid monad on the category of sets is suitable, by 6.5.1, hence the free plain operad monad on **Set**/$\mathbb{N}$ is also suitable, by 6.5.4. In particular it is cartesian, as claimed in 4.1.11.

In both examples the theorems were used to establish that T^+ and $\mathcal{E}^+$, or T_E^+ and $\mathcal{E}_E^+$, were cartesian (rather than suitable). We will use the full iterative strength in Section 7.1 to construct the 'opetopes'.

For technical purposes later on, we will need a refined version of these results. Wide pullbacks are defined on p. 364.

Proposition 6.5.6 *If $\mathcal{E}$ is a presheaf category and the functor T preserves wide pullbacks then the same is true of $\mathcal{E}^+$ and T^+ in Theorem 6.5.2, and of $\mathcal{E}_E^+$ and T_E^+ in Theorem 6.5.4. Moreover, if T is finitary then so are T^+ and T_E^+.*

6.6 Structured categories

Any monoidal category has an underlying plain multicategory. Here we meet 'T-structured categories', for any cartesian monad T, which bear the same relation to T-multicategories as strict monoidal categories do to plain multi-categories. At the end we briefly consider the non-strict case.

A strict monoidal category is a monoid in **Cat**, or, equivalently, a category in **Monoid**. This makes sense because the category **Monoid** is cartesian. More generally, if T is a cartesian monad on a cartesian category $\mathcal{E}$ then the category $\mathcal{E}^T$ of algebras is also cartesian (since the forgetful functor $\mathcal{E}^T \longrightarrow \mathcal{E}$ creates limits), so the following definition makes sense:

Definition 6.6.1 Let T be a cartesian monad on a cartesian category $\mathcal{E}$. Then a T-**structured category** is a category in $\mathcal{E}^T$, and we write T-**Struc** or $(\mathcal{E}, T)$-**Struc** for the category **Cat**$(\mathcal{E}^T)$ of T-structured categories.

A T-structured category is, incidentally, a generalized multicategory:

$$T\text{-}\mathbf{Struc} \cong (\mathcal{E}^T, \mathrm{id})\text{-}\mathbf{Multicat}$$

where id is the identity monad.

Example 6.6.2 For any cartesian category $\mathcal{E}$ we have

$$(\mathcal{E}, \mathrm{id})\text{-}\mathbf{Struc} \cong (\mathcal{E}, \mathrm{id})\text{-}\mathbf{Multicat} \cong \mathbf{Cat}(\mathcal{E}).$$

Example 6.6.3 If T is the free monoid monad on the category $\mathcal{E}$ of sets then T-**Struc** is the category **StrMonCat**$_\text{str}$ of strict monoidal categories and strict monoidal functors.

For an alternative definition, lift T to a monad $\mathbf{Cat}(T)$ on $\mathbf{Cat}(\mathcal{E})$, then define a T-structured category as an algebra for $\mathbf{Cat}(T)$. This is equivalent; more precisely, there is an isomorphism of categories

$$\mathbf{Cat}(\mathcal{E}^T) \cong \mathbf{Cat}(\mathcal{E})^{\mathbf{Cat}(T)}.$$

In the plain case $\mathbf{Cat}(T)$ is the free strict monoidal category monad on $\mathbf{Cat(Set)} = \mathbf{Cat}$, so algebras for $\mathbf{Cat}(T)$ are certainly the same as T-structured categories.

Example 6.6.4 Let T be the monad on **Set** corresponding to a strongly regular algebraic theory, as in 4.1.6. It makes sense to take models of such a theory in any category possessing finite products. A T-structured category is an algebra for $\mathbf{Cat}(T)$, which is merely a model of the theory in **Cat**.

Example 6.6.5 A specific instance is the monad $T = (1 + -)$ on **Set** corresponding to the theory of pointed sets (4.1.9). Then a T-structured category is a category A together with a functor from the terminal category 1 into A; in other words, it is a category A with a distinguished object.

Example 6.6.6 The principle described in 6.6.4 for finite product theories holds equally for finite limit theories. For instance, if T is the free plain operad monad on $\mathcal{E} = \mathbf{Set}^{\mathbb{N}}$ (as in 4.1.11) then a T-structured category is an operad in **Cat**, that is, a **Cat**-operad (p. 68). So we now have three descriptions of **Cat**-operads: as operads in **Cat**, as S-operads where S is the free strict monoidal category monad on **Cat** (4.2.17), and as T-structured categories.

Example 6.6.7 When T is the free strict ω-category monad on the category of globular sets (4.1.18), a T-structured category is what has been called a 'strict monoidal globular category' (Street, 1998, §1 or Batanin, 1998a, §2).

Example 6.6.8 Take the free category monad **fc** on the category $\mathcal{E}$ of directed graphs (Chapter 5). Then an **fc**-structured category is a category in $\mathcal{E}^{\mathbf{fc}} \cong \mathbf{Cat}$, that is, a strict double category.

Every strict monoidal category A has an underlying plain multicategory UA, and every plain multicategory C generates a free strict monoidal category FC, giving an adjunction $F \dashv U$ (Section 2.3). The same applies for generalized multicategories. Any T-structured category A has an underlying T-multicategory UA, whose graph is given by composing along the upper slopes

of the pullback diagram

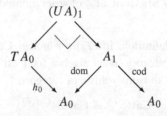

in which $h_0 : T A_0 \longrightarrow A_0$ is the T-algebra structure on A_0. Conversely, the free T-structured category FC on a T-multicategory C has graph

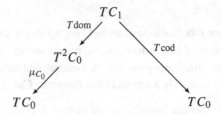

and the T-algebra structure on $(FC)_i = TC_i$ is μ_{C_i} ($i = 0, 1$). We then have the desired adjunction

$$T\text{-}\mathbf{Struc}$$
$$F \dashv U \qquad\qquad (6.9)$$
$$T\text{-}\mathbf{Multicat}.$$

Example 6.6.9 Consider once more the free monoid monad T on $\mathcal{E} = \mathbf{Set}$. Take the terminal plain multicategory 1, which has graph

$$\mathbb{N} \xleftarrow{\ 1\ } \mathbb{N} \xrightarrow{\ !\ } 1.$$

Then $F1$ is a strict monoidal category with graph

$$\mathbb{N} \xleftarrow{\ +\ } T\mathbb{N} \xrightarrow{\ T!\ } \mathbb{N}.$$

The objects of $F1$ are the natural numbers and a map $m \longrightarrow n$ in $F1$ is a sequence $(m_1, \ldots, m_n)$ of natural numbers such that $m_1 + \cdots + m_n = m$. So $F1$ is the strict monoidal category $\mathbb{D}$ of (possibly empty) finite totally ordered sets, with addition as tensor and 0 as unit. This is also suggested by diagram (2.9) (p. 83).

Example 6.6.10 Any object K of a cartesian category $\mathcal{K}$ generates a category DK in $\mathcal{K}$, the **discrete category** on K, uniquely determined by its underlying graph

$$K \xleftarrow{\ 1\ } K \xrightarrow{\ 1\ } K.$$

In particular, if $h = (TX \xrightarrow{\ h\ } X)$ is an algebra for some cartesian monad T then Dh is a T-structured category and UDh is a T-multicategory with graph

$$TX \xleftarrow{\ 1\ } TX \xrightarrow{\ h\ } X.$$

So UDh is the T-multicategory h^+ discussed in Example 4.2.22, and the triangle of functors

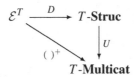

commutes up to natural isomorphism.

I have tried to emphasize that multicategories are truly different from monoidal categories (as well as providing a more natural language in many situations). This is borne out by the fact that the functor U : **StrMonCat**$_{\text{str}}$ ⟶ **Multicat** is far from an equivalence: it is faithful, but neither full (2.1.10) nor essentially surjective on objects (p. 61).

There is, however, a 'representation theorem' saying that every T-multicategory embeds fully in some T-structured category. In particular, every plain multicategory is a full sub-multicategory of the underlying multicategory of some strict monoidal category (2.1.4). I do not know of any use for the theorem; it does not reduce the study of T-multicategories to the study of T-structured categories any more than the Cayley Representation Theorem reduces the study of finite groups to the study of the symmetric groups. I therefore leave the proof as an exercise. The precise statement is as follows.

Definition 6.6.11 Let T be a cartesian monad on a cartesian category $\mathcal{E}$. A map $f : C \longrightarrow C'$ of T-multicategories is **full and faithful** if the square

$$
\begin{array}{ccc}
C_1 & \xrightarrow{\ (\text{dom},\text{cod})\ } & TC_0 \times C_0 \\
{\scriptstyle f_1}\downarrow & & \downarrow{\scriptstyle Tf_0 \times f_0} \\
C_1' & \xrightarrow[\ (\text{dom},\text{cod})\ ]{} & TC_0' \times C_0'
\end{array}
$$

is a pullback.

Proposition 6.6.12 *Let T be a cartesian monad on a cartesian category* $\mathcal{E}$, *and consider the adjunction of (6.9) (p. 204). For each T-multicategory C, the unit map* $C \longrightarrow UFC$ *is full and faithful.*   □

We finish with two miscellaneous thoughts.

First, we have been considering generalizations of plain multicategories, which are structures whose operations are 'many in, one out'. But in Section 2.3 we also considered PROs (and their symmetric cousins, PROPs), whose operations are 'many in, many out'. A PRO consists of a set S (the objects, or 'colours', often taken to have only one element) and a strict monoidal category whose underlying monoid of objects is the free monoid on S. The generalization to arbitrary cartesian monads T on cartesian categories $\mathcal{E}$ is clear: a T-**PRO** should be defined as a pair (S, A) where $S \in \mathcal{E}$ and A is a T-structured category whose underlying T-algebra of objects is the free T-algebra on S. We will not do anything with this definition, but see Section 7.6 for further discussion of 'many in, many out'.

Second, if strict monoidal categories generalize to T-structured categories, what do weak monoidal categories generalize to? One answer comes from realizing that the category $\mathbf{Cat}(\mathcal{E})$ has the structure of a strict 2-category and the monad $\mathbf{Cat}(T)$ the structure of a strict 2-monad. We can then define a **weak T-structured category** to be a weak algebra for this 2-monad, and indeed do the same in the lax case. In particular, if T is the free plain operad monad on $\mathbf{Set}^{\mathbb{N}}$ (6.6.6) then a weak T-structured category is like a $\mathbf{Cat}$-operad, but with the operadic composition only obeying associativity and unit laws up to coherent isomorphism. For example, let $P(n)$ be the category of Riemann surfaces whose boundaries are identified with the disjoint union of $(n + 1)$ copies of S^1, and define composition by gluing: then P forms a weak T-structured category. Compare 2.2.12 where, not having available the refined language of generalized multicategories, we had to quotient out by isomorphism and so lost (for instance) any information about automorphisms of the objects. We do not, however, pursue weak structured categories any further in this book.

6.7 Change of shape

To do higher-dimensional category theory we are going to want to move between n-categories and $(n + 1)$-categories and ω-categories, between globular and cubical and simplicial structures, and so on. In Chapter 9 we will see that a weak n-category can be defined as an algebra for a certain $T_{(n)}$-operad, where $T_{(n)}$ is the free strict n-category monad on the category of n-globular sets. So if we want to be able to relate n-categories to $(n + 1)$-categories, for instance, then we will need some way of relating $T_{(n)}$-operads to $T_{(n+1)}$-operads and some way of relating their algebras. In this section we set up the supporting theory: in other words, we show what happens to T-multicategories and their algebras as the monad T varies.

Formally, we expect the assignment $(\mathcal{E}, T) \longmapsto (\mathcal{E}, T)$-**Multicat** to be functorial in some way. We saw in Section 6.1 that there are notions of lax, colax, weak, and strict maps of monads, some of which are special cases of others, and we will soon see that a map $(\mathcal{E}, T) \longrightarrow (\mathcal{E}', T')$ of any one of these types induces a functor

$$(\mathcal{E}, T)\text{-}\mathbf{Multicat} \longrightarrow (\mathcal{E}', T')\text{-}\mathbf{Multicat}.$$

First we need some terminology. A lax map of monads $(Q, \psi):$ $(\mathcal{E}, T) \longrightarrow (\mathcal{E}', T')$ is **cartesian** if the functor Q is cartesian (but note that the natural transformation ψ need not be cartesian). Cartesian monads, cartesian lax maps of monads, and transformations form a sub-2-category $\mathbf{CartMnd}_{\mathrm{lax}}$ of $\mathbf{Mnd}_{\mathrm{lax}}$. Dually, a colax map of monads (P, ϕ) is **cartesian** if the functor P *and* the natural transformation ϕ are cartesian. Cartesian monads, cartesian colax maps of monads, and transformations form a sub-2-category $\mathbf{CartMnd}_{\mathrm{colax}}$ of $\mathbf{Mnd}_{\mathrm{colax}}$.

These definitions appear haphazard, with natural transformations required to be cartesian, or not, at random. I can justify this only pragmatically: they are the conditions required to make the following constructions work.

The main constructions are as follows. Let $(Q, \psi) : (\mathcal{E}, T) \longrightarrow (\mathcal{E}', T')$ be a cartesian lax map of cartesian monads. Then there is an induced functor

$$Q_* = (Q, \psi)_* : (\mathcal{E}, T)\text{-}\mathbf{Multicat} \longrightarrow (\mathcal{E}', T')\text{-}\mathbf{Multicat}$$

sending an $(\mathcal{E}, T)$-multicategory C to the $(\mathcal{E}', T')$-multicategory Q_*C whose underlying graph is given by composing along the upper slopes of the pullback diagram

$$
\begin{array}{ccc}
& (Q_*C)_1 & \\
\swarrow & \searrow & \\
T'QC_0 & & QC_1 \\
\psi_{C_0} \searrow & Q\mathrm{dom} \swarrow & \searrow Q\mathrm{cod} \\
& QTC_0 & QC_0 = (Q_*C)_0
\end{array}
$$

and whose composition and identities are defined in an evident way. Dually, let $(P, \phi) : (\mathcal{E}, T) \longrightarrow (\mathcal{E}', T')$ be a cartesian colax map of cartesian monads.

Then there is an induced functor

$$P_* = (P, \phi)_* : (\mathcal{E}, T)\text{-}\mathbf{Multicat} \longrightarrow (\mathcal{E}', T')\text{-}\mathbf{Multicat}$$

sending an $(\mathcal{E}, T)$-multicategory C to the $(\mathcal{E}', T')$-multicategory $P_* C$ with underlying graph

$$PC_1 = (P_* C)_1$$

$$\begin{array}{c} \quad {}^{P\mathrm{dom}} \swarrow \qquad \qquad \searrow {}^{P\mathrm{cod}} \\ PTC_0 \qquad \qquad \qquad \\ {}^{\phi_{C_0}} \swarrow \qquad \qquad \qquad \qquad \\ T'PC_0 \qquad \qquad \qquad PC_0 = (P_* C)_0. \end{array}$$

Note that $(Q_* C)_0 = Q(C_0)$ and $(P_* C)_0 = P(C_0)$, so if Q or P preserves all finite limits then Q_* or P_* restricts to a functor

$$(\mathcal{E}, T)\text{-}\mathbf{Operad} \longrightarrow (\mathcal{E}', T')\text{-}\mathbf{Operad}$$

between categories of operads.

After filling in all the details we obtain two maps of strict 2-categories,

$$\mathbf{CartMnd}_{\mathrm{lax}} \longrightarrow \mathbf{CAT}, \qquad \mathbf{CartMnd}_{\mathrm{colax}} \longrightarrow \mathbf{CAT},$$

both defined on objects by $(\mathcal{E}, T) \longmapsto (\mathcal{E}, T)\text{-}\mathbf{Multicat}$. The first is defined using pullbacks, so is only a weak functor; the second is strict. These dual functors agree where they intersect: the 2-categories $\mathbf{CartMnd}_{\mathrm{lax}}$ and $\mathbf{CartMnd}_{\mathrm{colax}}$ 'intersect' in the 2-category $\mathbf{CartMnd}_{\mathrm{wk}}$ of cartesian monads, cartesian weak maps of monads, and transformations, and the square

$$\begin{array}{ccc} \mathbf{CartMnd}_{\mathrm{wk}} & \lhook\joinrel\longrightarrow & \mathbf{CartMnd}_{\mathrm{colax}} \\ \big\downarrow & & \big\downarrow \\ \mathbf{CartMnd}_{\mathrm{lax}} & \longrightarrow & \mathbf{CAT} \end{array}$$

commutes up to natural isomorphism. (A **cartesian weak map** of monads is a cartesian lax map (Q, ψ) in which ψ is a natural isomorphism; then ψ is automatically cartesian.)

Example 6.7.1 Let $Q : \mathcal{D} \longrightarrow \mathcal{D}'$ be a cartesian functor between cartesian categories. Then $(Q, \mathrm{id}) : (\mathcal{D}, \mathrm{id}) \longrightarrow (\mathcal{D}', \mathrm{id})$ is a strict map of monads and so induces (unambiguously) a functor between the categories of multicategories,

$$\left((\mathcal{D}, \mathrm{id})\text{-}\mathbf{Multicat} \xrightarrow{\ Q_*\ } (\mathcal{D}', \mathrm{id})\text{-}\mathbf{Multicat} \right) = \left(\mathbf{Cat}(\mathcal{D}) \xrightarrow{\ Q_*\ } \mathbf{Cat}(\mathcal{D}') \right).$$

This is the usual induced functor between categories of internal categories.

There is another way in which the two processes are compatible, involving adjunctions. Suppose we have a diagram

$$(\mathcal{E}, T)$$

$$(P,\phi) \Big\uparrow \dashv \Big\downarrow (Q,\psi) \qquad\qquad (6.10)$$

$$(\mathcal{E}', T')$$

in which (P, ϕ) is a cartesian colax map of cartesian monads, (Q, ψ) is a cartesian lax map, and there is an adjunction of functors $P \dashv Q$ under which ϕ and ψ are mates (Section 6.1). Then, as may be checked, there arises an adjunction between the functors P_* and Q_* constructed above:

$$(\mathcal{E}, T)\text{-}\mathbf{Multicat}$$

$$P_* \Big\uparrow \dashv \Big\downarrow Q_* \qquad\qquad (6.11)$$

$$(\mathcal{E}', T')\text{-}\mathbf{Multicat}.$$

Example 6.7.2 Let T be a cartesian monad on a cartesian category $\mathcal{E}$. Then there is a diagram

$$(\mathcal{E}^T, \mathrm{id})$$

$$(F,v) \Big\uparrow \dashv \Big\downarrow (U,\varepsilon)$$

$$(\mathcal{E}, T)$$

of the form (6.10), in which F and U are the free and forgetful functors and v and ε are certain canonical natural transformations. This gives rise by the process just described to an adjunction

$$(\mathcal{E}, T)\text{-}\mathbf{Struc}$$

$$F_* \Big\uparrow \dashv \Big\downarrow U_*$$

$$(\mathcal{E}, T)\text{-}\mathbf{Multicat},$$

none other than the adjunction that was the subject of the previous section.

Example 6.7.3 Let $(P, \phi) : (\mathcal{E}, T) \longrightarrow (\mathcal{E}', T')$ be a cartesian colax map of cartesian monads, and suppose that the functor P has a right adjoint Q. Then P_* has a right adjoint too: for taking the mate $\psi = \overline{\phi}$ gives Q the structure of a cartesian lax map of monads, leading to an adjunction $P_* \dashv Q_*$.

We have considered change of shape for T-multicategories; let us now do the same for T-algebras and T-structured categories, concentrating on lax rather than colax maps. If $(Q, \psi) : (\mathcal{E}, T) \longrightarrow (\mathcal{E}', T')$ is a cartesian lax map of cartesian monads then the induced functor $\mathcal{E}^T \longrightarrow \mathcal{E}'^{T'}$ is also cartesian, so by Example 6.7.1 induces in turn a functor $\mathbf{Cat}(\mathcal{E}^T) \longrightarrow \mathbf{Cat}(\mathcal{E}'^{T'})$ – in other words, induces a functor on structured categories,

$$Q_* : (\mathcal{E}, T)\text{-}\mathbf{Struc} \longrightarrow (\mathcal{E}', T')\text{-}\mathbf{Struc}.$$

The change-of-shape processes for algebras, structured categories and multicategories are compatible: for any cartesian lax map $(Q, \psi) : (\mathcal{E}, T) \longrightarrow (\mathcal{E}', T')$ of cartesian monads, the diagram

$$
\begin{array}{ccc}
\mathcal{E}^T & \xrightarrow{\ D\ } (\mathcal{E}, T)\text{-}\mathbf{Struc} \xrightarrow{\ U_*\ } (\mathcal{E}, T)\text{-}\mathbf{Multicat} \\
Q_* \downarrow \qquad\qquad\quad Q_* \downarrow \qquad\qquad\qquad Q_* \downarrow \\
\mathcal{E}'^{T'} \xrightarrow[\ D\ ] {} (\mathcal{E}', T')\text{-}\mathbf{Struc} \xrightarrow[\ U_*\ ]{} (\mathcal{E}', T')\text{-}\mathbf{Multicat}
\end{array}
$$

commutes up to natural isomorphism. Here D is the discrete category functor defined in 6.6.10, and the commutativity of the left-hand square can be calculated directly. The U_*s are the forgetful functors, which, as we saw in 6.7.2, are induced by lax maps (U, ε). To see that the right-hand square commutes, it is enough to see that the square of lax maps

$$
\begin{array}{ccc}
(\mathcal{E}^T, \mathrm{id}) & \xrightarrow{(U,\varepsilon)} & (\mathcal{E}, T) \\
(Q_*, \mathrm{id}) \downarrow & & \downarrow (Q, \psi) \\
(\mathcal{E}'^{T'}, \mathrm{id}) & \xrightarrow[(U,\varepsilon)]{} & (\mathcal{E}', T')
\end{array}
$$

commutes, and this is straightforward.

Finally, we answer the question posed in the introduction to the section: how do algebras for multicategories behave under change of shape? An algebra for an $(\mathcal{E}, T)$-multicategory C is an object X over C_0 acted on by C, so if $(Q, \psi) : (\mathcal{E}, T) \longrightarrow (\mathcal{E}', T')$ is a cartesian lax or colax map then we might hope that QX, an object over QC_0, would be acted on by Q_*C. In other words, we might hope for a functor from $\mathbf{Alg}(C)$ to $\mathbf{Alg}(Q_*C)$. Such a functor does indeed exist, in both the lax and colax cases.

First take a cartesian lax map of cartesian monads, $(Q, \psi) : (\mathcal{E}, T) \longrightarrow (\mathcal{E}', T')$. For any T-multicategory C there is an induced lax map of monads

$$(\mathcal{E}/C_0, T_C) \longrightarrow (\mathcal{E}'/QC_0, T'_{Q_*C}),$$

comprising the functor $\mathcal{E}/C_0 \longrightarrow \mathcal{E}'/QC_0$ induced by Q and a natural transformation ψ that may easily be determined. This in turn induces a functor on categories of algebras:

$$\mathbf{Alg}(C) \longrightarrow \mathbf{Alg}(Q_*C),$$

$$\begin{pmatrix} T_C X \\ \downarrow h \\ X \end{pmatrix} \longmapsto \begin{pmatrix} T'_{Q_*C} QX \\ \downarrow \psi^C_X \\ QT_C X \\ \downarrow Qh \\ QX \end{pmatrix}.$$

Now take a cartesian colax map of cartesian monads, $(P, \phi) : (\mathcal{E}, T) \longrightarrow (\mathcal{E}', T')$. For any T-multicategory C there is an induced *weak* map of monads

$$(\mathcal{E}/C_0, T_C) \longrightarrow (\mathcal{E}'/PC_0, T'_{P_*C}),$$

which amounts to saying that if $X = (X \xrightarrow{p} C_0) \in \mathcal{E}/C_0$ then $T'_{P_*C} PX \cong PT_C X$ canonically; and indeed, we have a diagram

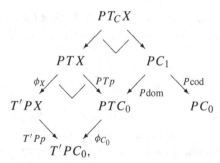

giving the isomorphism required.

Example 6.7.4 In Section 2.3 we considered the adjunction between plain multicategories and strict monoidal categories, and in 6.7.2 we saw that it is induced by certain maps of monads:

$$\begin{array}{ccc} (\mathbf{Monoid}, \mathrm{id}) & & \mathbf{StrMonCat}_{\mathrm{str}} \\ (F,\nu) \Big\uparrow \dashv \Big\downarrow (U,\varepsilon) & \longmapsto & F_* \Big\uparrow \dashv \Big\downarrow U_* \\ (\mathbf{Set}, \text{free monoid}) & & \mathbf{Multicat}. \end{array}$$

The lax map (U, ε) induces a functor $\mathbf{Alg}(A) \longrightarrow \mathbf{Alg}(U_* A)$ for each strict monoidal category A. When A is regarded as a $(\mathbf{Monoid}, \mathrm{id})$-multicategory, an A-algebra is a lax monoidal functor $A \longrightarrow \mathbf{Set}$. This is the same thing as an algebra for the underlying multicategory $U_* A$ (2.1.13), and the induced functor is in fact an isomorphism.

Conversely, the colax map (F, ν) induces a functor $\mathbf{Alg}(C) \longrightarrow \mathbf{Alg}(F_* C)$ for each multicategory C, whose explicit form is left as an exercise.

It is no coincidence that the functors induced by U in this example are isomorphisms. Roughly speaking, this is because U is monadic:

Proposition 6.7.5 *Let C be a cartesian category, let S and S' be cartesian monads on C, and let $\lambda : S \circ S' \longrightarrow S' \circ S$ be a cartesian distributive law. Write*

$$
\begin{array}{ccc}
C^S & \xrightarrow{\ \tilde{S}\ } & C^S \\
U \downarrow & \nearrow \psi & \downarrow U \\
C & \xrightarrow[\ S' \circ S\ ]{} & C
\end{array}
$$

for the induced lax map of monads, as in 6.1.5. Then for any $\tilde{S}$-multicategory C, there is an isomorphism of categories $\mathbf{Alg}(C) \cong \mathbf{Alg}((U, \psi)_* C)$.

The monads $\tilde{S}$ and $S' \circ S$ are cartesian (6.1.3, 6.1.4), so it does make sense to talk about multicategories for them.

Proof We just prove this in the case where C is an $\tilde{S}$-operad O, since that is all we will need later and the proof is a little easier.

As well as the lax map shown, we have a strict map of monads

$$
\begin{array}{ccc}
C^S & \xrightarrow{\ \tilde{S}\ } & C^S \\
U \downarrow & \parallel & \downarrow U \\
C & \xrightarrow[\ S'\ ]{} & C.
\end{array}
$$

So we have an S'-operad $(U, \mathrm{id})_* O$, hence a monad $S'_{(U,\mathrm{id})_* O}$ on C, of which $\tilde{S}_O$ is a lift to C^S. By Lemma 6.1.4, there is a corresponding distributive law

$$
S \circ S'_{(U,\mathrm{id})_* O} \longrightarrow S'_{(U,\mathrm{id})_* O} \circ S.
$$

This gives $S'_{(U,\mathrm{id})_*} O \circ S$ the structure of a monad on $\mathcal{C}$; but it can be checked that this monad is exactly $(S' \circ S)_{(U,\psi)_* O}$, so by Lemma 6.1.5,

$$\mathbf{Alg}(O) = (\mathcal{C}^S)^{\widetilde{S}_O} \cong \mathcal{C}^{S'_{(U,\mathrm{id})_*} O \circ S} \cong \mathcal{C}^{(S' \circ S)_{(U,\psi)_* O}} = \mathbf{Alg}((U, \psi)_* O),$$

as required. ☐

The forgetful functor in Example 6.7.4 is the case $\mathcal{C} = \mathbf{Set}$, $S = $ (free monoid), $S' = \mathrm{id}$. We will use the proposition when comparing definitions of weak 2-category in Section 9.4.

6.8 Enrichment

We finish with a brief look at a topic too large to fit in this book. Its slogan is 'what can we enrich in?'

Take, for example, **Ab**-categories: categories enriched in (or 'over') abelian groups. The simplest definition of an **Ab**-category is as a class C_0 of objects together with an abelian group $C(a, b)$ for each $a, b \in C_0$, a bilinear composition function

$$C(a, b), C(b, c) \longrightarrow C(a, c)$$

for each $a, b, c \in C_0$, and an identity $1_a \in C(a, a)$ for each $a \in C_0$, satisfying associativity and identity axioms. You *could* express composition as a linear map out of a tensor product, but this would be an irrelevant elaboration; put another way, the first definition of **Ab**-category can be understood by someone who knows what a multilinear map is but has not yet learned about tensor products.

More generally, if V is a plain multicategory then there is an evident definition of V-enriched category. Classically one enriches categories in monoidal categories (as in Section 1.3), but this is an unnaturally narrow setting; in the terminology of Section 3.3, representability of the multicategory is an irrelevance.

More generally still, suppose we have some type of categorical structure – 'widgets', say. Then the question is: for what types of structure V can we make a definition of 'V-enriched widget'? In the previous paragraphs widgets were categories, and we saw that V could be a plain multicategory. (In fact, that is not all V can be, as we will soon discover.) In my (1999a) paper, the question is answered when widgets are T-multicategories for almost any cartesian monad T. The general definition of enriched T-multicategory is short, simple, and given below, but unwinding its implications takes more space than we have; hence the following sketch.

Let T be a monad on a category $\mathcal{E}$, and suppose that both T and $\mathcal{E}$ are suitable (Section 6.5), so that there is a cartesian free T-multicategory monad T^+ on the category $\mathcal{E}^+$ of T-graphs. For any object C_0 of $\mathcal{E}$, there is a unique T-multicategory structure on the T-graph

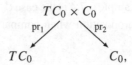

$$
\begin{array}{ccc}
 & TC_0 \times C_0 & \\
 {\scriptstyle \mathrm{pr}_1}\swarrow & & \searrow{\scriptstyle \mathrm{pr}_2} \\
 TC_0 & & C_0,
\end{array}
$$

and we write this T-multicategory as IC_0, the **indiscrete** T-multicategory on C_0. Then IC_0 is a T^+-algebra, so by 4.2.22 gives rise to a T^+-multicategory $(IC_0)^+$ whose domain map is the identity.

Definition 6.8.1 Let T be a suitable monad on a suitable category $\mathcal{E}$ and let V be a T^+-multicategory. A **V-enriched T-multicategory** is an object C_0 of $\mathcal{E}$ together with a map $(IC_0)^+ \longrightarrow V$ of T^+-multicategories.

Example 6.8.2 The most basic example is when T is the identity monad on the category $\mathcal{E}$ of sets. Then $\mathcal{E}^+$ is the category of directed graphs, T^+ is the free category monad **fc**, and T^+-multicategories are the **fc**-multicategories of Chapter 5. We therefore have a notion of 'V-enriched category' for any **fc**-multicategory V. In the special case that V is a monoidal category (5.1.8), we recover the standard definition of enrichment. More generally, if V is a bicategory (5.1.5) then we recover the less well-known definition of category enriched in a bicategory (Walters, 1981), and if V is a plain multicategory (5.1.6) then we recover the definition of category enriched in a plain multicategory. For more on enriched categories in this broad sense, see my (2002).

Example 6.8.3 The next most basic example is when T is the free monoid monad on the category $\mathcal{E}$ of sets. Theorem 7.1.3 tells us that any symmetric monoidal category gives rise canonically to what is there called a T_2-multicategory. By definition, T_2 is the free T-operad monad and T^+ the free T-multicategory monad, so a T_2-multicategory is a special kind of T^+-multicategory. Hence any symmetric monoidal category gives rise canonically to a T^+-multicategory, giving us a definition of plain multicategory enriched in a symmetric monoidal category, and in particular, of plain operad in a symmetric monoidal category. These are exactly the usual definitions (pp. 64, 68).

Example 6.8.4 Borcherds (1997) introduced certain structures called 'relaxed multilinear categories' in his definition of vertex algebras over a vertex group, and Soibelman (1997, 1999) defined the same structures independently in his work on quantum affine algebras. As explained by Borcherds, they can be regarded as categorical structures in which the maps have singularities whose severity is measured by trees. They also arise completely naturally in the theory of enrichment: if T is the free monoid monad on the category $\mathcal{E}$ of sets, as in the previous example, then there is a certain canonical T^{+}-multicategory V such that V-enriched T-multicategories are precisely relaxed multilinear categories. See Leinster (1999a, Ch. 4) for details.

Example 6.8.5 In the next chapter we introduce the sequence $(T_n)_{n\in\mathbb{N}}$ of 'opetopic' monads. By definition, T_n is the free T_{n-1}-operad monad, and this means that there is a notion of T_{n-1}-multicategory enriched in a T_n-multicategory. More vaguely, a T_n-multicategory is naturally regarded as an $(n+1)$-dimensional structure, so n-dimensional structures can be enriched in $(n+1)$-dimensional structures.

Notes

Most parts of this chapter have appeared before (1998a, §4), (2000b, Ch. 3). The thought that an operad is a cartesian monad equipped with a cartesian natural transformation down to the free monoid monad (Section 6.2) is closely related to Kelly's idea of a 'club' (1974, 1992). See Snydal (1999a, 1999b) for more on the relaxed multicategory definition of vertex algebra.

Chapter 7
Opetopes

John Dee [...] summoned angels of dubious celestial provenance by invoking names like Zizop, Zchis, Esiasch, Od and Iaod.

Eco (1995)

Operads lead inescapably into geometry. In this chapter we see that as soon as the notion of generalized operad is formulated, the notion of opetope is unavoidable. Opetopes are something like simplices: they are a completely canonical family of polytopes, as pervasive in higher-dimensional algebra as simplices are in geometry.

In Section 7.1 opetopes are defined and their geometric representation explained. Intertwined with the definition of opetope is the definition of a certain sequence $(T_n)_{n \in \mathbb{N}}$ of cartesian monads; we also look at T_n-multicategories and their relation to symmetric multicategories. Section 7.2 is formally about T_n-structured categories (in the sense of Section 6.6); translated into geometry, this means diagrams of opetopes pasted together.

As we shall see, there is a category of n-dimensional pasting diagrams for each natural number n. When $n = 1$ this is the category $\mathbb{D}$ of finite totally ordered sets (the augmented simplex category); when $n = 2$ it is a category of trees, as found in parts of quantum algebra. In Section 7.3 we analyse the category of trees in some detail. Some of this is analogous to some of the standard analysis of $\mathbb{D}$: see for instance Mac Lane (1971, VII.5), where the monics, epics, primitive face and degeneracy maps, and standard factorization properties are all worked out, and the universal role of $\mathbb{D}$ (as the free monoidal category containing a monoid) is established.

Opetopes were invented by Baez and Dolan (1997) so that they could frame a definition of weak n-category. The strategy is simple: opetopes together with face maps form a category, a presheaf on that category is called an opetopic set, and a weak n-category is an opetopic set with certain properties. This is like both the definition of Kan complex (a simplicial set with horn-filling properties) and the definition of representable multicategory (a multicategory with universality properties, 3.3.1). Opetopic sets are discussed in Section 7.4, and opetopic definitions of weak n-category in Section 7.5.

We finish (Section 7.6) with a short section on the 'many in, many out' approach to higher-dimensional category theory. Opetopes arise from the concept of operations with many inputs and a single output. We could start instead with the concept of operations having both many inputs and many outputs, and try to construct shapes analogous to opetopes. But this turns out to be impossible, as we see. (Formally, 3-computads do not form a presheaf category.) This does not mean that it is hopeless to try to develop a many-in, many-out framework for higher categorical structures, but it does mean that such a framework would be qualitatively different from the simplicial, cubical, globular, and opetopic frameworks, in each of which there is a genuine category of shapes.

7.1 Opetopes

The following table shows some of the types of generalized operad that we have met. As usual, T is a cartesian monad on a cartesian category $\mathcal{E}$.

$\mathcal{E}$	T	T-operads
Set	identity	monoids
Set	free monoid	plain operads
$\mathbf{Set}^{\mathbb{N}} \simeq \mathbf{Set}/\mathbb{N}$	free plain operad	(see 4.2.14)

Each row is generated automatically from the last by taking T to be 'free operad of the type in the last row'. Technicalities aside, it is clear that this table can be continued indefinitely.

This gives an absolutely fundamental infinite sequence of categories $\mathcal{E}$, monads T, and types of generalized operad. The bulk of this chapter consists of working out what they look like. It is plain from the first few rows that there will be some geometrical content – for instance, in the third row, the operations in a T-operad are indexed by trees (Example 4.2.14).

The table can be compared with that of Ginzburg and Kapranov (1994, p. 204). They include columns marked 'Geometry' (whose entries are 'vector bundles', 'manifolds' and '?(moduli spaces)'), 'Linear Physics' and 'Nonlinear Physics'. Their table, like the one above, has only three rows. Here we show how to continue forever – in our columns, at least.

To do this formally we need to recall the results on free operads in Section 6.5. There we met the concept of 'suitable' categories and monads and saw that they provided a good context for free operads. Specifically, let T be a suitable monad on a suitable category $\mathcal{E}$ with a terminal object. Then Theorem 6.5.4 tells us that the forgetful functor

$$(\mathcal{E}, T)\text{-}\mathbf{Operad} \longrightarrow \mathcal{E}/T1$$

sending a T-operad to its underlying T-graph has a left adjoint, that the induced monad ('free T-operad') on $\mathcal{E}/T1$ is suitable, and that the category $\mathcal{E}/T1$ is suitable. Trivially, $\mathcal{E}/T1$ has a terminal object. Also, Theorem 6.5.1 tells us that the category of sets and the identity monad on it are both suitable. This makes possible:

Definition 7.1.1 For each $n \in \mathbb{N}$, the suitable category $\mathcal{E}_n$ and the suitable monad T_n on $\mathcal{E}_n$ are defined inductively by

- $\mathcal{E}_0 = \mathbf{Set}$ and $T_0 = \mathrm{id}$
- $\mathcal{E}_{n+1} = \mathcal{E}_n/T_n 1$ and $T_{n+1} = $ (free T_n-operad).

In fact, there is for each $n \in \mathbb{N}$ a set O_n such that $\mathcal{E}_n$ is canonically isomorphic to $\mathbf{Set}/O_n$. (Recall also from p. 29 that $\mathbf{Set}/O_n \simeq \mathbf{Set}^{O_n}$.) First, $O_0 = 1$. Now suppose, inductively, that $n \geq 0$ and $\mathcal{E}_n \cong \mathbf{Set}/O_n$. The terminal object of $\mathbf{Set}/O_n$ is $(O_n \xrightarrow{1} O_n)$, so if we put

$$\begin{pmatrix} O_{n+1} \\ \downarrow t \\ O_n \end{pmatrix} = T_n \begin{pmatrix} O_n \\ \downarrow 1 \\ O_n \end{pmatrix}$$

then we have

$$\mathcal{E}_{n+1} = \frac{\mathcal{E}_n}{T_n 1} \cong \frac{\mathbf{Set}/O_n}{O_{n+1} \xrightarrow{t} O_n} \cong \frac{\mathbf{Set}}{O_{n+1}}.$$

This gives an infinite sequence of sets and functions

$$\cdots \xrightarrow{t} O_{n+1} \xrightarrow{t} O_n \xrightarrow{t} \cdots \xrightarrow{t} O_1 \xrightarrow{t} O_0.$$

The first few steps of the iteration are:

n	O_n	$\mathcal{E}_n$	T_n	T_n-operads
0	1	**Set**	identity	monoids
1	1	**Set**	free monoid	plain operads
2	$\mathbb{N}$	**Set**/$\mathbb{N}$	free plain operad	(see 4.2.14 and below)
3	{trees}	**Set**/{trees}		

We call O_n the set of n-**dimensional opetopes**, or n-**opetopes**. 'Opetope' is pronounced with three syllables, as in 'OPEration polyTOPE'; opetopes encode fundamental information about operations and are naturally represented as polytopes of the corresponding dimension. Let us see how this works in low dimensions.

The unique 0-opetope is drawn as a point:

$$O_0 = \{\ \bullet\ \}.$$

The monad T_0 on $\mathbf{Set}/O_0 \cong \mathbf{Set}$ is the identity. A T_0-multicategory is an ordinary category, with objects a depicted as labelled points (0-opetopes) and maps θ as labelled arrows, as usual:

$$\underset{a}{\bullet} \ , \qquad \overset{\theta}{\underset{a \qquad\qquad b}{\bullet\longrightarrow\!\!\!\bullet}} \ .$$

In particular, a T_0-operad is a monoid, the elements θ of which are drawn as

$$\overset{\theta}{\bullet\!\longrightarrow\!\!\bullet} \ .$$

The underlying graph structure of a T_0-operad is formally a set over $T_0 1 = O_1$, that is, a family of sets indexed by the elements of O_1. Of course, O_1 has only one element and the 'underlying graph structure' of a monoid is just a set. But since we want to view the elements of a T_0-operad as labels on arrows, we choose to draw the unique element of O_1 as an arrow:

$$O_1 = \{\ \bullet\!\longrightarrow\!\!\bullet\ \}.$$

So the unique 1-opetope is drawn as a 1-dimensional polytope.

Next we have the monad $T_1 =$ (free monoid) on the category $\mathbf{Set}/O_1 \cong \mathbf{Set}$. The free monoid $T_1 E$ on a set E has elements of the form

$$\bullet\overset{a_1}{\longrightarrow\!\!\bullet}\overset{a_2}{\longrightarrow} \quad \cdots \quad \overset{a_n}{\longrightarrow\!\!\bullet}$$

($n \geq 0$, $a_i \in E$). A T_1-multicategory is a plain multicategory, and with the diagrams we are using it is natural to draw an object a as labelling an edge,

$$\bullet\overset{a}{\longrightarrow\!\!\bullet} \ ,$$

and a map $\theta : a_1, \ldots, a_n \longrightarrow a$ as labelling a 2-dimensional region,

This is an alternative to our usual picture,

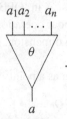

$$
\begin{array}{c}
a_1 a_2 \quad a_n \\
\mid\mid \quad \cdots \mid \\
\theta \\
\mid \\
a
\end{array}
$$

Sometimes we reduce clutter by omitting arrows:

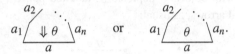

$$
a_1 \diagup \begin{array}{c} a_2 \\ \Downarrow \theta \end{array} \diagdown a_n \quad \text{or} \quad a_1 \diagup \begin{array}{c} a_2 \\ \theta \end{array} \diagdown a_n .
$$

In this paradigm, composition is drawn as

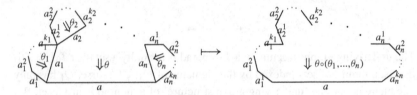

and the function assigning identities as

$$
\underline{\qquad a \qquad} \quad \longmapsto \quad \underset{a}{\overset{a}{\triangle}}{}^{\Downarrow 1_a} .
$$

A T_1-*operad* (plain operad) looks just the same except that the edges are no longer labelled: so an n-ary operation $\theta \in P(n)$ of an operad P is drawn as

with n 'input' edges along the top and one 'output' edge along the bottom. The underlying T_1-graph of a T_1-operad is a set over $T_1 1 = O_2 \cong \mathbb{N}$, so we draw 2-opetopes as 2-dimensional polytopes:

$$
O_2 = \left\{ \begin{array}{ccccc} \bigcirc\!\!\Downarrow , & \triangle\Downarrow , & \triangle\Downarrow , & \square\Downarrow , & \pentagon\Downarrow , & \cdots \end{array} \right\}.
$$

Next we have the monad $T_2 = $ (free plain operad) on the category

$$
\mathbf{Set}/O_2 \cong \mathbf{Set}/\mathbb{N} \simeq \mathbf{Set}^{\mathbb{N}}.
$$

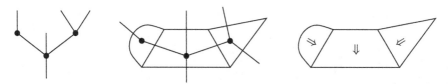

Fig. 7-A. How a tree corresponds to a diagram of pasted-together 2-opetopes

An object E of $\mathbf{Set}/O_2$ is a family $(E(\omega))_{\omega \in O_2}$ of sets, with the elements of $E(\omega)$ regarded as potential labels to be stuck on the 2-opetope ω. The free operad $T_2 E$ on E is formed by pasting together labelled 2-opetopes, with output edges joined to input edges. For instance, if $a_1 \in E(3)$, $a_2 \in E(1)$ and $a_3 \in E(2)$ then

 (7.1)

is an element of $(T_2 E)(4)$, and if $b_1 \in E(5)$, $b_2 \in E(0)$, $b_3, b_4 \in E(3)$ and $b_5 \in E(1)$ then

 (7.2)

is an element of $(T_2 E)(8)$. Previously we represented operations in a free operad as trees with labelled vertices, so instead of (7.1) we drew

(Section 2.3 and 4.1.11). These pictures are dual to one another, as Fig. 7-A demonstrates; we return to this correspondence in Section 7.3.

We described T_2-multicategories in Example 4.2.14 in terms of trees; we now describe them in terms of opetopes. The objects of a T_2-multicategory C form a family $(C_0(\omega))_{\omega \in O_2}$ of sets; put another way, they form a graded set $(C_0(n))_{n \in \mathbb{N}}$ with, for instance, $a \in C_0(3)$ drawn as

Arrows look like

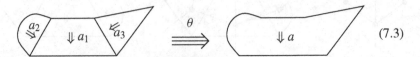

(7.3)

$(a_1 \in C_0(3), a_2 \in C_0(1), a_3 \in C_0(2), a \in C_0(4))$. The 2-opetope in the codomain always has the same number of input edges as the diagram in the domain (four, in this case); here it is drawn irregularly to make the equality self-evident. Visualize the whole of (7.3) as a 3-dimensional polytope with one flat bottom face labelled by a and three curved top faces labelled by a_1, a_2 and a_3 respectively, and with a label θ in the middle. On a sheet of paper we must settle for 2-dimensional representations such as the one above. Composition takes a diagram of arrows such as

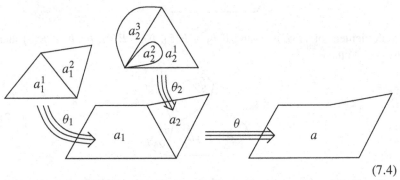

(7.4)

and produces a single arrow

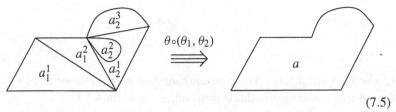

(7.5)

The identity on an object $a \in C_0(n)$ looks like

$$\langle a \rangle \stackrel{1_a}{\Longrightarrow} \langle a \rangle ,$$

where there are n input edges in both the domain and the codomain.

In particular, a T_2-operad consists of a collection of operations such as

together with composition and identities as above. So the elements of $O_3 = T_2(1)$ – the 3-opetopes – are thought of as 3-dimensional polytopes, for instance

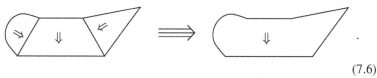

$$(7.6)$$

The function $t : O_3 \longrightarrow O_2$ is 'target'; it sends the 3-opetope above to the 2-opetope with four input edges.

This gives a systematic way of portraying opetopes and T_n-multicategories for arbitrary n. In Section 7.4 we will see that each n-opetope does indeed give rise to an n-dimensional topological space.

Some crude examples of T_n-multicategories are provided by symmetric structures. First, any commutative monoid $(A, +, 0)$ gives rise to a T_n-operad for every $n \in \mathbb{N}$: an operation of shape $\omega \in O_n$ is just an element of A (regardless of what ω is), composition is $+$, and the identity is 0. We have already seen this in the case $n = 1$ (plain operads, 2.2.7). Formally, let **CommMon** be the category of commutative monoids and $\Delta : \textbf{Set} \longrightarrow \textbf{Set}/O_n$ the functor sending a set A to $(A \times O_n \xrightarrow{\text{pr}_2} O_n)$; then we have

Theorem 7.1.2 *For each $n \in \mathbb{N}$ there is a canonical functor*

$$\textbf{CommMon} \longrightarrow T_n\textbf{-Operad}$$

making the diagram

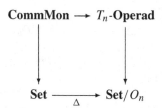

commute, where the vertical arrows are the forgetful functors.

Proof This follows from T_n being a finitary familially representable monad on a slice of **Set**: see Example C.4.2. $\qquad\square$

More interestingly, any symmetric multicategory A gives rise to a T_n-multicategory C for every $n \in \mathbb{N}$, canonically up to isomorphism. The objects of C of shape $\omega \in O_n$ are simply the objects of A (regardless of ω). The arrows of C are arrows of A: for example, if $n = 2$ and $\widehat{a}, \widetilde{a}, \ldots$ are objects of A then an arrow

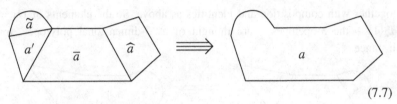

$$\tag{7.7}$$

in C might be defined to be an arrow

$$a', \widehat{a}, \overline{a}, \widetilde{a} \longrightarrow a$$

in A – or indeed, the same but with the four domain objects ordered differently. There is no canonical ordering of the 2-opetopes making up a given 2-pasting diagram; more precisely, there is no method of ordering that is stable under substitution of the kind shown in diagrams (7.4) and (7.5). This is why we needed to start with a *symmetric* multicategory: composition in C cannot be defined without permuting the lists of objects. It is also why C is canonical only up to isomorphism. The precise statement of the construction is:

Theorem 7.1.3 *For each $n \in \mathbb{N}$ there is a functor*

$$\textbf{SymMulticat} \longrightarrow T_n\textbf{-Multicat},$$

canonical up to isomorphism, making the diagram

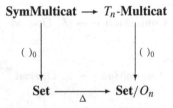

commute, where both vertical arrows are the functors assigning to a multicategory its object-of-objects.

Proof Again this follows from T_n being a finitary familially representable monad on a slice of **Set**: Example C.4.9. □

Symmetric structures have a ghostly presence throughout this book, hovering just beyond our world of cartesian monads and generalized multicategories.

Much of the time we are concerned with labelled cell diagrams such as the domain of (7.7), and exercise full sensitivity to their geometric configuration. By passing to a symmetric multicategory we destroy all the geometry.

In this sense, symmetric multicategories are the ultimate in crudeness. There are various situations in mathematics where symmetric multicategories (or symmetric monoidal categories) are customarily used, but generalized multicategories provide a more sensitive and more general approach. Operads in a symmetric monoidal category and categories enriched in a (symmetric or not) monoidal category are two examples; the more thoughtful approaches replace symmetric monoidal categories by T_2-multicategories and **fc**-multicategories, respectively (Section 6.8). There are also entire approaches to higher-dimensional category theory based on symmetric structures, as discussed in Section 7.5; crude does not mean ineffective.

7.2 Categories of pasting diagrams

The principal thing that you can do in a higher-dimensional category is to take a diagram of cells and form its composite. Not just any old diagram will do: it must, for instance, be connected and have the cells oriented compatibly. For example, the acceptable diagrams of 1-cells are those of the form

$$\bullet\!\!-\!\!\!\longrightarrow\!\!\bullet\!\!-\!\!\!\longrightarrow \quad \cdots \quad \longrightarrow\!\!\bullet \tag{7.8}$$

where the number of arrows is a non-negative integer. Let us call a composable diagram of pasted-together n-cells an 'n-pasting diagram'. For $n \geq 2$, the class of n-pasting diagrams depends on the shape of cells that you have chosen to use in your theory of higher-dimensional categories: globular, cubical, simplicial, opetopic, ...

Pasting diagrams play an important role in any theory of higher categories. What distinguishes the opetopic theory is that pasting diagrams are the same thing as cell shapes of one dimension higher. For instance, the 1-pasting diagram (7.8) with k arrows corresponds to the 2-opetope

with k arrows along the top; we draw them differently, but there is a natural identification. So we may conveniently *define* an **(opetopic) n-pasting diagram** to be an $(n + 1)$-opetope, for any $n \in \mathbb{N}$. In this section we show how, for each n, the n-pasting diagrams form a category $\mathbf{Pd}_n$ – and actually, rather more than just a category.

The formal method is as follows. So far we have looked at T_n-operads and T_n-multicategories; now we look at T_n-structured categories. $\mathbf{PD}_n$ is defined as the free T_n-structured category on the terminal T_n-multicategory, and $\mathbf{Pd}_n$ as the underlying category of $\mathbf{PD}_n$. So, we begin by recalling what T-structured categories are in general and what they look like when $T = T_n$; then we look at free structured categories; finally, we arrive at the category of n-pasting diagrams. The next section, 7.3, is a detailed examination of the case $n = 2$: it turns out that $\mathbf{Pd}_2$ is a category of trees.

Recall from Section 6.6 that if T is a cartesian monad on a cartesian category $\mathcal{E}$ then a T-structured category is an internal category in $\mathcal{E}^T$. Alternatively, note that T lifts naturally to a monad $\mathbf{Cat}(T)$ on $\mathbf{Cat}(\mathcal{E})$, and a T-structured category is then a $\mathbf{Cat}(T)$-algebra.

A T_0-structured category is a category in $\mathbf{Set}^{T_0} \cong \mathbf{Set}$, that is, a category.

A T_1-category is a category in $\mathbf{Set}^{T_1} \cong \mathbf{Monoid}$, that is, a strict monoidal category. (Alternatively, $\mathbf{Cat}(T_1)$ is the free strict monoidal category monad on $\mathbf{Cat}(\mathbf{Set}) \cong \mathbf{Cat}$, and a T_1-structured category is a $\mathbf{Cat}(T_1)$-algebra.)

A T_2-structured category is a category in $(\mathbf{Set}/\mathbb{N})^{T_2} \cong \mathbf{Operad}$. Alternatively, it is an operad in $\mathbf{Cat}$, or '$\mathbf{Cat}$-operad', as we saw in 6.6.6. Diagrammatically, a T_2-structured category A consists of

- a set $A_0(k)$ for each $k \in \mathbb{N}$, with $a \in A_0(k)$ drawn as a label on the kth 2-opetope,

- a set $A(a, b)$ for each $k \in \mathbb{N}$ and $a, b \in A_0(k)$, with $\theta \in A(a, b)$ drawn as

- a function defining composition or 'gluing' of objects,

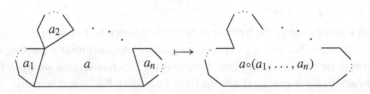

and an identity or 'unit' object,

- a function defining composition of arrows,

$$(a \xrightarrow{\theta} b \xrightarrow{\phi} c) \qquad \longmapsto \qquad (a \xrightarrow{\phi\theta} c),$$

and an identity arrow $(a \xrightarrow{1_a} a)$ on each object a

- a function defining gluing of arrows:

$$a \xrightarrow{\theta} b, \; a_1 \xrightarrow{\theta_1} b_1, \; \ldots, \; a_k \xrightarrow{\theta_k} b_k$$

give rise to

$$a \circ (a_1, \ldots, a_k) \xrightarrow{\theta * (\theta_1, \ldots, \theta_k)} b \circ (b_1, \ldots, b_k),$$

all satisfying the usual kinds of axioms. The sets A_0 and $A_1 = \coprod_{a,b \in C_0} A(a,b)$ both have the structure of plain operads: thus, A can be viewed as a category in **Operad**. Regrouping the data, there is for each n a category $A(n)$ whose object-set is $A_0(n)$: thus, A can also be viewed as an operad in **Cat**.

An analogous diagrammatic description applies to T_n-structured categories for any $n \in \mathbb{N}$.

Next recall from Section 6.6 that there is an adjunction between T-structured categories and T-multicategories, which for $T = T_n$ will be denoted

$$T_n\text{-}\mathbf{Struc}$$
$$F_n \Big\uparrow \dashv \Big\downarrow U_n$$
$$T_n\text{-}\mathbf{Multicat}.$$

Also recall from Section 7.1 that a T_n-multicategory consists of a set of objects labelling n-opetopes, a set of maps whose domains are labelled n-pasting diagrams and whose codomains are labelled single n-opetopes, and functions defining composition and identities. Now, let us see what this adjunction looks like.

The functor U_n 'forgets how to tensor but remembers multilinear maps'. When $n = 0$ it is the identity, when $n = 1$ it sends a strict monoidal category to its underlying plain multicategory, and when $n = 2$ it sends a **Cat**-operad A to the T_2-multicategory whose objects are the same as those of A and

whose maps

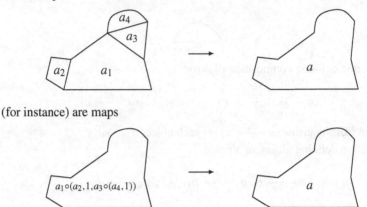

(for instance) are maps

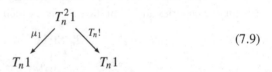

in A.

The free functor F_n is formal pasting: if C is a T_n-multicategory then the objects (respectively, arrows) of $F_n C$ are the formal pastings of objects (respectively, arrows) of C. Trivially, F_0 is the identity. We described F_1 on p. 83, in a different diagrammatic style.

Definition 7.2.1 Let $n \geq 0$. The **structured category of n-pasting diagrams**, **PD**$_n$, is defined by **PD**$_n = F_n 1$. In other words, **PD**$_n$ is the free T_n-structured category on the terminal T_n-multicategory.

So **PD**$_n$ is an internal category in $(\mathbf{Set}/O_n)^{T_n}$; its underlying graph is

$$
\begin{array}{ccc}
 & T_n^2 1 & \\
{\scriptstyle \mu_1}\swarrow & & \searrow{\scriptstyle T_n !} \\
T_n 1 & & T_n 1
\end{array}
\tag{7.9}
$$

where the T_n-algebra structures on $T_n 1$ and $T_n^2 1$ are both components of the multiplication μ of the monad T_n.

A T_0-structured category is just a category, and **PD**$_0$ is the terminal category (whose object is viewed as the unique 0-pasting diagram $\bullet$).

A T_1-structured category is a strict monoidal category, and we have already seen in Example 6.6.9 that **PD**$_1$ is $\mathbb{D}$, the strict monoidal category of (possibly empty) finite totally ordered sets $\mathbf{n} = \{1, \ldots, n\}$; addition is tensor and $\mathbf{0}$ is the unit. The diagram above is in this case

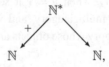

where $\mathbb{N}^*$ is the set of finite sequences of natural numbers and the right-hand map sends $(m_1, \ldots, m_n)$ to n.

The **Cat**-operad $\mathbf{PD}_2$ is described in detail in the next section.

We have been looking at $\mathbf{PD}_n$, the T_n-structured category of n-pasting diagrams, but sometimes it is useful to forget the more sophisticated structure and pass to the mere category of n-pasting diagrams. Formally, we have cartesian forgetful functors

$$(\mathbf{Set}/O_n)^{T_n} \longrightarrow \mathbf{Set}/O_n \longrightarrow \mathbf{Set}$$

and these induce a forgetful functor

$$T_n\text{-}\mathbf{Struc} = \mathbf{Cat}((\mathbf{Set}/O_n)^{T_n}) \longrightarrow \mathbf{Cat}(\mathbf{Set}) = \mathbf{Cat},$$

making possible the following definition.

Definition 7.2.2 Let $n \geq 0$. The **category of n-pasting diagrams**, $\mathbf{Pd}_n$, is the image of $\mathbf{PD}_n$ under the forgetful functor $T_n\text{-}\mathbf{Struc} \longrightarrow \mathbf{Cat}$.

For example, $\mathbf{Pd}_0$ is the terminal category, $\mathbf{Pd}_1$ is the category $\mathbb{D}$ of finite totally ordered sets, and $\mathbf{Pd}_2$ can be regarded as a category of trees (see below). The objects of $\mathbf{Pd}_n$ really are the n-pasting diagrams: for by (7.9), the object-set of $\mathbf{Pd}_n$ is the underlying set of $T_n 1 \in \mathbf{Set}/O_n$, which is the set O_{n+1} of $(n + 1)$-opetopes or n-pasting diagrams.

In Section 7.4 we will consider the category of all opetopes; beware that this is quite different from the categories $\mathbf{Pd}_n$ of n-pasting diagrams.

It is instructive to contemplate the situation for arbitrary cartesian $\mathcal{E}$ and T. We have functors

$$T\text{-}\mathbf{Struc} = \mathbf{Cat}(\mathcal{E}^T) \xrightarrow{\;\;V\;\;} \mathbf{Cat}(\mathcal{E})$$

$$F \uparrow \dashv \downarrow U$$

$$T\text{-}\mathbf{Multicat},$$

where V is forgetful, and so we have an internal category $VF1$ in $\mathcal{E}$. The object-of-objects of $VF1$ is $T1$, which may be thought of as the object of T-pasting diagrams; hence $VF1$ may be thought of as the (internal) category of T-pasting diagrams. In some situations (such as when $\mathcal{E} = \mathcal{E}_n \cong \mathbf{Set}/O_n$) there is an obvious cartesian 'forgetful' functor $\mathcal{E} \longrightarrow \mathbf{Set}$, and then there is an induced functor $\mathbf{Cat}(\mathcal{E}) \longrightarrow \mathbf{Cat}$, giving a genuine category of T-pasting diagrams. For instance, if K is a set and T is the monad $K + (-)$ on $\mathcal{E} = \mathbf{Set}$ then this is the category $\mathbb{P}_K$ defined on p. 364.

7.3 A category of trees

We have seen that for each $n \geq 0$ there is a category $\mathbf{Pd}_n$ whose objects are n-pasting diagrams. We have also seen that 2-pasting diagrams correspond naturally to trees (Fig. 7-A). Hence $\mathbf{Tr} = \mathbf{Pd}_2$ is a category whose objects are trees. Here we describe it in some detail. The definition of a map between trees is perfectly natural but takes some getting used to; we approach it slowly.

First recall what trees themselves are. By definition (2.3.3), $\mathbf{tr}$ is the free plain operad on the terminal object of $\mathbf{Set}^{\mathbb{N}}$, and an n-leafed tree is an element of $\mathbf{tr}(n)$. As we saw, the sets $\mathbf{tr}(n)$ also admit the following recursive description:

- $| \in \mathbf{tr}(1)$
- if $n, k_1, \ldots, k_n \in \mathbb{N}$ and $\tau_1 \in \mathbf{tr}(k_1), \ldots, \tau_n \in \mathbf{tr}(k_n)$ then $(\tau_1, \ldots, \tau_n) \in \mathbf{tr}(k_1 + \cdots + k_n)$.

For the purposes of this text we will need no further description of what a tree is. But it is also possible, as you might expect, to describe a tree as a graph of a certain kind, and this alternative, 'concrete', description can be comforting. To say exactly *what* kind of graph is more delicate than meets the eye, and in papers on operads is often done only vaguely. I have therefore put a graph-theoretic definition of tree, and a proof of its equivalence to our usual one, in Appendix E.

In preparation for looking at maps in $\mathbf{Pd}_2$ – maps between trees – let us look again at maps in $\mathbf{Pd}_1 = \mathbb{D}$. An object of $\mathbb{D}$ is a natural number. A map is, as observed in the previous section, a finite sequence $(m_1, \ldots, m_n)$ of natural numbers; the domain of such a map is $m_1 + \cdots + m_n$ and the codomain is n. If we view natural numbers as finite sequences of $\bullet$s then what a map does is to take a finite sequence of $\bullet$s (the domain), partition it into a finite number of (possibly empty) segments, and replace each segment by a single $\bullet$ (giving the codomain). For example, Fig. 7-B illustrates the map

$$(0, 3, 1, 0, 2) : 6 \longrightarrow 5$$

in two different ways: in (a) as a partition, and in (b) as a function.

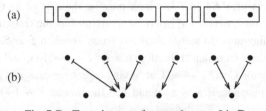

Fig. 7-B. Two pictures of a map $6 \longrightarrow 5$ in $\mathbb{D}$

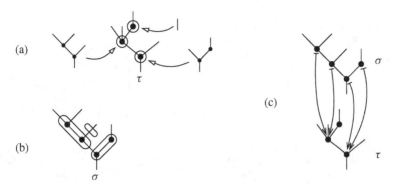

(a)

(b)

(c)

σ

τ

σ

τ

Fig. 7-C. Three pictures of a map in **Tr**(4)

Ordinarily $\mathbb{D} = \mathbf{Pd}_1$ is described as the category of finite totally ordered sets, but our new description leads smoothly into a description of the category $\mathbf{Tr} = \mathbf{Pd}_2$ of trees. **Tr** is the disjoint union $\coprod_{n \in \mathbb{N}} \mathbf{Tr}(n)$. An object of $\mathbf{Tr}(n)$ is an n-leafed tree. The set of maps in $\mathbf{Tr}(n)$ is

$$(T_2^2 1)(n) = (T_2(\mathbf{tr}))(n),$$

that is, a map is an n-leafed tree τ in which each k-ary vertex v has assigned to it a k-leafed tree σ_v; the domain of the map is the tree obtained by gluing the σ_vs together in the way dictated by the shape of τ, and the codomain is τ itself. Put another way, what a map does is to take a tree σ (the domain), partition it into a finite number of (possibly trivial) subtrees, and replace each of these subtrees by the corolla

with the same number of leaves, to give the codomain τ. Fig. 7-C depicts a certain map $\sigma \longrightarrow \tau$ in $\mathbf{Tr}(4)$ in three different ways: in (a) as a 4-leafed tree τ with a k-leafed tree σ_v assigned to each k-ary vertex v, in (b) as a 4-leafed tree σ partitioned into subtrees σ_v, and in (c) as something looking more like a function. We will return to the third point of view later; for now, just observe that there is an induced function from the vertices of σ to the vertices of τ, in which the inverse image of a vertex v of τ is the set of vertices of σ_v.

In some texts a map of trees is described as something that 'contracts some internal edges'. (Here an **internal edge** is an edge that is not the root or a leaf; maps of trees keep the root and leaves fixed. To 'contract' an internal edge means to shrink it down to a vertex.) With one important caveat, this is what our maps of trees do: for in a map $\sigma \longrightarrow \tau$, the replacement of each

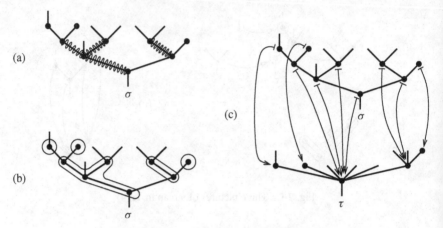

(a)

(c)

σ

(b)

σ

τ

Fig. 7-D. Three pictures of an epic in **Tr**(6)

partitioning subtree σ_v by the corolla with the same number of leaves amounts
to the contraction of all the internal edges of σ_v. For example, Fig. 7-D(a)
shows a tree σ with some of its edges marked for contraction, and Figs. 7-D(b)
and 7-D(c) show the corresponding maps $\sigma \longrightarrow \tau$ in two different styles (as
in Figs. 7-C(b) and (c)); so τ is the tree obtained by contracting the marked
edges of σ.

The caveat is that some of the σ_vs may be the trivial tree, and these are replaced by the 1-leafed corolla ⍦. This does *not* amount to the contraction of
internal edges: it is, rather, the addition of a vertex to the middle of a (possibly
external) edge. Any map of trees can be viewed as a combination of contractions of internal edges and additions of vertices to existing edges. For example,
the map illustrated in Fig. 7-C contracts two internal edges and adds a vertex
to one edge.

Analogously, any map in the category $\mathbb{D}$ of finite totally ordered sets can be
viewed as a combination of merging adjacent •s and adding new •s (Fig. 7-B);
this amounts to the factorization of any map as a surjection followed by an injection. So those who define their maps between trees to be just contractions of
internal edges are doing something analogous to considering only the surjective maps in the augmented simplex category $\mathbb{D}$. Indeed, the full subcategory
of **Tr** consisting of just those trees in which every vertex has exactly one edge
coming up out of it is isomorphic to $\mathbb{D}$; if we take only maps made out of contractions of internal edges then we obtain the subcategory of $\mathbb{D}$ consisting of
surjections only.

Fig. 7-E. (a) The category of 3-leafed stable trees, and (b) its classifying space

We will come back soon to this issue of surjections and injections in **Tr**, with more precision.

Some further understanding of the category of trees can be gained by considering just those trees in which each vertex has at least two branches coming up out of it. I will call these 'stable trees', following Kontsevich and Manin (1994, Definition 6.6.1). Formally, **StTr**(n) is the full subcategory of **Tr**(n) with objects defined by the recursive clauses

- $\mid\ \in$ **StTr**(1)
- if $n \geq 2$, $k_1, \ldots, k_n \in \mathbb{N}$, and $\tau_1 \in$ **StTr**$(k_1), \ldots, \tau_n \in$ **StTr**(k_n) then $(\tau_1, \ldots, \tau_n) \in$ **StTr**$(k_1 + \cdots + k_n)$,

and an n-**leafed stable tree** is an object of **StTr**(n). Since a stable tree can contain no subtree of the form ↑, all maps between stable trees are 'surjections', that is, consist of just contractions of internal edges, without insertions of new vertices. It follows that each category **StTr**(n) is finite, and so its classifying space can be represented by a finite CW complex; this may explain why topologists often like their trees to be stable.

The first few categories **StTr**(n) are trivial:

$$\mathbf{StTr}(0) = \emptyset,$$
$$\mathbf{StTr}(1) = \{\,\mid\,\},$$
$$\mathbf{StTr}(2) = \left\{\ \mathrel{\text{\Large Y}}\ \right\},$$

where in each case there are no arrows except for identities. The cases $n = 3$, 4, and 5 are illustrated in Figs. 7-E(a), 7-F(a), and 7-G(a). Identity arrows are not shown, and the categories **StTr**(n) are ordered sets: all diagrams commute. Vertices are also omitted; since the trees are stable, this does not cause ambiguity. Parts (b) of the figures show the classifying spaces of these categories, solid polytopes of dimensions 1, 2, and 3. In the case of 5-leafed trees (Fig. 7-G) only about half of the category is shown, corresponding to the front faces of the polytope; the back faces and the terminal object of the category (the 5-leafed corolla), which sits at the centre of the polytope, are hidden. The whole polytope has six pentagonal faces, three square faces, and 3-fold rotational symmetry about the central vertical axis.

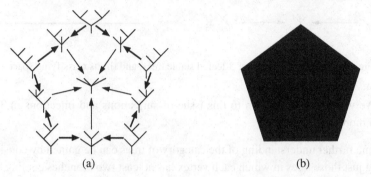

(a)   (b)

Fig. 7-F. (a) The category of 4-leafed stable trees, and (b) its classifying space

For $n \leq 5$, the classifying space $B(\mathbf{StTr}(n))$ is homeomorphic to the asso-
ciahedron K_n (Stasheff, 1963a, and 2.2.17 above), and it seems very likely that
this persists for all $n \in \mathbb{N}$. Indeed, the family of categories $(\mathbf{StTr}(n))_{n \in \mathbb{N}}$ forms
a sub-$\mathbf{Cat}$-operad $\mathbf{STTR}$ of $\mathbf{TR} = \mathbf{PD}_2$, and the classifying space functor B :
$\mathbf{Cat} \longrightarrow \mathbf{Top}$ preserves finite products, so there is a (non-symmetric) topological operad $B(\mathbf{STTR})$ whose nth part is the classifying space of $\mathbf{StTr}(n)$.
(To make B preserve finite products we must interpret $\mathbf{Top}$ as the category
of compactly generated or Kelley spaces: see Segal (1968, §1) and Gabriel
and Zisman (1967, III.2).) This operad $B(\mathbf{STTR})$ is presumably isomorphic
to Stasheff's operad $K = (K_n)_{n \in \mathbb{N}}$. A K-algebra is called an A_∞-**space**, and
should be thought of as an up-to-homotopy version of a topological semigroup;
the basic example is a loop space.

The categories $\mathbf{StTr}(n)$ also give rise to the notion of an A_∞-algebra (Stasheff, 1963b). For each $n \in \mathbb{N}$, there is a chain complex $P(n)$ whose degree k part
is the free abelian group on the set of n-leafed stable trees with $(n - k - 1)$
vertices. For instance,

$$P(4) = (\cdots \longrightarrow 0 \longrightarrow 0 \longrightarrow \mathbb{Z} \cdot L_2 \longrightarrow \mathbb{Z} \cdot L_1 \longrightarrow \mathbb{Z} \cdot L_0)$$

where the sets L_k are

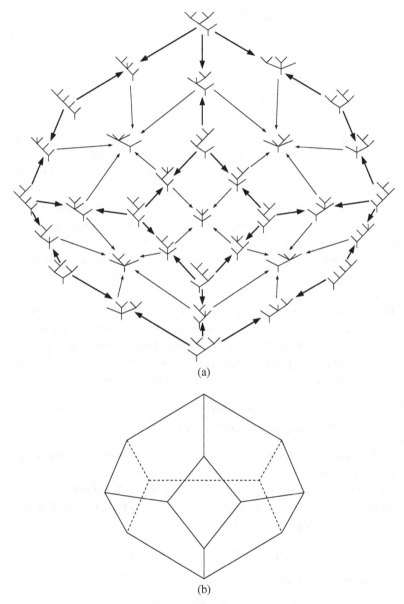

(a)

(b)

Fig. 7-G. (a) About half of the category of 5-leafed stable trees, and (b) the classifying space of the whole category

The differential d is defined by $d(\tau) = \sum \pm \sigma$, where if $\tau \in P(n)_k$ then the sum is over all $\sigma \in P(n)_{k-1}$ for which there exists a map $\sigma \longrightarrow \tau$. For instance,

$$d\left(\; \vcenter{\hbox{\includegraphics}} \;\right) = \pm \;\; \vcenter{\hbox{\includegraphics}} \;\; \pm \;\; \vcenter{\hbox{\includegraphics}} \;\; .$$

When the signs are chosen appropriately this defines an operad P of chain complexes. A P-algebra is called an A_∞-**algebra**, to be thought of as an up-to-homotopy differential graded non-unital algebra; the usual example is the singular chain complex of an A_∞-space. A P-category is called an A_∞-category (see 5.1.11), and consists of a collection of objects, a chain complex $\mathrm{Hom}(a, b)$ for each pair (a, b) of objects, maps defining binary composition, chain homotopies witnessing that this composition is associative up to homotopy, further homotopies witnessing that the previous homotopies obey the pentagon law up to homotopy, and so on. A_∞-categories are cousins of weak ω-categories, as we see in Section 10.2.

Finally, since the polytopes $K_n = B(\mathbf{StTr}(n))$ describe higher associativity conditions, they also arise in definitions of higher-dimensional category. For example, the pentagon K_4 occurs in the classical definition of bicategory (1.5.1), and the polyhedron K_5 occurs as the 'non-abelian 4-cocycle condition' in Gordon, Power and Street's definition of tricategory (1995).

We have already described the set $\mathbf{tr}(n)$ of n-leafed trees. Maps $\sigma \longrightarrow \tau$ between trees are described by induction on the structure of τ:

- if $\tau = |$ then there is only one map into τ; it has domain $|$ and we write it as $1_| : | \longrightarrow |$
- if $\tau = (\tau_1, \dots, \tau_n)$ for $\tau_1 \in \mathbf{tr}(k_1), \dots, \tau_n \in \mathbf{tr}(k_n)$ then a map $\sigma \longrightarrow \tau$ consists of trees $\rho \in \mathbf{tr}(n), \rho_1 \in \mathbf{tr}(k_1), \dots, \rho_n \in \mathbf{tr}(k_n)$ such that $\sigma = \rho \circ (\rho_1, \dots, \rho_n)$, together with maps

$$\rho_1 \xrightarrow{\;\theta_1\;} \tau_1, \quad \dots, \quad \rho_n \xrightarrow{\;\theta_n\;} \tau_n,$$

and we write this map as

$$\sigma = \rho \circ (\rho_1, \dots, \rho_n) \xrightarrow{\;!_\rho * (\theta_1, \dots, \theta_n)\;} (\tau_1, \dots, \tau_n) = \tau. \qquad (7.10)$$

It follows easily that the n-leafed corolla $\nu_n = (|, \dots, |)$ is the terminal object of $\mathbf{Tr}(n)$: the unique map from $\sigma \in \mathbf{tr}(n)$ to ν_n is $!_\sigma * (1_|, \dots, 1_|)$.

The rest of the structure of the **Cat**-operad **TR** can be described in a similarly explicit recursive fashion.

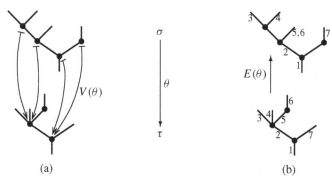

Fig. 7-H. The effect on (a) vertices and (b) edges of a certain map of 4-leafed trees

To make precise the intuition that a map of trees is a function of some sort, functors

$$V : \mathbf{Tr} \longrightarrow \mathbf{Set}, \qquad E : \mathbf{Tr}^{\mathrm{op}} \longrightarrow \mathbf{Set}$$

can be defined, encoding what happens on vertices and edges respectively. Both functors turn out to be faithful, which means that a map of trees is completely determined by its effect on either vertices or edges. The following account of V and E is just a sketch.

The more obvious of the two is the vertex functor V, defined on objects by

- $V(|) = \emptyset$
- $V((\tau_1, \ldots, \tau_n)) = 1 + V(\tau_1) + \cdots + V(\tau_n).$

The edge functor E can be defined by first defining a functor

$$E_n : \mathbf{Tr}(n)^{\mathrm{op}} \longrightarrow (n+1)/\mathbf{Set}$$

for each $n \in \mathbb{N}$, where $(n+1)/\mathbf{Set}$ is the category of sets equipped with $(n+1)$ ordered marked points. This definition is again by induction, the idea being that E_n associates with a tree its edge-set with the n input edges and the one output edge (root) distinguished. Fig. 7-H illustrates a map $\theta : \sigma \longrightarrow \tau$ in $\mathbf{Tr}(4)$; part (a) ($=$ Fig. 7-C(c)) shows its effect $V(\theta)$ on vertices; part (b) shows $E(\theta)$, taking $E(\tau) = \{1, \ldots, 7\}$ and labelling the image of $i \in \{1, \ldots, 7\}$ under $E(\theta)$ by an i on the edge $(E(\theta))(i)$ of σ.

A map of trees will be called surjective if it is built up from contractions of internal edges (the analogues of degeneracy maps in $\mathbb{D}$). Formally, the

surjective maps in **Tr** are defined by:

- $1_|$: | $\longrightarrow$ | is surjective
- with notation as in (7.10), $!_\rho * (\theta_1, \ldots, \theta_n)$ is surjective if and only if each θ_i is surjective and $\rho \neq$ |.

The crucial part is the last: the unique map $!_\rho$ from $\rho \in \mathbf{tr}(n)$ to the corolla v_n is made up of edge-contractions just as long as ρ is not the unit tree |.

Dually, a map of trees is **injective** if, informally, it is built up from adding vertices to the middle of edges (the analogues of face maps in $\mathbb{D}$). Formally,

- $1_|$: | $\longrightarrow$ | is injective
- with notation as above, $!_\rho * (\theta_1, \ldots, \theta_n)$ is injective if and only if each θ_i is injective and ρ is either v_n or | (the latter only being possible if $n = 1$).

The punchline is that the various possible notions of a map of trees being 'onto' (respectively, 'one-to-one') all coincide:

Proposition 7.3.1 *The following conditions on a map* $\theta : \sigma \longrightarrow \tau$ *in* **Tr** *are equivalent:*

a. θ is epic
b. θ is surjective
c. $V(\theta)$ is surjective
d. $E(\theta)$ is injective (sic).

Moreover, if each condition is replaced by its dual then the equivalence persists.

Proof Omitted. 'Moreover' is not just an application of formal duality, since surjectivity and injectivity are not formal duals.   □

We finish this section by re-considering briefly what we have done with trees, but this time with 2-pasting diagrams instead. This gives a very good impression of the category of n-pasting diagrams for arbitrary n.

The objects of the category $\mathbf{Pd}_2 = \mathbf{Tr}$ are the opetopic 2-pasting diagrams. A map $\theta : \sigma \longrightarrow \tau$ in $\mathbf{Pd}_2$ takes a 2-pasting diagram σ, partitions it into a finite number of sub-pasting diagrams, and replaces each of these sub-pasting diagrams by the 2-opetope

with the same number of input edges, to give the codomain τ. Another way to put this is that each of the 2-opetopes v making up the pasting diagram τ has assigned to it a 2-pasting diagram σ_v with the same number of input edges,

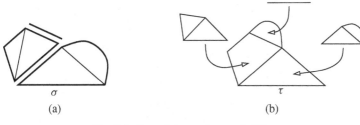

Fig. 7-I. Two pictures of a map in **Pd**$_2$

and when the σ_vs are pasted together according to the shape of τ, the result is the pasting diagram σ. Fig. 7-I shows a map in **Pd**$_2$, in (a) as a partition of σ and in (b) as a family (σ_v) indexed over the regions v of τ; these correspond precisely to the tree pictures in Figs. 7-C(b) and (a) respectively.

More generally, a map of n-pasting diagrams consists of the replacement of some sub-pasting diagrams by their bounding opetopes. When the sub-pasting diagrams are non-trivial, this amounts to the removal of some internal faces of codimension one. For $n = 2$, faces of codimension one are edges, and the trivial case is the replacement of the 2-pasting diagram

by the 2-opetope

In Fig. 7-I, there are two instances of edge-deletion and one instance of inflating the trivial 2-pasting diagram. This is the distinction between epics (degeneracy maps) and monics (face maps) in **Pd**$_2$, as we saw for trees.

The classical connections between trees and A_∞-structures can, of course, be phrased equally in terms of 2-pasting diagrams. This is probably the natural approach if we want to incorporate A_∞-structures into higher-dimensional algebra. Stable trees correspond to 2-pasting diagrams not containing any copies of either of the 2-opetopes

– in other words, those that can be drawn using only straight lines.

Because of the inversion of dimensions, the 'vertex functor' $V : \mathbf{Tr} = \mathbf{Pd}_2 \longrightarrow \mathbf{Set}$ assigns to a 2-pasting diagram its set of faces, or regions, or constituent 2-opetopes. (Compare C.4.6.) The 'edge functor' $E : \mathbf{Pd}_2{}^{\text{op}} \longrightarrow \mathbf{Set}$

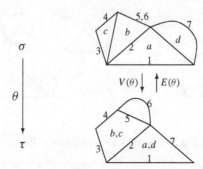

Fig. 7-J. The effects of V and E on a map of 2-pasting diagrams

is still aptly named. Nothing interesting happens on the *actual* vertices of 2-pasting diagrams. Fig. 7-J shows the effects of V and E on the map $\theta : \sigma \longrightarrow \tau$ of Fig. 7-I: the regions of σ are labelled a, b, c, d, and the image of the region a under the function $V(\theta)$ is also labelled a, and so on; similarly, the edges of τ are labelled $1, \ldots, 7$ and their images under $E(\theta)$ are labelled correspondingly. The same example was also shown in Fig. 7-H, using trees.

The pictures of 2-pasting diagrams can be taken seriously, that is, geometrically realized. This leads quickly into the geometric realization of opetopic sets, and so to the underlying ('singular') opetopic set of a topological space, one of the motivating examples of a weak ω-category. We come to this in the next section.

7.4 Opetopic sets

Opetopes were defined by Baez and Dolan in order to give a definition of weak n-category. Their definition has been subject to various modifications by various other people, all of the form 'a weak n-category is an opetopic set with certain properties'. The next two sections are a discussion of the general features of such definitions, not concentrating on any version in particular.

In this section I will explain what an opetopic set is. Again, there are various proposed definitions, most of which have been proved equivalent by Cheng (see the Notes). Rather than giving any particular one of them, I will list some properties satisfied by the category $\mathbb{O}$ of opetopes, an opetopic set being a presheaf on $\mathbb{O}$. Using this, I will show how every topological space has an underlying opetopic set, to be thought of as its 'singular opetopic set' or 'fundamental ω-groupoid'. This will motivate the definition of weak n-category in the next section.

Fig. 7-K. A 3-opetope with 21 sub-opetopes

Opetopic sets should be thought of as something like simplicial sets. A simplicial set is a presheaf on the category Δ of simplices; an opetopic set is a presheaf on the category $\mathbb{O}$ of opetopes. Actually, it might be more apposite to compare opetopic sets to presheaves on the category Δ_{inj} of non-empty finite totally ordered sets and order-preserving injections, rather than whole simplicial sets: we only consider face maps between opetopes, not degeneracies.

Here is an informal description of the category $\mathbb{O}$ of opetopes. The set of objects is the set $\bigsqcup_{n\in\mathbb{N}} O_n$ of all opetopes of all dimensions. A map $\omega' \longrightarrow \omega$ is an embedding of ω' as a face of ω. For example, there are four maps in $\mathbb{O}$ from the unique 1-opetope

$$\longrightarrow$$

into the 2-opetope

$$\omega = \diagup\!\!\Downarrow\!\!\diagdown \; ,$$

corresponding to the three input edges and one output edge of ω. There are also four maps from the unique 0-opetope $\bullet$ into ω, corresponding to its four vertices. Along with the identity 1_ω, that enumerates all of the maps into ω in $\mathbb{O}$. Similarly, there are 21 maps whose codomain is the 3-opetope ω illustrated in Fig. 7-K: seven maps from the unique 0-opetope, nine from the unique 1-opetope, one from the 2-opetope with two input edges, two from the 2-opetope with three input edges, one from the 2-opetope with six input edges, and one from ω itself (the identity 1_ω).

In order to prove results about the relation between opetopic sets and topological spaces, we will need to know some specific properties of the category $\mathbb{O}$. They are listed here, along with a few more properties that will not be needed but add detail to the picture. So, $\mathbb{O}$ is a small category such that:

a. the set of objects of $\mathbb{O}$ is a disjoint union of subsets $(O_n)_{n\in\mathbb{N}}$; we write $\dim(\omega) = n$ if $\omega \in O_n$

b. if $\omega' \xrightarrow{\rho} \omega$ is a map in $\mathbb{O}$ then $\dim(\omega') \leq \dim(\omega)$

c. if $\omega' \xrightarrow{\rho} \omega$ is a map in $\mathbb{O}$ with $\dim(\omega') = \dim(\omega)$ then $\rho = 1_\omega$
d. every map in $\mathbb{O}$ is monic
e. every map in $\mathbb{O}$ is a composite of maps of the form $\omega' \longrightarrow \omega$ with $\dim(\omega) = \dim(\omega') + 1$
f. if $\dim(\omega) = \dim(\omega') + 1$ then every map $\omega' \longrightarrow \omega$ can be classified as either a 'source embedding' or a 'target embedding'
g. if $\omega \in \mathbb{O}$ with $\dim(\omega) \geq 1$ then the set of pairs (ω', ρ) with $\dim(\omega) = \dim(\omega') + 1$ and $\rho : \omega' \longrightarrow \omega$ is finite, and there is exactly one such pair for which ρ is a target embedding.

Many of these properties can be compared to properties of Δ_{inj}. They also imply that every object of $\mathbb{O}$ is the codomain of only finitely many maps – in other words, has only a finite set of sub-opetopes.

An **opetopic set** is by definition a functor $X : \mathbb{O}^{\mathrm{op}} \longrightarrow \mathbf{Set}$. If $Y : \Delta^{\mathrm{op}} \longrightarrow \mathbf{Set}$ is a simplicial set and $[n] \in \Delta$ then an element y of $Y[n]$ is usually depicted as a label attached to an n-simplex, and the images of y under the various face maps as labels on the various faces of the n-simplex, as in

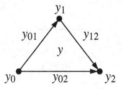

for $n = 2$. Similarly, an opetopic set can be thought of as a system of labelled opetopes: if $\omega \in O_n$ and $x \in X(\omega)$ then x is depicted as a label on the n-opetope ω, and the images $(X\rho)(x) \in X(\omega')$ of x under the various face maps $\rho : \omega' \longrightarrow \omega$ as labels on the faces of ω. For example, if ω is the 2-opetope with three input edges then $x \in X(\omega)$ is drawn as

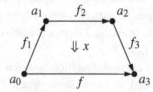

where a_0, a_1, a_2, a_3 are the elements of $X(\bullet)$ induced from x by the four different maps from the unique 0-opetope $\bullet$ to ω in $\mathbb{O}$, and similarly f_1, f_2, f_3, f for the unique 1-opetope. The elements of $\coprod_{\omega \in O_n} X(\omega)$ are called the n-**cells** of X, for $n \in \mathbb{N}$.

Opetopic sets can also be (informally) defined without reference to a category of opetopes. Thus, an opetopic set is a commutative diagram of sets and

functions

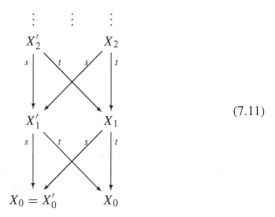

$$(7.11)$$

where for each $n \geq 1$, the set X'_n and the functions $s : X'_n \longrightarrow X'_{n-1}$ and $t : X'_n \longrightarrow X_{n-1}$ are defined from the sets $X_n, X'_{n-1}, X_{n-1}, \ldots, X'_1, X_1, X_0$ and the functions s, t between them in the way now explained. As usual for directed graphs or globular sets, elements of X_0 and X_1 are called **0-cells** and **1-cells** respectively, and drawn as labelled points or intervals. An element of X'_1 is called a 1-**pasting diagram in** X, and consists of a diagram

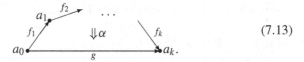

$$(7.12)$$

of cells in X ($k \in \mathbb{N}$). An element $\alpha \in X_2$ is called a **2-cell**; its source $s\alpha$ is a 1-pasting diagram in X, and its target $t\alpha$ a 1-cell. If $s\alpha$ is the 1-pasting diagram of (7.12) and $t\alpha$ is

$$a_0 \xrightarrow{\ g\ } a_k$$

then α is drawn as

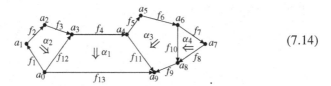

$$(7.13)$$

An element of X'_2 is a **2-pasting diagram in** X, that is, consists of a finite diagram of cells of the form (7.13) pasted together, a typical example being

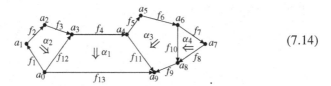

$$(7.14)$$

Note that the arrows go in compatible directions: for instance, the target or output edge f_{11} of α_3 is a source or input edge of α_1. The source of this element of X_2' is

$$a_0 \xrightarrow{\ f_1\ } \quad \cdots \quad \xrightarrow{\ f_9\ } a_9 \quad \in X_1',$$

and the target is $f_{13} \in X_1$. An element $\Gamma \in X_3$ is called a **3-cell**; if, for instance, $s\Gamma$ is the 2-pasting diagram (7.14) then $t\Gamma$ is of the form (7.13) with $k = 9$ and $g = f_{13}$, and we picture Γ as being a label on the evident 3-opetope (whose output face is itself labelled α and whose input faces are labelled $\alpha_1, \alpha_2, \alpha_3, \alpha_4$). And so it continues.

Perhaps the easiest example of an opetopic set is that arising from a topological space. We can define this rigorously using only a few of the properties of $\mathbb{O}$ listed above. But first we need to 'recall' two constructions, one from topology and one from category theory.

The topology is the cone construction. This is the canonical way of embedding a given space into a contractible space, and amounts formally to a functor **Cone** and a (monic) natural transformation ι as shown:

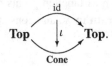

Given a space E, the contractible space **Cone**(E) is the pushout

$$
\begin{array}{ccc}
E & \xrightarrow{\ e \mapsto (e,1)\ } & E \times [0, 1] \\
\downarrow & & \downarrow \\
1 & \longrightarrow & \mathbf{Cone}(E)
\end{array}
$$

in **Top**, where 1 is the one-point space. If E is empty then **Cone**$(E) = 1$; otherwise **Cone**(E) is $E \times [0, 1]$ with all points of the form $(e, 1)$ identified. The inclusion $\iota_E : E \hookrightarrow \mathbf{Cone}(E)$ sends e to $(e, 0)$.

The category theory is the construction from any functor

$$J : \mathbb{C} \longrightarrow \mathcal{E}$$

of a pair of adjoint functors

$$\mathcal{E} \underset{F}{\overset{U}{\underset{\top}{\rightleftarrows}}} [\mathbb{C}^{\mathrm{op}}, \mathbf{Set}]. \tag{7.15}$$

Here $\mathbb{C}$ is any small category and $\mathcal{E}$ any category with small colimits. The right adjoint U is defined by

$$(UE)(C) = \mathcal{E}(JC, E)$$

($E \in \mathcal{E}, C \in \mathbb{C}$). The left adjoint F is the left Kan extension of J along the Yoneda embedding:

$$\mathbb{C} \xrightarrow{\text{Yoneda}} [\mathbb{C}^{\text{op}}, \textbf{Set}]$$

with J going down to $\mathcal{E}$ and F dotted back up.

Explicitly, F is given by the coend formula

$$FX = \int^{C \in \mathbb{C}} XC \times JC$$

($X \in [\mathbb{C}^{\text{op}}, \textbf{Set}]$). As the Yoneda embedding is full and faithful, this is a 'genuine' extension: $F(\mathbb{C}(-, C)) \cong JC$. The best-known example – and probably the best remedy for readers new to Kan extensions and coends – involves simplicial sets. Here

$$(\mathbb{C} \xrightarrow{J} \mathcal{E}) = (\Delta \xrightarrow{J} \textbf{Top})$$

where J sends $[n] \in \Delta$ to the standard n-simplex Δ^n. Then in the adjunction

$$\textbf{Top} \underset{F}{\overset{U}{\rightleftarrows_\top}} [\Delta^{\text{op}}, \textbf{Set}],$$

U sends a space to its underlying (singular) simplicial set and F is geometric realization. The coend formula above becomes the formula more familiar to topologists,

$$FX = (\coprod_{n \in \mathbb{N}} X_n \times \Delta^n)/\sim$$

(Segal, 1968, §1, Adams, 1978, p. 58). The isomorphism $F(\Delta(-, [n])) \cong \Delta^n$ asserts that the realization of the simplicial n-simplex $\Delta(-, [n])$ is the topological n-simplex Δ^n.

To define both the underlying opetopic set of a topological space and, conversely, the geometric realization of an opetopic set, it is therefore only necessary to define a functor $J : \mathbb{O} \longrightarrow \textbf{Top}$. We do this under the assumption that $\mathbb{O}$ is a category with properties (a)–(c) (p. 241). The idea is, of course, that J assigns to each n-opetope ω the topological space $J(\omega)$ looking like our usual picture of ω (and homeomorphic to the closed n-disc). In the underlying opetopic set UE of a space E, an element of $(UE)(\omega)$ is a continuous map from $J(\omega)$ into E; conversely, if X is an opetopic set then the space FX is formed by gluing together copies of the spaces $J(\omega)$ according to the recipe X.

For $n \in \mathbb{N}$, let $\mathbb{O}(n)$ be the full subcategory of $\mathbb{O}$ with object-set $\bigcup_{k \leq n} O_k$. We will construct for each n a functor $J_n : \mathbb{O}(n) \longrightarrow \textbf{Top}$ such that the diagram

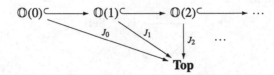

commutes. Since $\mathbb{O}$ is the colimit of the top row, this will induce a functor J of the form desired.

The unique 0-opetope • is drawn as a one-point space, so we define J_0 to have constant value 1. Let $n \in \mathbb{N}$ and suppose that we have defined J_n. By the mechanism described above, J_n induces a geometric realization functor

$$F_n : [\mathbb{O}(n)^{\mathrm{op}}, \textbf{Set}] \longrightarrow \textbf{Top}.$$

The functor J_{n+1} is defined as follows. Its value on the subcategory $\mathbb{O}(n)$ of $\mathbb{O}(n+1)$ is the same as that of J_n. Any object ω of $\mathbb{O}$ induces a functor

$$\mathbb{O}(-, \omega)|_{\mathbb{O}(n)} : \quad \mathbb{O}(n)^{\mathrm{op}} \quad \longrightarrow \quad \textbf{Set},$$
$$\chi \quad \longmapsto \quad \mathbb{O}(\chi, \omega),$$

and if $\omega \in O_{n+1}$ then we put

$$J_{n+1}(\omega) = \textbf{Cone}(F_n(\mathbb{O}(-, \omega)|_{\mathbb{O}(n)})).$$

This defines J_{n+1} on objects. By assumptions (b) and (c) on $\mathbb{O}$, the only remaining maps on which J_{n+1} needs to be defined are those of the form $\rho : \omega' \longrightarrow \omega$ where ω' is an object of $\mathbb{O}(n)$ and $\omega \in O_{n+1}$. This is done by taking $J_{n+1}(\rho)$ to be the composite

$$J_n(\omega') \xrightarrow{\sim} F_n(\mathbb{O}(-, \omega')|_{\mathbb{O}(n)}) \xrightarrow{F_n(\rho_*)} F_n(\mathbb{O}(-, \omega)|_{\mathbb{O}(n)}) \xhookrightarrow{\iota} J_{n+1}(\omega)$$

where the isomorphism is the one noted above and ρ_* is composition with ρ.

To see why this is sensible, consider the first inductive steps. The geometric realization functor

$$F_0 : [\mathbb{O}(0)^{\mathrm{op}}, \textbf{Set}] \cong \textbf{Set} \longrightarrow \textbf{Top}$$

realizes sets as discrete spaces. If ω is the unique 1-opetope then

$$\mathbb{O}(-,\omega)|_{\mathbb{O}(0)} : \mathbb{O}(0)^{\text{op}} \longrightarrow \textbf{Set}$$

has value 2, since there are two maps from the unique 0-opetope ω' into ω; hence

$$J_1(\omega) = \textbf{Cone}(\mathbb{O}(-,\omega)|_{\mathbb{O}(0)})$$

is the cone on the discrete 2-point space, which is homeomorphic to the unit interval $[0,1]$. The two maps $\omega' \rightrightarrows \omega$ are sent by J_1 to the two maps $1 \rightrightarrows [0,1]$ picking out the endpoints. For the next inductive step, $n = 1$, note that

$$F_1 : [\mathbb{O}(1)^{\text{op}}, \textbf{Set}] \longrightarrow \textbf{Top}$$

is the usual functor geometrically realizing directed graphs. If ω is the 2-opetope with r input edges then $\mathbb{O}(-,\omega)|_{\mathbb{O}(1)}$ is the directed graph

with $r + 1$ vertices and $r + 1$ edges, whose geometric realization is the circle S^1; hence $J_1(\omega)$ is $\textbf{Cone}(S^1)$, homeomorphic to the closed disc D^2.

It should certainly be true in general that if ω is an n-opetope then $J(\omega)$ is homeomorphic to the n-disc D^n, but we do not have enough information about $\mathbb{O}$ to prove that here.

A similar construction can be tried with strict ω-categories in place of spaces. The idea now is that every n-opetope gives rise to a strict n-category – hence to a strict ω-category in which the only cells of dimension greater than n are identities – and every map between opetopes gives rise to a strict functor between the resulting strict ω-categories. For example, the strict 2-category associated to the 2-opetope

is freely generated by this diagram, in other words, by 0-cells a_0, a_1, a_2, a_3, 1-cells $f_i : a_{i-1} \longrightarrow a_i$ ($i = 1, 2, 3$) and $f : a_0 \longrightarrow a_3$, and a 2-cell $f_3 \circ f_2 \circ f_1 \longrightarrow f$. However, the formal construction is more difficult than that for topological spaces because there is a question of orientation. I do not know whether properties (a)–(g) of $\mathbb{O}$ suffice to do the construction, but we would certainly need to know more about $\mathbb{O}$ in order to say anything really useful.

7.5 Weak n-categories: a sketch

The opetopic sets arising from topological spaces and from strict ω-categories should all be weak ω-categories. 'Should' means that if someone proposes to define a weak ω-category as an opetopic set with certain properties then these two families of examples must surely be included – or if not, their concept of weak ω-category is very different from mine. So let us now look at the properties we might ask of an opetopic set in order for it to qualify as a weak ω-category.

The situation is roughly like that in Section 3.3, where we characterized the (plain) multicategories that arise from monoidal categories and so were able to re-define a monoidal category as a multicategory with properties. The main observation was that if $a_1, \dots, a_k$ are objects of a monoidal category A then in the underlying multicategory C, the canonical map

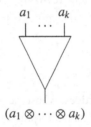

$$(a_1 \otimes \cdots \otimes a_k)$$

is 'universal' as a map with domain $a_1, \dots, a_k$, and this determines the tensor $(a_1 \otimes \cdots \otimes a_k)$ up to isomorphism. It follows that a monoidal category can be re-defined as a multicategory in which every sequence of objects is the domain of some 'universal' map. Actually, there is a choice of notions of 'universality': in Section 3.3 we spoke of both universal maps and the more general pre-universal maps, and if we use pre-universals then we must add to this re-definition the requirement that the composite of pre-universal maps is pre-universal.

Similarly, a weak ω-category can be thought of as an opetopic set in which there are 'enough universals'. Consider, for example, the opetopic set X arising from a *strict* ω-category A. The 0- and 1-cells of X are the same as the 0- and 1-cells of A, and a 2-cell in X of the form (7.13) (p. 243) is a 2-cell

$$a_0 \xrightarrow[g]{\overset{f_k \circ \cdots \circ f_1}{\Downarrow}} a_k$$

in A. Now, given such a string of 1-cells $f_1, \ldots, f_k$, there is a distinguished 2-cell ε in X of this form: the one corresponding to the identity 2-cell

in A. We would like to pin down some universal property of the 2-cells of X arising in this way. The obvious approach is to say something like 'every 2-cell of X of the form (7.13) factors uniquely through ε', but to say 'factors' we need to know about composition of *2-cells*, even though at present we are only trying to discuss composition of *1-cells*. . . .

This problem of 'downwards induction' means that it is easier to define weak n-category, for finite n, than weak ω-category.

Here, then, is a sketch of a definition of weak n-category, roughly that proposed by Baez and Dolan in (1997). (A summary can also be found as Definition **X** in my (2001b).) Let $n \in \mathbb{N}$. In a moment I will say what it means for a cell of an opetopic set to be 'pre-universal' (or 'universal' in the terminology of the sources just cited), a condition depending on n. A **weak n-category** is an opetopic set X such that

- every pasting diagram is the source of a pre-universal cell
- the composite of pre-universal cells is pre-universal.

The first condition means that if $n \geq 1$ and Φ is an n-pasting diagram in X then there exists a pre-universal $(n + 1)$-cell ε with source Φ:

$$\Phi \xrightarrow{\ \varepsilon\ } \phi.$$

(In the notation of diagram (7.11), p. 243, we have $\varepsilon \in X_{n+1}$, $\Phi = s\varepsilon \in X'_n$, and $\phi = t\varepsilon \in X_n$.) Think of ϕ as a – or, with a pinch of salt, 'the' – composite of the cells making up the pasting diagram Φ, and ε as asserting that ϕ is a composite of Φ. This suggests the correct meaning of the second condition: that if

$$\Psi \xrightarrow{\ \varepsilon\ } \psi$$

is such that the $(n + 1)$-cell ε and all of the n-cells making up the n-pasting diagram Ψ are pre-universal, then the n-cell ψ is also pre-universal.

What should it mean for a cell to be pre-universal? For a start, we define pre-universality in such a way that all cells of dimension greater than n in a weak n-category are trivial. This means that if $k \geq n$ then every k-pasting diagram Φ is the source of precisely one $(k + 1)$-cell, whose target is to be thought of as the composite of Φ. Taking $k = n$, we obtain a composition of n-cells, obeying strict laws.

Now, an n-cell

$$\Phi \xrightarrow{\ \varepsilon\ } \phi$$

is pre-universal if and only if every n-cell

$$\Phi \xrightarrow{\ \alpha\ } \psi$$

with source Φ factors uniquely through ε – in other words, there is a unique n-cell $\bar{\alpha}$ such that α is the composite

$$\Phi \xrightarrow{\ \varepsilon\ } \phi \xrightarrow{\ \bar{\alpha}\ } \psi.$$

Equivalently, ε is pre-universal if for every $(n-1)$-cell ψ parallel to ϕ, composition with ε induces a bijection

$$X(\phi, \psi) \xrightarrow{\ \sim\ } X(\Phi, \psi) \tag{7.16}$$

of hom-sets. We think of ϕ as a composite of Φ, or as 'the' composite if it is understood that it is only defined up to isomorphism.

At the next level down, we want to say that an $(n-1)$-cell $\Phi \xrightarrow{\ \varepsilon\ } \phi$ is pre-universal if for every $(n-2)$-cell ψ parallel to ϕ, composition with ε induces an equivalence (7.16) of hom-categories. The difficulties are that we must first put a category structure on the domain and codomain, and, more significantly, that 'composition' with ε is not actually a functor, since composition of $(n-1)$-cells is only defined up to isomorphism. At this point we really need to set up some more language, and this would carry us beyond the scope of this informal account, so I refer the curious reader to the texts cited in the Notes.

The case of weak 2-categories with only one 0-cell is the familiar one of monoidal categories as representable multicategories (Section 3.3). An opetopic set X with only one 0-cell and trivial above dimension 2 is a set C_0 (the 1-cells of X) together with a set $C(a_1, \ldots, a_n; a)$ for each sequence $a_1, \ldots, a_n, a$ of elements of C_0. If X is a weak 2-category then there is, as explained above, a composition for 2-cells, obeying strict associativity and unit laws; this gives a multicategory C with the indicated hom-sets. A 2-cell of X is pre-universal if and only if the corresponding arrow of C is pre-universal; the axiom that every 1-pasting diagram of X is the source of a pre-universal 2-cell is the axiom that every sequence of objects of C is the source of a pre-universal arrow; the only other axiom is that the composite of pre-universals is pre-universal. So a one-object weak 2-category is nothing other than a representable multicategory, which in turn is the same thing as a monoidal category (Theorem 3.3.4). See Cheng (2003d) for details.

Although the approach described above is modelled on that of Baez and Dolan (1997), if you look at their paper then you will see immediately that it has features quite different from any that we have considered. Here is a description of what they actually do.

The most striking feature is that they work throughout with *symmetric* multicategories, and that these play essentially the same role as our generalized multicategories. (In fact they call symmetric multicategories 'typed operads', and symmetric operads '(untyped) operads'; this is like the 'coloured operad' terminology discussed on p. 68.) The most important thing that they do with symmetric multicategories is this: given a symmetric multicategory C, they construct a new symmetric multicategory C^+, the **slice** of C, such that

$$\mathbf{Alg}(C^+) \cong \text{(symmetric multicategories over } C \text{ with the same object-set).}$$

(7.17)

The left-hand side is the category of algebras for C as a *symmetric* multicategory (p. 80). An object of the category on the right-hand side is a map $D \xrightarrow{\;f\;} C$ where D is a symmetric multicategory with $D_0 = C_0$ and f is a map of symmetric multicategories with $f_0 = 1_{C_0}$; a map is a commutative triangle.

To describe C^+ explicitly it is helpful to use the following terminology: a **reduction law** in a multicategory D is an equation stating what the composite of some family of arrows in D is. So reduction laws in D correspond to trees of arrows in D, but we think of a reduction law as also knowing what the composite of this tree is; thus, the collection of reduction laws in D describes the composition in D completely (indeed, very redundantly). Here 'tree' must be understood in the symmetric sense: branches are allowed to cross over, but the topologically-obvious identifications are made so that any tree is equivalent to a 'combed tree' – one where all the crossings are at the top. Now:

- an object of C^+ is an arrow of C
- an arrow of C^+ is a reduction law in C
- a reduction law in C^+ is a way of assembling a family of reduction laws in C to form a new reduction law.

So if $\theta_1, \ldots, \theta_n, \theta$ are arrows of C then an arrow $\theta_1, \ldots, \theta_n \longrightarrow \theta$ in C^+ is a tree with n vertices, totally ordered, such that if θ_i is written at the ith vertex then the evident composite can be formed in C and is equal to θ.

Example 7.5.1 Consider the terminal symmetric multicategory 1, whose algebras are the commutative monoids (2.2.2). Then according to the explicit description, the objects of 1^+ are the natural numbers; an arrow $k_1, \ldots,$

$k_n \longrightarrow k$ in 1^+ is a k-leafed tree with n vertices, totally ordered, such that the ith vertex has k_i branches coming up out of it; composition is by substituting trees into vertices of other trees. So the arrows $k_1, \ldots, k_n \longrightarrow k$ describe the ways of combining one k_1-ary operation, one k_2-ary operation, ..., and one k_n-ary operation of a generic symmetric multicategory in order to obtain a k-ary operation. Hence 1^+ is the symmetric multicategory whose algebras are symmetric operads; we met it as $\mathcal{O}'$ in 2.2.23.

Example 7.5.2 Let I be the initial symmetric operad, which has only one object and one arrow (the identity). The explicit description says that I^+ also has only one object, in other words, is an operad. The reduction laws of I say that if you take n copies of the identity arrow, permute them somehow, then compose them, then you obtain the identity arrow. So an n-ary operation of the operad I^+ is an element of the symmetric group S_n, and I^+ is the operad $\mathbf{S}$ of symmetries (2.2.20).

Baez and Dolan define an n-**dimensional opetope** to be an object of $I^{+\cdots+}$, where there are n +s above the I. For $n = 0$ and $n = 1$ this looks fine: there is only one opetope in each of these dimensions. But a 2-dimensional opetope is an object of I^{++}, that is, an arrow of I^+, and the set of such is the disjoint union of all the symmetric groups – not the expected answer, $\mathbb{N}$. In their world, there is not just one 2-dimensional opetope looking like

but six, corresponding to the 3! permutations of the three input edges. Similarly, as observed by Cheng (2003a, 1.3), there are 311 040 copies of the 3-dimensional opetope shown in Fig. 7-K (p. 241). (Verifying this is a useful exercise for understanding the slice construction.) We come back to this discrepancy later.

Their next step is to define opetopic set. We omit this; it requires a little more multicategory theory. Then, as sketched above, they define what it means for a cell of an opetopic set to be universal, hence what it means for an opetopic set to be a weak n-category.

Actually, they do more. Their definition of opetope uses I as its starting point, but it is possible to start with any other symmetric multicategory C instead, to obtain a definition of 'n-dimensional C-opetope'. There is a corresponding notion of 'C-opetopic set', and of what it means for a C-opetopic set to be an 'n-coherent C-algebra' (in the case $C = I$, a weak n-category). For example, let C be the terminal operad 1. Then an n-coherent 1-algebra is what they call a 'stable n-category', or might also be called a symmetric monoidal

n-category, as in the periodic table (p. 17). A 0-coherent C-algebra is just a C-algebra, so a stable 0-category is a commutative monoid, as it should be.

We have discussed elsewhere (p. 224) the difference between symmetric and generalized multicategories. In short, the generalized multicategory approach allows the geometry to be represented faithfully, whereas the symmetric approach destroys the geometry and squashes everything into one dimension; but symmetric multicategories are perhaps easier to get one's hands on. Cheng offers the following analogy (2003a, 1.4): she does not like tidying her desk. As she works away and produces more and more pages of notes, each page gets put into its natural position on the desktop, with notes on related topics side-by-side, the sheet currently being written on at the front, and so on. If she is forced to clear up her desk then she must destroy this natural configuration and stack the papers in some arbitrary order; but having put them into a single stack, they are much easier to carry around.

Despite symmetric multicategories not falling readily into our scheme of generalized multicategories, there are many points of similarity between Baez and Dolan's constructions and ours.

Slicing is an example. Their construction of the slice C^+ of a symmetric multicategory C proceeds in two stages. First they show how to slice a multicategory by one of its algebras: given a symmetric multicategory D and a D-algebra X, they construct a symmetric multicategory D/X such that

$$\mathbf{Alg}(D/X) \simeq \mathbf{Alg}(D)/X. \qquad (7.18)$$

Then they show how to construct, for any set S, a symmetric multicategory $\mathbf{Mti}_S$ such that

$$\mathbf{Alg}(\mathbf{Mti}_S) \simeq \mathbf{SymMulticat}_S$$

where the right-hand side is the category whose objects are the symmetric multicategories with object-set S and whose maps are those leaving the object-set fixed. (They write D/X as X^+ and leave $\mathbf{Mti}_S$ nameless. We wrote $\mathbf{Mti}_S$ as $\mathcal{O}'_S$ in 2.2.23.) The slice multicategory C^+ of C is then defined by

$$C^+ = \mathbf{Mti}_{C_0}/C, \qquad (7.19)$$

so that

$$\mathbf{Alg}(C^+) \simeq \mathbf{Alg}(\mathbf{Mti}_{C_0})/C \simeq \mathbf{SymMulticat}_{C_0}/C,$$

as required. These constructions are mirrored in our world of generalized multicategories. First, we saw in Proposition 6.3.3 how to slice a multicategory by an algebra: if T is a cartesian monad on a cartesian category $\mathcal{E}$, D is a

T-multicategory, and X is a D-algebra, then there is a T multicategory D/X satisfying (7.18). Second, assuming that $\mathcal{E}$ and T are suitable, Theorem 6.5.4 tells us that for any $S \in \mathcal{E}$ there is a cartesian monad T_S^+ on a cartesian category $\mathcal{E}_S^+$ such that

$$\mathbf{Alg}(T_S^+) \simeq (\mathcal{E}, T)\text{-}\mathbf{Multicat}_S,$$

and so if we take $\mathbf{Mti}_S$ to be the terminal T_S^+-multicategory then

$$\mathbf{Alg}(\mathbf{Mti}_S) \simeq (\mathcal{E}, T)\text{-}\mathbf{Multicat}_S$$

by 4.3.8. So we can define the T_S^+-multicategory C^+ by the same formula (7.19) as in the symmetric case, to reach the analogous conclusion

$$\mathbf{Alg}(C^+) \simeq \mathbf{Alg}(\mathbf{Mti}_{C_0})/C \simeq (\mathcal{E}, T)\text{-}\mathbf{Multicat}_{C_0}/C.$$

This neatly illustrates the contrast between the two approaches. With generalized multicategories, slicing raises the dimension: if C is a T-multicategory then C^+ is a $T_{C_0}^+$-multicategory. When we draw the pictures this dimension-shift seems perfectly natural. On the other hand, if C is a symmetric multicategory then C^+ is again a symmetric multicategory, and this is certainly convenient. Note also that there is a flexibility in the generalized approach lacking in the symmetric approach, revealed if we attempt to drop the restriction fixing the object-set. That is, suppose that given a (T- or symmetric) multicategory C we wish to define a new multicategory C^+ whose algebras are *all* multicategories over C. This is easy for generalized multicategories: we just replace $\mathcal{E}_S^+$ and T_S^+ by $\mathcal{E}^+$ and T^+ in the construction above, using Theorem 6.5.2, and C^+ is a T^+-multicategory. But it is impossible in the symmetric theory, as there is no symmetric multicategory whose algebras are all symmetric multicategories.

Another point where the two approaches proceed in analogous ways is in the construction of opetopes. As discussed, Baez and Dolan considered so-called C-opetopes for any symmetric multicategory C, not just the case $C = I$ that generates the ordinary opetopes. The analogous point for generalized multicategories is that the opetopic categories $\mathcal{E}_n$ and monads T_n were generated from $\mathcal{E}_0 = \mathbf{Set}$ and $T_0 = \mathrm{id}$ (Definition 7.1.1), but we could equally well have started from any other suitable category $\mathcal{E}$ and suitable monad T on $\mathcal{E}$. This would give a new sequence $(O_{T,n})_{n \in \mathbb{N}}$ of objects of $\mathcal{E}$ – the objects of 'n-dimensional T-opetopes'.

We now come to the discrepancy between Baez and Dolan's opetopes and ours, mentioned on p. 252. A satisfactory resolution has been found by Cheng (2003a, 2003b), and can be explained roughly as follows. The problem in Baez and Dolan's approach is that the slicing process results in a loss of information.

For example, as we observed, the symmetric multicategory I^{++} has six objects drawn as

but formally it contains nothing to say that they are in any way related. That information about symmetries has been discarded in the slicing process. With each new slicing another layer of information about symmetries is lost, so that even a simple 3-dimensional opetope is reproduced into hundreds of thousands of apparently unrelated copies.

The remedy is to work with structures slightly more sophisticated than symmetric multicategories, capable of holding just a little more information. For the short while that we discuss them, let us call them **enhanced symmetric multicategories**; the definition can be found in Cheng (2003a, 2.1) or (2003b, 1.1). An enhanced symmetric multicategory is like a symmetric multicategory, but as well as having the usual objects and arrows, it also has **morphisms** between objects, making objects and morphisms into a category. The morphisms are *not* the same as the arrows, although naturally there are axioms relating them; note in particular that whereas an arrow has a finite sequence of objects as its domain, a morphism has only one. An enhanced symmetric multicategory in which the category of objects and morphisms is discrete is an ordinary symmetric multicategory. (This explains Cheng's terminology: for her, enhanced symmetric multicategories are merely 'symmetric multicategories', and symmetric multicategories are 'object-discrete symmetric multicategories'. Compare also 4.2.18 above.)

The virtue of enhanced symmetric multicategories is that like symmetric multicategories, they can be sliced, but unlike symmetric multicategories, slicing retains all the information about symmetries. So in the enhanced version of I^{++}, all six objects shaped like the 2-dimensional opetope above are isomorphic – in other words, there is essentially only one of them. In fact, Cheng has proved (2003b) that there is a natural one-to-one correspondence between her modified version of Baez and Dolan's opetopes and the opetopes defined in the present text. They also correspond to the multitopes of Hermida, Makkai and Power: see the Notes at the end of this chapter.

7.6 Many in, many out

Multicategories and, more specifically, opetopes, are based on the idea of operations taking many inputs and one output. We could, however, decide to allow multiple outputs too, so that 2-opetopes are replaced by shapes like

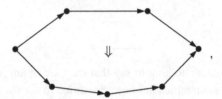

,

3-opetopes by shapes looking something like soccer balls, and so on.

The idea of doing higher category theory with cells shaped like this is attractive, since they encompass all the other shapes in common use: globular, simplicial, cubical and opetopic. In other words, these shapes are intended to represent all possible composable diagrams of cells in an n- or ω-category, and so we are aiming for a universal, canonical, totally unbiased approach, and that seems very healthy.

There is, however, a fundamental obstruction. It turns out that it does not make sense to talk about 'these shapes', at least in dimensions higher than 2. They are simply not well-defined.

To make these statements precise we define an analogue of opetopic set with multiple outputs as well as multiple inputs, then show that these structures, unlike opetopic sets, do not form a presheaf category. So there is no category of many-in, many-out shapes analogous to the category $\mathbb{O}$ of opetopes. Actually, it is not quite opetopic sets of which we define an analogue, but rather n-dimensional versions in which all cells have dimension at most n. We need only go up to dimension 3 to encounter the obstruction. (It is no coincidence that the obstruction to simple-minded coherence for n-categories also appears in dimension 3.) These n-dimensional analogues of opetopic sets are the n-computads of Street (1976, 1996); let us define them for $n \leq 3$.

The category 0-**Cptd** of **0-computads** is **Set**.

A **1-computad** is a directed graph; if $\mathbb{C}(1)$ is the category $(0 \rightrightarrows 1)$ then

$$1\text{-}\mathbf{Cptd} = [\mathbb{C}(1)^{\mathrm{op}}, \mathbf{Set}].$$

We have the familiar adjunction

$$\mathbf{Cat} \underset{F_1}{\overset{U_1}{\underset{\top}{\rightleftarrows}}} 1\text{-}\mathbf{Cptd}$$

(induced, if you like, by the functor $\mathbb{C}(1) \longrightarrow \mathbf{Cat}$ sending 0 to the terminal category and 1 to the category consisting of a single arrow, using the mechanism described on p. 244). There is also a functor $P_1 : 1\text{-}\mathbf{Cptd} \longrightarrow \mathbf{Set}$ sending a directed graph to the set of parallel pairs of edges in it.

A **2-computad** is a 1-computad X together with a set X_2 and a function $\xi : X_2 \longrightarrow P_1 U_1 F_1(X)$. A map $(X, X_2, \xi) \longrightarrow (X', X_2', \xi')$ is a pair $(X \longrightarrow X', X_2 \longrightarrow X_2')$ of maps making the obvious square commute. Concretely, a 2-computad consists of a collection X_0 of 0-cells, a collection X_1 of 1-cells between 0-cells, and a collection of X_2 of 2-cells α of the form

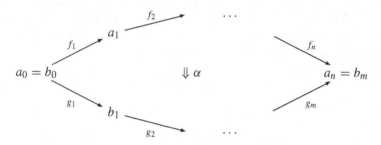

where $n, m \in \mathbb{N}$, $a_i, b_j \in X_0$, and $f_i, g_j \in X_1$. (The possibility exists that $n = m = 0$: this will prove critical.) Hence

$$\text{2-Cptd} \simeq [\mathbb{C}(2)^{\mathrm{op}}, \text{Set}]$$

where $\mathbb{C}(2)$ is the evident category with one object for each pair (n, m) of natural numbers and then two further objects (one for each of dimensions 1 and 0). There is an adjunction

$$\textbf{Str-2-Cat} \underset{F_2}{\overset{U_2}{\underset{\top}{\rightleftarrows}}} \textbf{2-Cptd}$$

between strict 2-categories and 2-computads (induced, if you like, by the evident functor $\mathbb{C}(2) \longrightarrow \textbf{Str-2-Cat}$). There is also a functor $P_2 : \textbf{2-Cptd} \longrightarrow \textbf{Set}$ sending a 2-computad to the set of parallel pairs of 2-cells in it.

Similarly, a **3-computad** is a 2-computad X together with a set X_3 and a function $X_3 \longrightarrow P_2 U_2 F_2(X)$, and maps are defined in the obvious way. But there is no category $\mathbb{C}(3)$ of 'computopes of dimension ≤ 3':

Proposition 7.6.1 *The category of 3-computads is not a presheaf category.*

The following proof is due to Makkai and Zawadowski (private communication) and to Carboni and Johnstone (2002).

Proof The category 3-**Cptd** is a certain comma category, the Artin gluing (p. 374) of the functor

$$P_2 U_2 F_2 : \textbf{2-Cptd} \longrightarrow \textbf{Set}.$$

Proposition C.3.4 tells us that 3-**Cptd** is a presheaf category only if $P_2 U_2 F_2$ preserves wide pullbacks. We show that it does not even preserve ordinary pullbacks (which by Carboni and Johnstone (1995, 4.4(ii)) implies that 3-**Cptd**, far from being a presheaf category, is not even locally cartesian closed).

Let S be the full subcategory of 2-**Cptd** consisting of those 2-computads with only one 0-cell and no 1-cells. Plainly $S \simeq$ **Set**. The inclusion $S \hookrightarrow$ 2-**Cptd** preserves pullbacks, so it suffices to show that the functor $S \longrightarrow$ **Set** obtained by restricting $P_2 U_2 F_2$ to S does not preserve pullbacks.

If $X \in S$ and $C \in$ **Str**-2-**Cat** then a map $X \longrightarrow U_2(C)$ consists of an object c of C together with a 2-cell

of C for each 2-cell of X. Example 1.2.3 tells us that a strict 2-category with only one 0-cell and one 1-cell is just a commutative monoid, so $U_2 F_2(X)$ is the 2-computad with one 0-cell, one 1-cell, and 2-cells corresponding to the elements of the free commutative monoid $M(X_2)$ on the set X_2 of 2-cells of X. Since any two 2-cells of $U_2 F_2(X)$ are parallel, we now just have to show that the endofunctor

$$A \longmapsto M(A) \times M(A)$$

of **Set** does not preserve pullbacks. This follows from the fact that M itself does not preserve pullbacks, which we proved in Example 4.1.5.   □

Notes

Opetopes were defined by Baez and Dolan (1997). Hermida, Makkai and Power defined their own version (2000, 2001, 2002, 1998), calling them multitopes, as did I (1998a, 4.1), re-using the name 'opetope'. After tweaking Baez and Dolan's definition (see p. 254), Cheng proved all three notions equivalent (2003a, 2003b). On ordering the opetopes making up a pasting diagram, she writes:

> The three different theories arise, essentially, from three different ways of tackling this issue. Baez and Dolan propose listing them in every order possible, giving one description *for each* ordering. Hermida, Makkai and Power propose picking one order at random. Leinster proposes *not* picking any order.

(Unpublished document, but compare Cheng, 2003a, 1.3.) Burroni has also considered 'ω-multigraphs', which are similar to opetopic sets (1991, 1993). Variants of Baez and Dolan's definition of weak n-category have been proposed by Hermida, Makkai and Power (2000, 2001, 2002, 1998), Makkai (1999), and Cheng (2003a, 2003c, 2003e).

Categories of trees have been used in a variety of situations: see, for instance, Borcherds (1997), Ginzburg and Kapranov (1994), Kontsevich and Manin (1994), and Soibelman (1997, 1999). Some authors consider only stable trees, or consider all trees but only surjective maps between them. Our maps of trees are the identity on leaf-sets, but some authors allow the leaves to be moved; doubtless this is natural for certain parts of higher-dimensional algebra.

For more on A_∞-spaces, A_∞-algebras, and A_∞-categories see, for example, the book of Markl, Shnider and Stasheff (2002).

The explanation (Section 7.6) of computads as being like opetopic sets is a historical inversion; computads were introduced by Street in (1976). (His original computads are what we call 2-computads here.) The question of whether higher-dimensional computads form a presheaf category has been wreathed in confusion: Schanuel observed correctly that 2-computads form a presheaf category (unpublished), Carboni and Johnstone claimed incorrectly that this extended to n-computads for all $n \in \mathbb{N}$ (1995, 4.6), Makkai and Zawadowski proved that 3-computads do *not* form a presheaf category (unpublished), and Carboni and Johnstone published a variant of their proof in a corrigendum (2002); meanwhile, Batanin found a notion of generalized computad and claimed that they form a presheaf category (1998b), before retracting and refining that claim (2002b) in the light of Makkai and Zawadowski's argument. It should be added that defining 'n-computad' for $n > 3$ is much easier if one makes Carboni and Johnstone's error, and is otherwise not straightforward.

There is a good deal more to be said about 'many in, many out' than I have said here. See, for instance, the work of Szabo (1975) and Koslowski (2003) on polycategories, and of Gan (2002) on dioperads; see also the remarks on PROs and PROPs herein.

PART III

n-categories

Chapter 8
Globular operads

Once Theofilos was [...] painting a mural in a Mytilene baker's shop [...] As was his habit, he had depicted the loaves of bread upright in their trays, like heraldic emblems on an out-thrust shield – so that no one could be in any doubt that they were loaves of bread and very fine ones, too. The irate baker pointed out that in real life loaves thus placed would have fallen to the floor. 'No,' replied Theofilos – surely with that calm, implacable self-certainty which carried him throughout what most people would call a miserable life – 'only real loaves fall down. Painted ones stay where you put them.'

The Athenian (1980)

In the next two chapters we explore one possible definition of weak ω-category. Its formal shape is very simple: we take the category $\mathcal{E}$ of globular sets and the free strict ω-category monad T on $\mathcal{E}$, construct a certain T-operad L, and define a weak ω-category to be an L-algebra.

In order to see that this is a *reasonable* definition, and to get a feel for the concepts involved, we proceed at a leisurely pace. The present chapter is devoted to contemplation of the monad T (Section 8.1), of T-operads (= globular operads, Section 8.2), and of their algebras (Section 8.3). In Chapter 9 we define the particular globular operad L and look at its algebras – that is, at weak ω-categories. To keep the explanation in this chapter uncluttered, the discussion of the finite-dimensional case (n-categories) is also deferred to Chapter 9; we stick to ω-categories here.

Globular operads are an absolutely typical example of generalized operads. Pictorially, they are typical in that $T1$ is a family of shapes (globular pasting diagrams), and an operation in a T-operad is naturally drawn as an arrow with data fitting one of these shapes as its input. (Compare plain operads, where the 'shapes' are mere finite sequences.) Technically, globular operads are typical in that T is, in the terminology of Appendix C, a finitary familially representable

263

monad on a presheaf category. Hence the explanations contained in this chapter can easily be adapted to many other species of generalized operad.

8.1 The free strict ω-category monad

In Section 1.4 we defined a strict ω-category as a globular set equipped with extra structure, and a strict ω-functor as a map of globular sets preserving that structure. There is consequently a forgetful functor from the category **Str-ω-Cat** of strict ω-categories and strict ω-functors to the category $[\mathbb{G}^{op}, \mathbf{Set}]$ of globular sets. In Appendix F it is shown that this forgetful functor has a left adjoint, that the adjunction is monadic, and that the induced monad (T, μ, η) on $[\mathbb{G}^{op}, \mathbf{Set}]$ is cartesian.

To understand this in pictorial terms, we start by considering $T1$, the free strict ω-category on the terminal globular set

$$1 = (\cdots \rightrightarrows 1 \rightrightarrows \cdots \rightrightarrows 1).$$

The free strict ω-category functor takes a globular set and creates all possible formal composites in it. A typical element of $(T1)(2)$ looks like

$$(8.1)$$

where each k-cell drawn represents the unique member of $1(k)$. Note that although this picture contains four dots representing 0-cells of 1, they actually all represent the same 0-cell; of course, 1 only *has* one 0-cell. The same goes for the 1- and 2-cells. So we have drawn a flattened-out version of the true, twisted, picture. We call an element of $(T1)(m)$ a **(globular) m-pasting diagram** and write $T1 = \mathbf{pd}$.

Since the theory of strict ω-categories includes identities, there is for each $m \geq 2$ an element of $\mathbf{pd}(m)$ looking like (8.1). Although the pictures look the same, they are regarded as *different* pasting diagrams for different values of m; the sets $\mathbf{pd}(m)$ and $\mathbf{pd}(m')$ are considered disjoint when $m \neq m'$. When it comes to understanding the definition of weak n-category, this point will be crucial.

In the globular set 1, all cells are endomorphisms – in other words, the source and target maps are equal. It follows that the same is true in the globular set $\mathbf{pd}$. We write $\partial : \mathbf{pd}(m + 1) \longrightarrow \mathbf{pd}(m)$ instead of s or t, and call ∂ the

boundary operator. For instance, the boundary of the 2-pasting diagram (8.1) is the 1-pasting diagram

It is easy to describe the globular set **pd** explicitly: writing ( )* for the free monoid functor on **Set**, we have $\mathbf{pd}(0) = 1$ and $\mathbf{pd}(m + 1) = \mathbf{pd}(m)^*$. That is, an $(m + 1)$-pasting diagram is a sequence of m-pasting diagrams. For example, the 2-pasting diagram depicted in (8.1) is the sequence

$$(\bullet \longrightarrow \!\!\!\!\!\to \bullet \longrightarrow \!\!\!\!\!\to \bullet \longrightarrow \!\!\!\!\!\to \bullet, \; \bullet, \; \bullet \longrightarrow \!\!\!\!\!\to \bullet)$$

of 1-pasting diagrams, so if we write the unique element of **pd**(0) as • then (8.1) is the double sequence

$$((\bullet, \bullet, \bullet), (), (\bullet)) \in \mathbf{pd}(2).$$

The boundary map $\partial : \mathbf{pd}(m + 1) \longrightarrow \mathbf{pd}(m)$ is defined inductively by

$$\left(\mathbf{pd}(m + 1) \xrightarrow{\ \partial\ } \mathbf{pd}(m)\right) = \left(\mathbf{pd}(m) \xrightarrow{\ \partial\ } \mathbf{pd}(m - 1)\right)^*$$

$(m \geq 1)$. The correctness of this description of **pd** follows from the results of Appendix F.

Having described **pd** as a globular set, we turn to its strict ω-category structure: how pasting diagrams may be composed.

Typical binary compositions are

$$(8.2)$$

illustrating the composition function $\circ_1 : \mathbf{pd}(2) \times_{\mathbf{pd}(1)} \mathbf{pd}(2) \longrightarrow \mathbf{pd}(2)$, and

$$(8.3)$$

illustrating the composition function $\circ_0 : \mathbf{pd}(3) \times_{\mathbf{pd}(0)} \mathbf{pd}(3) \longrightarrow \mathbf{pd}(3)$.

These compositions are possible because the boundaries match: in (8.2), the 1-dimensional boundaries of the two pasting diagrams on the left-hand side are equal, and similarly for the 0-dimensional boundaries (8.3) – indeed, this is inevitable as there is only one 0-pasting diagram.

(The arguments on the left-hand side of (8.2) are stacked vertically rather than horizontally just to make the picture more compelling; strictly speaking we should have written

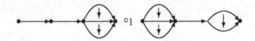

instead. The same applies to (8.3).)

A typical nullary composition (identity) is

   (8.4)

illustrating the identity function $i : \mathbf{pd}(1) \longrightarrow \mathbf{pd}(2)$ in the strict ω-category $\mathbf{pd}$. (Recall the remarks above on degenerate pasting diagrams.) So the left-hand side of (8.4) is a 1-pasting diagram $\pi \in \mathbf{pd}(1)$, and the right-hand side is 2-pasting diagram $1_\pi = i(\pi) \in \mathbf{pd}(2)$. We have $\partial(1_\pi) = \pi$.

We will need to consider not just binary and nullary composition in $\mathbf{pd}$, but composition 'indexed' by arbitrary shapes, in the sense now explained. The first binary composition (8.2) above is indexed by the 2-pasting diagram

, in that we were composing one 2-cell with another by joining along their bounding 1-cells. The composition can be represented as

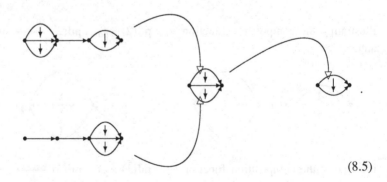

(8.5)

In general, the ways of composing pasting diagrams are indexed by pasting diagrams themselves: for instance,

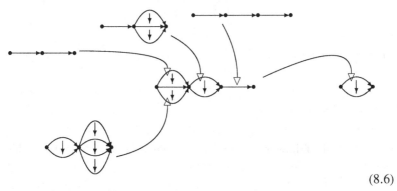

$$(8.6)$$

represents the composition

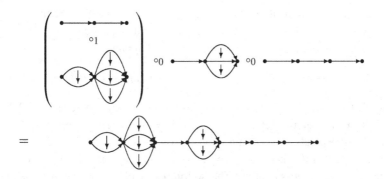

(with the same pictorial convention on positioning the arguments as previously).

This describes the free strict ω-category **pd** on the terminal globular set. Before progressing to free strict ω-categories in general, let us pause to consider an alternative way of representing pasting diagrams, due to Batanin (1998a): as trees.

First a warning: these are not the same trees as appear elsewhere in this text (Section 7.3, for instance). Not only is there a formal difference, but also the two kinds of trees play very different roles. The exact connection remains unclear, but for the purposes of understanding what is written here they can be regarded as entirely different species.

The idea is that, for instance, the 2-pasting diagram

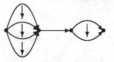

can be portrayed as the tree

(8.7)

according to the following method. The pasting diagram is three 1-cells long, so the tree begins life as

Then the first column is three 2-cells high, the second none, and the third one, so it grows to (8.7). Finally, there are no 3-cells so it stops there.

Formally, an *m*-**stage level tree** ($m \in \mathbb{N}$) is a diagram

$$\tau(m) \longrightarrow \tau(m-1) \longrightarrow \cdots \longrightarrow \tau(1) \longrightarrow \tau(0) = 1$$

in the skeletal category $\mathbb{D}$ of finite (possibly empty) totally ordered sets (1.2.2); we write $\mathbf{lt}(m)$ for the set of all *m*-stage level trees. The element of $\mathbf{lt}(2)$ in (8.7) corresponds to a certain diagram $4 \longrightarrow 3 \longrightarrow 1$ in $\mathbb{D}$, for example. (Note that if τ is an *m*-stage tree with $\tau(m) = 0$ then the height of the picture of τ will be less than m.) The **boundary** $\partial\tau$ of an *m*-stage tree τ is the $(m-1)$-stage tree obtained by removing all the nodes at height m, or formally, truncating

$$\tau(m) \longrightarrow \tau(m-1) \longrightarrow \cdots \longrightarrow \tau(1) \longrightarrow \tau(0)$$

to

$$\tau(m-1) \longrightarrow \cdots \longrightarrow \tau(1) \longrightarrow \tau(0).$$

This defines a diagram

$$\cdots \xrightarrow{\partial} \mathbf{lt}(m) \xrightarrow{\partial} \mathbf{lt}(m-1) \xrightarrow{\partial} \cdots \xrightarrow{\partial} \mathbf{lt}(0) \qquad (8.8)$$

of sets and functions, hence a globular set $\mathbf{lt}$ with $s = t = \partial$.

Proposition 8.1.1 *There is an isomorphism of globular sets* $\mathbf{pd} \cong \mathbf{lt}$.

Proof There is one 0-stage tree, and informally it is clear that an $(m + 1)$-stage tree amounts to a finite sequence of m-stage trees (placed side by side and with a root node adjoined). Formally, take an $(m + 1)$-stage tree τ, write $\tau(1) = \{1, \ldots, r\}$, and for $1 \leq i \leq r$ and $0 \leq p \leq m$, let

$$\tau_i(p) = \{j \in \tau(p + 1) \mid \partial^p(j) = i\}.$$

Then we have a finite sequence $(\tau_1, \ldots, \tau_r)$ of m-stage trees and so, inductively, a finite sequence of m-pasting diagrams, which is an $(m + 1)$-pasting diagram. It is easy to check that this defines an isomorphism. ☐

Composition and identities in the strict ω-category **pd** can also be expressed in the pictorial language of trees, in a simple way: see Batanin (1998a) or Leinster (2001a, Ch. II).

Let us now see what T does to an arbitrary globular set X. An m-cell of $T X$ is a formal pasting-together of cells of X of dimension at most m: for instance, a typical element of $(T X)(2)$ looks like

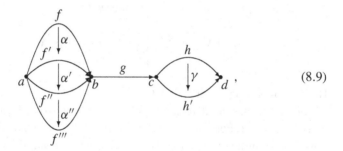

$$(8.9)$$

where $a, b, c, d \in X(0)$, $f, f', f'', f''', g, h, h' \in X(1)$, $\alpha, \alpha', \alpha'', \gamma \in X(2)$, and $s(\alpha) = f, t(\alpha) = f'$, and so on.

We can describe the functor T explicitly; in the terminology of Appendix C, we give a 'familial representation'.

First we associate with each pasting diagram π the globular set $\widehat{\pi}$ that 'looks like π'. If π is the unique 0-pasting diagram then

$$\widehat{\pi} = (\cdots \rightrightarrows \emptyset \rightrightarrows \emptyset \rightrightarrows 1).$$

Inductively, suppose that $m \geq 0$ and $\pi \in \mathbf{pd}(m + 1)$: then $\pi = (\pi_1, \ldots, \pi_r)$ for some $r \in \mathbb{N}$ and $\pi_1, \ldots, \pi_r \in \mathbf{pd}(m)$, and we put

$$\widehat{\pi} = (\cdots \rightrightarrows \coprod_{i=1}^{r} \widehat{\pi}_i(1) \rightrightarrows \coprod_{i=1}^{r} \widehat{\pi}_i(0) \rightrightarrows \{0, 1, \ldots, r\}). \qquad (8.10)$$

The source and target maps in all but the bottom dimension are the evident disjoint unions, and in the bottom dimension they are defined at $x \in \widehat{\pi}_i(0)$ by

$$s(x) = i - 1, \qquad t(x) = i.$$

For example, if π is the 2-pasting diagram (8.1) then $\widehat{\pi}$ is of the form

$$\cdots \; \Longrightarrow \; \emptyset \; \Longrightarrow \; 4 \; \Longrightarrow \; 7 \; \Longrightarrow \; 4$$

where '4' means a 4-element set, etc. This reflects the fact that the picture (8.1) contains four 0-cells, seven 1-cells, four 2-cells, and no higher cells.

The case of degenerate pasting diagrams deserves attention. If $\pi \in \mathbf{pd}(m)$ then $1_\pi \in \mathbf{pd}(m + 1)$ is represented by the same picture as π; formally, $\widehat{1_\pi} = \widehat{\pi}$. In fact, if $\sigma \in \mathbf{pd}(m + 1)$ then $\widehat{\sigma}(m + 1) = \emptyset$ if and only if $\sigma = 1_{\partial\sigma}$.

An m-cell of TX is meant to be an 'm-pasting diagram labelled by cells of X', that is, an m-pasting diagram π together with a map $\widehat{\pi} \longrightarrow X$ of globular sets, which suggests that there is an isomorphism

$$(TX)(m) \cong \coprod_{\pi \in \mathbf{pd}(m)} [\mathbb{G}^{\mathrm{op}}, \mathbf{Set}](\widehat{\pi}, X). \qquad (8.11)$$

This is proved as Proposition F.2.3.

For each m-pasting diagram π ($m \geq 1$) there are source and target inclusions $\partial\pi \rightrightarrows \widehat{\pi}$. For instance, when π is the 2-pasting diagram of (8.1), these embed $\widehat{\partial\pi}$ (a string of three 1-cells) as the top and bottom edges of $\widehat{\pi}$. The formal definition is straightforward and left as an exercise. Given a globular set X, these embeddings induce functions $(TX)(m) \rightrightarrows (TX)(m - 1)$ for each m, so that TX becomes a globular set. So we now have the desired explicit description of the free strict ω-category functor T.

Finally, T is not just a functor but a monad. The multiplication turns a pasting diagram of pasting diagrams of cells of some globular set X into a single pasting diagram of cells of X by 'erasing the joins'; compare (8.6). The unit realizes a single cell of X as a (trivial) pasting diagram of cells of X.

We could also try to describe the multiplication and unit explicitly in terms of the family $(\widehat{\pi})_{m \in \mathbb{N}, \pi \in \mathbf{pd}(m)}$ 'representing' T. This can be done, but takes appreciable effort and seems to be both very complicated and not especially illuminating; as discussed in the Notes to Appendix C, the full theory of familially representable monads on presheaf categories is currently beyond us. But for the purposes of this chapter, we have all the description of T that we need.

8.2 Globular operads

A **globular operad** is a T-operad. The purpose of this section is to describe globular operads pictorially. The more general T-multicategories are not mentioned until the end of Chapter 10, and there only briefly.

A globular operad P is a T-graph

$$
\begin{array}{ccc}
& P & \\
{}^{d}\diagup & & \diagdown \\
\mathbf{pd} = T1 & & 1
\end{array}
\qquad (8.12)
$$

equipped with composition and identity operations satisfying associativity and identity axioms. (In a standard abuse of language, we use P to mean either the whole operad or just the globular set at the apex of the diagram.) We consider each part of this description in turn.

A T-graph (8.12) whose object-of-objects is 1 will be called a **collection**. So a collection is merely a globular set over **pd**, and consists of a set $P(\pi)$ for each $m \in \mathbb{N}$ and m-pasting diagram π, together with a pair of functions $P(\pi) \overset{s}{\underset{t}{\rightrightarrows}} P(\partial\pi)$ (when $m \geq 1$) satisfying the usual globularity equations. (Formally, 1.1.7 tells us that a presheaf on $\mathbb{G}$ over **pd** is the same thing as a presheaf on the category of elements of **pd**, whose objects are pasting diagrams and whose arrows are generated by those of the form $\partial\pi \overset{\sigma}{\underset{\tau}{\rightrightarrows}} \pi$ subject to the duals of the globularity equations.)

If we were discussing plain rather than globular operads then $\mathbf{pd} = T1$ would be replaced by $\mathbb{N}$, and a collection would be a sequence $(P(k))_{k \in \mathbb{N}}$ of sets. In that context we think of an element θ of $P(k)$ as an operation of arity k (even though it does not actually act on anything before a P-algebra is specified) and draw it as

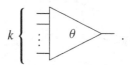

Similarly, if P is a (globular) collection and π an m-pasting diagram for some $m \in \mathbb{N}$, we think of an element of $P(\pi)$ as an 'operation of arity π' and draw it as an arrow whose input is (a picture of) π and whose output is a single m-cell. For instance, if $m = 2$ and

$$
\pi =
$$

then $\theta \in P(\pi)$ is drawn as

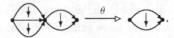

So θ is thought of as an operation capable of taking data shaped like the pasting diagram π as input and producing a single 2-cell as output. This is a figurative description but, as we shall see, becomes literal when an algebra for P is present.

Composition in a globular operad P is a map comp : $P \circ P \longrightarrow P$ of collections. The collection $P \circ P$ is the composite down the left-hand diagonal of the diagram

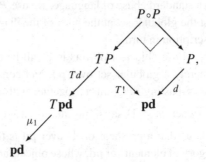

and a typical element of $(P \circ P)(2)$ is depicted as

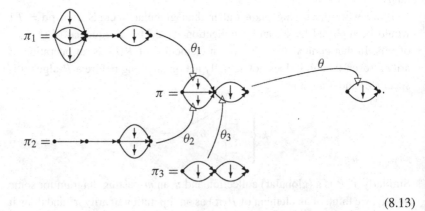

(8.13)

Here $\theta_1 \in P(\pi_1)$, $\theta_2 \in P(\pi_2)$, $\theta_3 \in P(\pi_3)$, $\theta \in P(\pi)$, and it is meant to be implicit that θ_1, θ_2, and θ_3 match on their sources and targets: $t\theta_1 = s\theta_2$ and $tt\theta_1 = ss\theta_3$. The left-hand half of the diagram (containing the θ_is) is an element of the fibre over π of the map $T! : (TP)(2) \longrightarrow \mathbf{pd}(2)$, and the right-hand half (θ) is an element of the fibre over π of the map $d : P(2) \longrightarrow$ **pd**(2) (that is, an element of $P(\pi)$), so the whole diagram is a 2-cell of $P \circ P$. More precisely, it is an element of $(P \circ P)(\pi \circ (\pi_1, \pi_2, \pi_3))$, where

$$\pi \circ (\pi_1, \pi_2, \pi_3) =$$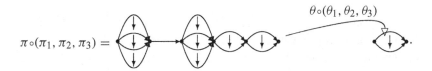

is the composite of π with π_1, π_2, π_3 in the ω-category **pd**. So, the composition function comp of the globular operad P sends the data assembled in (8.13) to an element $\theta \circ (\theta_1, \theta_2, \theta_3) \in P(\pi \circ (\pi_1, \pi_2, \pi_3))$, which may be drawn as

$$\pi \circ (\pi_1, \pi_2, \pi_3) =$$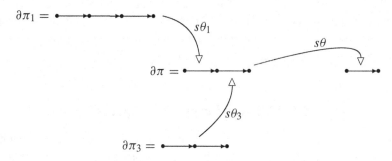

(The 'linear' notation $\pi \circ (\pi_1, \pi_2, \pi_3)$ and $\theta \circ (\theta_1, \theta_2, \theta_3)$ should not be taken too seriously: there is evidently no canonical order in which to put the π_is.)

That comp : $P \circ P \longrightarrow P$ is a map of globular sets says that composition is compatible with source and target. In the example above,

$$s(\theta \circ (\theta_1, \theta_2, \theta_3)) = s\theta \circ (s\theta_1, s\theta_3),$$
$$t(\theta \circ (\theta_1, \theta_2, \theta_3)) = t\theta \circ (t\theta_2, t\theta_3),$$

where the composite

$$s\theta \circ (s\theta_1, s\theta_3) \in P(\bullet \!\longrightarrow\!\bullet \quad \bullet\!\longrightarrow\!\bullet \quad \bullet\!\longrightarrow\!\bullet \quad \bullet\!\longrightarrow\!\bullet \quad \bullet\!\longrightarrow\!\bullet)$$

is as shown:

and $t\theta \circ (t\theta_2, t\theta_3)$ similarly.

To see what identities in a globular operad P are, let $\iota_m \in \mathbf{pd}(m)$ be the m-pasting diagram looking like a single m-cell. (Formally, ι_0 is the unique element of $\mathbf{pd}(0)$ and

$$\iota_{m+1} = (\iota_m) \in (\mathbf{pd}(m))^* = \mathbf{pd}(m+1),$$

using the description of $\mathbf{pd}$ on p. 265). Then the map $1 \xrightarrow{\eta_1} \mathbf{pd}$ sends the unique m-cell of 1 to ι_m, so the identities function $\mathrm{ids} : 1 \longrightarrow P$ consists of an element $1_m \in P(\iota_m)$ for each $m \in \mathbb{N}$. For instance, the 2-dimensional identity operation 1_2 of P is drawn as

$$\bigcirc\!\!\!\downarrow\!\!\!\triangleright \xrightarrow{\quad 1_2 \quad} \bigcirc\!\!\!\downarrow\!\!\!\triangleright\,.$$

That ids is a map of globular sets says that $s(1_m) = 1_{m-1} = t(1_m)$ for all $m \geq 1$.

Finally, the composition and identities in P are required to obey associativity and identity laws. Together these say that there is only one way of composing any 'tree' of operations of the operad: for instance, if

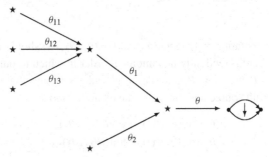

is a diagram of the same general kind as (8.13), with each $\star$ representing a 2-pasting diagram, then

$$\theta\circ(\theta_1\circ(\theta_{11},\theta_{12},\theta_{13}),\theta_2) = (\theta\circ(\theta_1,\theta_2))\circ(\theta_{11},\theta_{12},\theta_{13},1_2).$$

We have now unwound all of the data and axioms for a globular operad. Although it may seem complicated on first reading, it is summed up simply: a globular operad is a collection of operations together with a unique composite for any family of operations that might plausibly be composed.

8.3 Algebras for globular operads

We now confirm what was suggested implicitly in the previous section: that if P is a globular operad then a P-algebra structure on a globular set X consists of a function

$$\overline{\theta} : \{\text{labellings of } \pi \text{ by cells of } X\} \longrightarrow X(m)$$

for each number m, m-pasting diagram π, and operation $\theta \in P(\pi)$, satisfying sensible axioms.

So, fix a globular operad P. According to the general definition, an algebra for P is an algebra for the monad T_P on the category of globular sets, which is defined on objects $X \in [\mathbb{G}^{\mathrm{op}}, \mathbf{Set}]$ by

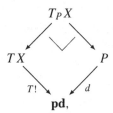

$$
\begin{array}{c}
T_P X \\
T X \qquad \qquad P \\
T! \qquad \qquad d \\
\mathbf{pd},
\end{array}
$$

that is, by

$$(T_P X)(m) \cong \coprod_{\pi \in \mathbf{pd}(m)} P(\pi) \times [\mathbb{G}^{\mathrm{op}}, \mathbf{Set}](\widehat{\pi}, X).$$

So a P-algebra is a globular set X together with a function

$$h_\pi : P(\pi) \times [\mathbb{G}^{\mathrm{op}}, \mathbf{Set}](\widehat{\pi}, X) \longrightarrow X(m)$$

for $m \in \mathbb{N}$ and $\pi \in P(m)$, satisfying axioms. Writing $h_\pi(\theta, -)$ as $\overline{\theta}$ and re-calling that a map $\widehat{\pi} \longrightarrow X$ of globular sets is a 'labelling of π by cells of X', we see that this is exactly the description above. For example, suppose that

$$\pi = \qquad \in \mathbf{pd}(2), \qquad\qquad (8.14)$$

that $\theta \in P(\pi)$, and that

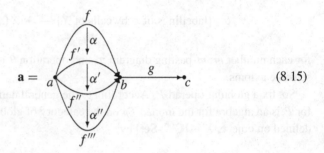

$$\mathbf{a} = \qquad (8.15)$$

is a diagram of cells in X: then $\bar{\theta}$ assigns to this diagram a 2-cell $\bar{\theta}(\mathbf{a})$ of X.

What are the axioms? First, $h : T_P X \longrightarrow X$ must be a map of globular sets, which says that $\overline{s(\theta)} = s \circ \bar{\theta}$ and $\overline{t(\theta)} = t \circ \bar{\theta}$. So in our example, $\bar{\theta}(\mathbf{a})$ is a 2-cell of the form

where

$$k = \overline{s(\theta)}(\underset{d}{\bullet} \xrightarrow{f} \underset{b}{\bullet} \xrightarrow{g} \underset{c}{\bullet}),$$
$$k' = \overline{t(\theta)}(\underset{d}{\bullet} \xrightarrow{f''} \underset{b}{\bullet} \xrightarrow{g} \underset{c}{\bullet}),$$
$$d = \overline{ss(\theta)}(\underset{a}{\bullet}),$$
$$e = \overline{tt(\theta)}(\underset{c}{\bullet}).$$

Second, $h : T_P X \longrightarrow X$ must obey the usual axioms for an algebra for a monad. These say that composition in the operad is interpreted in the model (algebra) as ordinary composition of functions, and identities similarly.

An example for composition: take 2-pasting diagrams

$$\pi_1 = \quad , \qquad \pi_2 = \quad ,$$

$$\pi = \quad$$

and write

$$\pi \circ (\pi_1, \pi_2) = \quad.$$

Let

$$\theta_1 \in P(\pi_1), \qquad \theta_2 \in P(\pi_2),$$
$$\theta \in P(\pi)$$

be operations of P satisfying $tt(\theta_1) = ss(\theta_2) \in P(\bullet)$. Let

$$\mathbf{a}_1 = \quad, \qquad \mathbf{a}_2 = \quad,$$

$$\mathbf{a} = \quad$$

be diagrams of cells in X (from which the 1-cell labels have been omitted). Then there is a composite operation $\theta \circ (\theta_1, \theta_2) \in P(\pi \circ (\pi_1, \pi_2))$, and the composition-compatibility axiom on the algebra X says that

$$\overline{\theta \circ (\theta_1, \theta_2)}(\mathbf{a}) = \overline{\theta}\left(\quad \overline{\theta_1}(\mathbf{a}_1) \quad \overline{\theta_2}(\mathbf{a}_2) \quad \right).$$

An example for identities: if α is a 2-cell of X then $\overline{1_2}(\alpha) = \alpha$. In general, the m-pasting diagram ι_m (defined on p. 274) satisfies $\widehat{\iota_m} \cong \mathbb{G}(-, m)$, and the identity axiom says that

$$\overline{1_m} : [\mathbb{G}^{\mathrm{op}}, \mathbf{Set}](\widehat{\iota_m}, X) \longrightarrow X(m)$$

is the canonical (Yoneda) isomorphism.

We meet a non-trivial example of a globular operad in the next chapter. Its algebras are, by definition, the weak ω-categories. As a trivial example for now, the terminal globular operad P is characterized by $P(\pi)$ having exactly one element for each pasting diagram π, and for the general reasons given in 4.3.8, a P-algebra is exactly a T-algebra, that is, a *strict* ω-category.

Notes

Globular operads were introduced by Batanin (1998a). He studied them in the wider context of 'monoidal globular categories', and considered 'operads in $\mathcal{C}$' for any monoidal globular category $\mathcal{C}$. There is a particular monoidal globular category *Span* such that operads in *Span* are exactly the globular operads of this chapter, which are the only kind of operads we need in order to define weak ω-categories.

The realization that Batanin's operads in *Span* are just T-operads, for T the free strict ω-category monad on globular sets, was first recorded in my paper of (1998a), and subsequently explained in more detail in my (2001a) and (2000b).

Chapter 9

A definition of weak n-category

Vico took it for granted that the first language of humanity was in the
form of hieroglyphics; that is, of metaphors and animated figures [...]
He had intimated war with just 'five real words': a frog, a mouse, a bird,
a ploughshare, and a bow.

Eco (1995)

Algebraic structures are often defined in a way that suggests conflict: gener-
ators vs. relations, operations vs. equations, composition vs. coherence. For
example, in the definition of bicategory one equips a 2-globular set first with
various composition operations, then with coherence isomorphisms to ensure
that some of the derived compositions are, in fact, essentially the same. One
imagines the two sides pulling against each other: more operations make the
structure bigger and wilder, more equations or coherence cells make it smaller
and more tame.

With this picture in mind, the most obvious way to go about defining weak
n-category is to set up a family of higher-dimensional composition operations
subject to a family of higher-dimensional coherence constraints. This is the
strategy in Batanin's and Penon's proposed definitions, both of which we discuss in Chapter 10. But it is not our strategy in this chapter.

In the definition proposed here, no distinction is made between composition
and coherence. They are seen as two aspects of a single idea, 'contraction', not
as opposing forces. This unified approach is in many ways more simple and
graceful: one idea instead of two.

Contractions are explained in Section 9.1. A map of globular sets may
have the property of being contractible, which viewed topologically means
something like being injective on homotopy; if so, it admits at least one contraction, which is something like a homotopy lifting. These definitions lead
to definitions of contractibility of, and contraction on, a globular operad. By

considering some low-dimensional situations, we see how contraction alone generates a natural theory of weak ω-categories.

In Section 9.2 weak ω-categories are defined formally as algebras for the initial globular operad equipped with a contraction. We look at some examples, including the fundamental ω-groupoid of a topological space.

The finite-dimensional case, weak n-categories, is a shade less easy than the infinite-dimensional case because we have to take care in the top dimension. In Section 9.3 we define weak n-categories and look at various ways of constructing weak n_1-categories from weak n_2-categories, for different (possibly infinite) values of n_1 and n_2. For instance, if a and b are 0-cells of an n-category X then there is a 'hom-$(n-1)$-category' $X(a, b)$.

The first test for a proposed definition of n-category is that it does something sensible when $n \leq 2$. We show in Section 9.4 that ours passes: weak 0-categories are sets, weak 1-categories are categories, and weak 2-categories are unbiased bicategories.

9.1 Contractions

Given the language of globular operads, we need one more concept in order to express our definition of weak ω-category: contractions. First we define contraction on a map of globular sets, then we define contraction on a globular operad, then we see what this has to do with the theory of weak ω-categories.

Definition 9.1.1 Let X be a globular set and let $m \in \mathbb{N}$. Two cells $\alpha^-, \alpha^+ \in X(m)$ are **parallel** if $m = 0$ or if $m \geq 1$, $s(\alpha^-) = s(\alpha^+)$, and $t(\alpha^-) = t(\alpha^+)$.

Definition 9.1.2 Let $q : X \longrightarrow Y$ be a map of globular sets. For $m \geq 1$ and $\phi \in Y(m)$, write (Fig. 9-A)

$$\mathrm{Par}_q(\phi) = \{(\theta^-, \theta^+) \in X(m-1) \times X(m-1) \mid \theta^- \text{ and } \theta^+ \text{ are parallel},$$
$$q(\theta^-) = s(\phi), \; q(\theta^+) = t(\phi)\}.$$

A **contraction** κ on q is a family of functions

$$\left(\mathrm{Par}_q(\phi) \xrightarrow{\;\kappa_\phi\;} X(m)\right)_{m \geq 1, \phi \in Y(m)}$$

such that for all $m \geq 1$, $\phi \in Y(m)$, and $(\theta^-, \theta^+) \in \mathrm{Par}_q(\phi)$,

$$s(\kappa_\phi(\theta^-, \theta^+)) = \theta^-, \quad t(\kappa_\phi(\theta^-, \theta^+)) = \theta^+, \quad q(\kappa_\phi(\theta^-, \theta^+)) = \phi.$$

So a map admits a contraction when it is 'injective on homotopy' in some oriented sense.

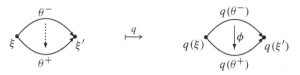

Fig. 9-A. Effect of a contraction κ, shown for $m = 2$. The dotted arrow is $\kappa_\phi(\theta^-, \theta^+)$

Definition 9.1.3 A **contraction** on a collection $(P \xrightarrow{d} T1)$ is a contraction on the map d. A **contraction** on a globular operad is a contraction on its underlying collection. A map, collection or operad is **contractible** if it admits a contraction.

Explicitly, if π is a pasting diagram then $\mathrm{Par}_d(\pi)$ is the set of parallel pairs (θ^-, θ^+) of elements of $P(\partial\pi)$. A contraction assigns to each pasting diagram π and parallel $\theta^-, \theta^+ \in P(\partial\pi)$ an element of $P(\pi)$ with source θ^- and target θ^+. We usually write $\mathrm{Par}_P(\pi)$ instead of $\mathrm{Par}_d(\pi)$; then a contraction κ on P consists of a function

$$\kappa_\pi : \mathrm{Par}_P(\pi) \longrightarrow P(\pi)$$

for each pasting diagram π, satisfying the source and target axioms.

To understand what contractions have to do with weak ω-categories, let us forget these definitions for a while and ask: what should the operad for weak ω-categories be? In other words, how should we pick a globular operad L in such a way that L-algebras might reasonably be called weak ω-categories? The elements of $L(\pi)$ are all the possible ways of composing a labelled diagram of shape π in an arbitrary weak ω-category, so the question is: what such ways should there be?

There are many sensible answers. Even for weak 2-categories there is an infinite family of definitions, all equivalent (Chapter 3). We saw, for instance, that we could choose to start with 100 different specified ways of composing diagrams of shape

just as long as we made them all coherently isomorphic. This particular choice seems bizarre, and in choosing our theory L we try to make it in some sense canonical.

So let us decide what operations to put into the theory of weak ω-categories, starting at the bottom dimension and working our way up.

There should not be any operations in dimension 0 except for the identity: the only way of obtaining a 0-cell in a weak ω-category is to start with that

same 0-cell and do nothing. So if • denotes the unique 0-pasting diagram then
we want $L(\bullet) = 1$.

In dimension 1, let us be unbiased and specify for each $k \in \mathbb{N}$ a single way
of composing a string

of k 1-cells. These specified compositions can be built up to make more complex ones: if we denote the specified k-fold composition of 1-cells in a weak
ω-category by

$$a_0 \xrightarrow{f_1} \quad \cdots \quad \xrightarrow{f_k} a_k \quad \longmapsto \quad a_0 \xrightarrow{(f_k \circ \cdots \circ f_1)} a_k$$

(as we did for unbiased bicategories) then an example of a built-up operation
is

$$a_0 \xrightarrow{f_1} a_1 \xrightarrow{f_2} a_2 \xrightarrow{f_3} a_3 \xrightarrow{f_4} a_4 \quad \longmapsto \quad a_0 \xrightarrow{((f_4 \circ f_3 \circ f_2) \circ f_1)} a_4 .$$

There should be no other ways of obtaining a 1-cell in a weak ω-category. So
if χ_k is the 1-pasting diagram made up of k 1-cells then we want $L(\chi_k)$ to be
the set $\mathbf{tr}(k)$ of k-leafed trees (2.3.3).

What operations in a weak ω-category result in a 2-cell? First, ordinary
composition makes a 2-cell from diagrams of shapes such as

$$\rho = \quad , \quad \sigma = \quad .$$

Choosing to be unbiased again, we specify an operation $\theta \in L(\rho)$ such that
$s(\theta) = t(\theta) = \mathrm{id}$. When acting in a weak ω-category, θ takes as input a diagram of 2-cells α_i and produces as output a single 2-cell as shown:

$$a \quad \Downarrow\alpha \Downarrow\beta \Downarrow\gamma \quad b \quad \longmapsto \quad a \quad \Downarrow \quad b . \tag{9.1}$$

Similarly, we specify one operation of arity σ whose source and target are
both the specified element of $L(\chi_2)$ (which is ordinary binary composition of
1-cells):

$$a \quad \Downarrow\alpha \quad b \quad \Downarrow\alpha' \quad c \quad \longmapsto \quad a \quad \Downarrow \quad c . \tag{9.2}$$

Second, there are coherence 2-cells. For instance, a string of three 1-cells gives rise to an associativity 2-cell:

$$\bullet \xrightarrow{f_1} \bullet \xrightarrow{f_2} \bullet \xrightarrow{f_3} \bullet \qquad \longmapsto \qquad \overset{((f_3 \circ f_2) \circ f_1)}{\underset{(f_3 \circ (f_2 \circ f_1))}{\Longleftrightarrow}} \qquad (9.3)$$

and similarly in more complex cases:

$$\bullet \xrightarrow{f_1} \bullet \xrightarrow{f_2} \bullet \xrightarrow{f_3} \bullet \xrightarrow{f_4} \bullet \qquad \longmapsto \qquad \overset{(1 \circ (f_4 \circ f_3 \circ f_2) \circ 1 \circ f_1)}{\underset{(f_4 \circ (f_3 \circ (f_2 \circ 1 \circ f_1)))}{\Longleftrightarrow}}. \qquad (9.4)$$

Being unbiased once more, we *specify* operations giving 2-cells in each of these two ways; we do not, for instance, insist that the coherence cell on the right-hand side of (9.4) should be equal to some composite of associativity and identity coherence cells.

Here comes the crucial point. It looks as if these two kinds of operations resulting in 2-cells, composition and coherence, are quite different – complementary, even. But they can actually be regarded as two instances of the same thought, and that makes matters much simpler.

First recall from Section 8.1 that any 1-dimensional picture of a pasting diagram can also be taken to represent a (degenerate) element of $\mathbf{pd}(2)$. In particular, the left-hand sides of (9.3) and (9.4) can be regarded as degenerate elements of $\mathbf{pd}(2)$: they are 1_{χ_3} and 1_{χ_4} respectively (see (8.4)). Thus, each of (9.1)–(9.4) portrays an element of $L(\pi)$ for some $\pi \in \mathbf{pd}(2)$.

Now note that four times over, we have taken a 2-pasting diagram π and elements $\theta^-, \theta^+ \in L(\partial\pi)$ and decreed that $L(\pi)$ should contain a specified element θ with source θ^- and target θ^+. In the first two cases this is obvious; in the third (9.3) we took $\pi = 1_{\chi_3}$ (hence $\partial\pi = \chi_3$) and the evident two elements $\theta^-, \theta^+ \in L(\chi_3)$; the last is similar.

Which pairs (θ^-, θ^+) of 1-dimensional operations should we use to generate the 2-dimensional operations? The simplest possible answer, and the properly unbiased one, is 'all of them'. So the principle is:

Let $\pi \in \mathbf{pd}(2)$ and $\theta^-, \theta^+ \in L(\partial\pi)$. Then there is a specified element $\theta \in L(\pi)$ satisfying $s(\theta) = \theta^-$ and $t(\theta) = \theta^+$.

'Specified' means that we take these operations θ as primitive and generate the new 2-dimensional operations in L freely using its operadic structure, just as one dimension down we took the k-fold composition operations ($k \in \mathbb{N}$) as primitive and generated from them derived 1-dimensional operations, indexed by trees. In dimension 2 the combinatorial situation is more difficult, and I will not attempt an explicit description of $L(\pi)$ for $\pi \in \mathbf{pd}(2)$. A categorical description takes its place; in the next section, L is defined as the universal operad containing elements specified in this way.

Because our principle takes in both composition and coherence at once, it produces operations traditionally regarded as a hybrid of the two. For instance, there is a specified operation of the form

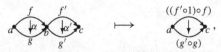

where traditionally operations such as this would be built up from horizontal composition (9.2) and coherence cells. The spirit of our definition is that composition and coherence are *not* separate entities: they are two sides of the same coin.

What we have said for dimension 2 applies equally in all dimensions, with just one small refinement: there can only be a θ satisfying $s(\theta) = \theta^-$ and $t(\theta) = \theta^+$ if θ^- and θ^+ are parallel. (This is trivial in dimension 2 because $L(0) = 1$.) So the general principle is:

> Let $n \in \mathbb{N}$, let $\pi \in \mathbf{pd}(n)$, and let θ^-, θ^+ be parallel elements of $L(\partial\pi)$. Then there is a specified element $\theta \in L(\pi)$ satisfying $s(\theta) = \theta^-$ and $t(\theta) = \theta^+$.

In other words:

> L is equipped with a contraction.

Everything that we did in choosing L is encapsulated in this statement. We started from the identity element of $L(0)$, part of the operad structure of L. Then we chose to specify k-fold compositions of 1-cells, which amounted to applying the contraction with $\pi = \chi_k$. Then the operad structure of L gave derived 1-dimensional operations such as $(f_1, f_2, f_3) \longmapsto ((f_3 \circ f_2) \circ f_1)$. Then we chose to specify one operation of arity π with any given source and target, for each 2-pasting diagram π. Then the operad structure of L gave derived 2-dimensional operations. All in all, we chose L to be the universal operad equipped with a contraction.

9.2 Weak ω-categories

We have decided that the operad for weak ω-categories ought to come with a specified contraction, and that it ought to be 'universal', 'minimal', or 'freely generated' as such. Precisely, it ought to be initial in the category of operads equipped with a contraction.

Definition 9.2.1 The category **OC** of (globular) **operads-with-contraction** has as objects all pairs (P, κ) where P is a globular operad and κ a contraction on P, and as maps $(P, \kappa) \longrightarrow (P', \kappa')$ all operad maps $f : P \longrightarrow P'$ preserving contractions:

$$f(\kappa_\pi(\theta^-, \theta^+)) = \kappa'_\pi(f\theta^-, f\theta^+)$$

whenever $m \geq 1$, $\pi \in \mathbf{pd}(m)$, and $(\theta^-, \theta^+) \in \mathrm{Par}_P(\pi)$.

Proposition 9.2.2 *The category* **OC** *has an initial object.*

Proof Appendix G. □

We write (L, λ) for the initial object of **OC**. In the previous section we described L explicitly in low dimensions and saw informally how to construct it in higher dimensions: if L is known up to and including dimension $n - 1$, then $L(n)$ is obtained by first closing under contraction then closing under n-dimensional operadic composition and identities.

Definition 9.2.3 A **weak ω-category** is an L-algebra.

We write **Wk-ω-Cat** for **Alg**(L). Since the algebras construction is functorial (p. 156), this category is determined uniquely up to isomorphism.

Observe that the maps in **Wk-ω-Cat** preserve the operations from L – that is, the weak ω-category structure – *strictly*. In this text we do not go as far as a definition of *weak ω-functor*, nor do we reach a definition of (weak) equivalence of weak ω-categories. These are serious omissions, and the reader may feel cheated that we are stopping when we have barely begun, but that is the state of the art.

Definition 9.2.3 is just one of many proposed definitions of weak ω-category. In my (2001b), it is 'Definition **L1**'. Its place among other definitions is discussed in Chapter 10.

Example 9.2.4 If P is a contractible operad then any contraction κ on P gives rise to a unique map $f : (L, \lambda) \longrightarrow (P, \kappa)$ in **OC**, whose underlying map $f : L \longrightarrow P$ of operads induces a functor

$$\mathbf{Alg}(P) \longrightarrow \mathbf{Wk\text{-}\omega\text{-}Cat}. \tag{9.5}$$

In other words, any algebra for an operad-with-contraction is canonically a weak ω-category. All the examples below of weak ω-categories are constructed in this way.

Functor (9.5) is always faithful (being the identity on underlying globular sets) and is full if for for each pasting diagram π, the function $f_\pi : L(\pi) \longrightarrow P(\pi)$ is surjective (exercise).

Example 9.2.5 The algebras for the terminal globular operad 1 are the strict ω-categories (by 4.3.8), so the unique map $L \longrightarrow 1$ induces a functor

$$\textbf{Str-}\omega\textbf{-Cat} \longrightarrow \textbf{Wk-}\omega\textbf{-Cat} \qquad (9.6)$$

– a strict ω-category is a special weak ω-category.

The operad 1 admits a unique contraction, and thus becomes the *terminal* object of **OC**. Contractibility of L implies that $L(\pi)$ is non-empty for each pasting diagram π, so by the observations of the previous example, functor (9.6) is full and faithful. We would expect this: if C and D are strict ω-categories then there ought to be a single notion of 'strict map $C \longrightarrow D$', independent of whether C and D are regarded as strict or as weak ω-categories.

Example 9.2.6 A directed graph is **indiscrete** if for all objects x and y, there is exactly one edge from x to y. Such a graph has a unique category structure. The ω-categorical analogue of this observation is that any **contractible** globular set X – one for which the unique map $X \longrightarrow 1$ is contractible – has a weak ω-category structure. To see the analogy, note that contractibility of a globular set says that any two parallel n-cells x and y have at least one $(n + 1)$-cell $f : x \longrightarrow y$ between them, and indiscreteness of a graph says that any two objects x and y have a map $f : x \longrightarrow y$ between them and any two parallel arrows have an equality between them.

A contractible globular set X acquires the structure of a weak ω-category as follows. Recall from Section 6.4 that there is an endomorphism operad $\textbf{End}(X)$, and that if P is any operad then a P-algebra structure on X is just an operad map $P \longrightarrow \textbf{End}(X)$. In particular, X is canonically an $\textbf{End}(X)$-algebra. So, by 9.2.4, it is enough to prove that contractibility of the globular set X implies contractibility of the operad $\textbf{End}(X)$.

It follows from the definition of **End** that for any globular set X and pasting diagram $\pi \in \textbf{pd}(n)$, an element of $(\textbf{End}(X))(\pi)$ is a sequence of functions

$$(f_n, f_{n-1}^-, f_{n-1}^+, f_{n-2}^-, f_{n-2}^+, \ldots, f_0^-, f_0^+)$$

making the diagram

$$
\begin{array}{ccccccccc}
(TX)(\pi) & \overset{s}{\underset{t}{\rightrightarrows}} & (TX)(\partial\pi) & \overset{s}{\underset{t}{\rightrightarrows}} & (TX)(\partial^2\pi) & \overset{s}{\underset{}{\rightrightarrows}} & \cdots & \overset{s}{\underset{t}{\rightrightarrows}} & (TX)(\partial^n\pi) \\
\downarrow{\scriptstyle f_n} & & \downarrow{\scriptstyle f_{n-1}^-}\;\downarrow{\scriptstyle f_{n-1}^+} & & \downarrow{\scriptstyle f_{n-2}^-}\;\downarrow{\scriptstyle f_{n-2}^+} & & & & \downarrow{\scriptstyle f_0^-}\;\downarrow{\scriptstyle f_0^+} \\
X(n) & \overset{s}{\underset{t}{\rightrightarrows}} & X(n-1) & \overset{s}{\underset{t}{\rightrightarrows}} & X(n-2) & \overset{s}{\underset{t}{\rightrightarrows}} & \cdots & \overset{s}{\underset{t}{\rightrightarrows}} & X(0)
\end{array}
$$

$$(9.7)$$

commute serially – that is,

$$s \circ f_i^- = f_{i-1}^- \circ s, \qquad t \circ f_i^+ = f_{i-1}^+ \circ t$$

for all $i \in \{1, \ldots, n\}$, interpreting both f_n^- and f_n^+ as f_n. (Compare Batanin, 1998a, Prop. 7.2.) Contractibility of $\mathbf{End}(X)$ says that given all of such a serially commutative diagram except for f_n, there exists a function f_n completing it. This holds if X is contractible.

Different contractions on X induce different contractions on $\mathbf{End}(X)$, hence different weak ω-category structures on X. They should all be 'equivalent', and if X is non-empty then the weak ω-category X should be equivalent to 1, but we do not attempt to make this precise.

Our final example of a weak ω-category is one of the principal motivations for the subject.

Example 9.2.7 Any topological space S gives rise to a weak ω-category $\Pi_\omega S$, its **fundamental ω-groupoid**. Indeed, there is a product-preserving functor

$$\Pi_\omega : \mathbf{Top} \longrightarrow \mathbf{Wk}\text{-}\omega\text{-}\mathbf{Cat}.$$

To show this, we first establish the relationship between spaces and globular sets. As in 1.4.6, the Euclidean discs D^n define a functor $\mathbb{G} \longrightarrow \mathbf{Top}$, and by the mechanism described on p. 244, this induces a pair of adjoint functors

$$\mathbf{Top} \overset{\Pi_\omega}{\underset{|\cdot|}{\overset{\top}{\rightleftarrows}}} [\mathbb{G}^{\mathrm{op}}, \mathbf{Set}].$$

The right adjoint is given on a space S by

$$(\Pi_\omega S)(n) = \mathbf{Top}(D^n, S),$$

so $\Pi_\omega S$ is just like the singular simplicial set of S, but with discs in place of simplices. The left adjoint is **geometric realization**, given on a globular set X by the coend formula

$$|X| = \int^{n \in \mathbb{G}} X(n) \times D^n.$$

For example, if π is an n-pasting diagram then $|\hat{\pi}|$ is the space resembling the usual picture of π; it is a finite CW-complex made up of cells of dimension at most n, and is contractible, as can be proved by induction.

To give the globular set $\Pi_\omega S$ the structure of a weak ω-category we define a contractible operad P (independently of S) and show that $\Pi_\omega S$ is naturally a P-algebra. By 9.2.4, this suffices.

Our operad P is analogous to the universal operad for iterated loop spaces (2.2.15). If π is an n-pasting diagram then an element of $P(\pi)$ is a map from D^n to $|\hat{\pi}|$ respecting the boundaries; for instance, if π is the 1-pasting diagram χ_k consisting of k arrows in a row then an element of $P(\pi)$ is an endpoint-preserving map $[0, 1] \longrightarrow [0, k]$. In general, an element of $P(\pi)$ is a sequence of maps

$$\theta = (\theta_n, \theta_{n-1}^-, \theta_{n-1}^+, \theta_{n-2}^-, \theta_{n-2}^+, \ldots, \theta_0^-, \theta_0^+)$$

making the diagram

commute serially (as in (9.7)). The maps along the top are induced by the source and target inclusions $\widehat{\partial^{i+1}\pi} \rightrightarrows \widehat{\partial^i \pi}$ described on p. 270; they are all injective, so θ_n determines the whole of θ. The obvious restriction maps $P(\pi) \rightrightarrows P(\partial\pi)$ make P into a collection.

We have to show that the collection P is contractible, has the structure of an operad, and acts on the globular set $\Pi_\omega S$ for any space S. Contractibility amounts to the condition that if π is an n-pasting diagram then any continuous map from the unit $(n-1)$-sphere into $|\hat{\pi}|$ extends to the whole unit n-ball D^n, which holds because $|\hat{\pi}|$ is contractible. An action of P on $\Pi_\omega S$ consists of a function

$$\bar{\theta} : [\mathbb{G}^{\mathrm{op}}, \mathbf{Set}](\hat{\pi}, \Pi_\omega S) \longrightarrow (\Pi_\omega S)(n)$$

for each $n \in \mathbb{N}$, $\pi \in \mathbf{pd}(n)$, and $\theta \in P(\pi)$, satisfying axioms involving the operad structure of P (yet to be defined). By the adjunction above, such a function can equally be written as

$$\bar{\theta} : \mathbf{Top}(|\hat{\pi}|, S) \longrightarrow \mathbf{Top}(D^n, S),$$

and we take $\bar{\theta}$ to be composition with θ_n.

All that remains is to endow the collection P with the structure of an operad and check some axioms. As suggested by the similarity between the diagrams in this example and the last, P can be constructed as an endomorphism operad. By the Yoneda Lemma, an element of $P(\pi)$ is a sequence of natural transformations

$$(\alpha_n, \alpha_{n-1}^-, \alpha_{n-1}^+, \ldots, \alpha_0^-, \alpha_0^+)$$

making the diagram

$$\mathbf{Top}(|\widehat{\pi}|, -) \rightrightarrows \mathbf{Top}(|\widehat{\partial\pi}|, -) \rightrightarrows \cdots \rightrightarrows \mathbf{Top}(|\widehat{\partial^n\pi}|, -)$$

$$\downarrow{\alpha_n} \qquad \alpha_{n-1}^- \Big\downarrow\Big\downarrow\alpha_{n-1}^+ \qquad \qquad \alpha_0^- \Big\downarrow\Big\downarrow\alpha_0^+$$

$$\mathbf{Top}(D^n, -) \rightrightarrows \mathbf{Top}(D^{n-1}, -) \rightrightarrows \cdots \rightrightarrows \mathbf{Top}(D^0, -)$$

commute serially. But for any $m \in \mathbb{N}$ and $\rho \in \mathbf{pd}(m)$ there are isomorphisms

$$\mathbf{Top}(|\widehat{\rho}|, S) \cong [\mathbb{G}^{\mathrm{op}}, \mathbf{Set}](\widehat{\rho}, \Pi_\omega S) \cong (T(\Pi_\omega S))(\rho)$$

and

$$\mathbf{Top}(D^m, S) \cong (\Pi_\omega S)(m),$$

both natural in $S \in \mathbf{Top}$, so an element of $P(\pi)$ is a sequence of families of functions

$$\left((f_{S,n})_{S\in\mathbf{Top}}, (f_{S,n-1}^-)_{S\in\mathbf{Top}}, (f_{S,n-1}^+)_{S\in\mathbf{Top}}, \ldots, (f_{S,0}^-)_{S\in\mathbf{Top}}, (f_{S,0}^+)_{S\in\mathbf{Top}}\right) \tag{9.8}$$

natural in S and making the diagram

$$(T(\Pi_\omega S))(\pi) \overset{s}{\underset{t}{\rightrightarrows}} (T(\Pi_\omega S))(\partial\pi) \overset{s}{\underset{t}{\rightrightarrows}} \cdots \overset{s}{\underset{t}{\rightrightarrows}} (T(\Pi_\omega S))(\partial^n\pi)$$

$$\downarrow{f_{S,n}} \qquad f_{S,n-1}^- \Big\downarrow\Big\downarrow f_{S,n-1}^+ \qquad\qquad f_{S,0}^- \Big\downarrow\Big\downarrow f_{S,0}^+$$

$$(\Pi_\omega S)(n) \overset{s}{\underset{t}{\rightrightarrows}} (\Pi_\omega S)(n-1) \overset{s}{\underset{t}{\rightrightarrows}} \cdots \overset{s}{\underset{t}{\rightrightarrows}} (\Pi_\omega S)(0)$$

commute serially for each $S \in \mathbf{Top}$. This now looks like (9.7), an operation in an endomorphism operad. Indeed, composition with T induces a cartesian monad T_* on the category

$$[\mathbf{Top}, [\mathbb{G}^{\mathrm{op}}, \mathbf{Set}]],$$

and if $R \in \mathbf{Top}$ then evaluation at R induces a strict map of monads

$$\mathrm{ev}_R : ([\mathbf{Top}, [\mathbb{G}^{\mathrm{op}}, \mathbf{Set}]], T_*) \longrightarrow ([\mathbb{G}^{\mathrm{op}}, \mathbf{Set}], T)$$

hence a functor

$$(\mathrm{ev}_R)_* : T_*\text{-}\mathbf{Operad} \longrightarrow T\text{-}\mathbf{Operad}.$$

We therefore have a T_*-operad $\mathbf{End}(\Pi_\omega)$ and a globular operad

$$P' = (\mathrm{ev}_\emptyset)_*(\mathbf{End}(\Pi_\omega)),$$

and a few calculations reveal that an element of $P'(\pi)$ consists of data as in (9.8). So $P'(\pi) \cong P(\pi)$, giving P the structure of an operad. Further checks show that the operad structure is compatible with the action on $\Pi_\omega S$ described above.

9.3 Weak n-categories

We now imitate what we did for ω-categories to obtain a definition of weak n-category. This is straightforward except for one subtlety. We then explore the relationship between weak n-categories for different values of n, and in particular we see how a weak n-category can be regarded as a weak ω-category trivial above dimension n.

First recall from Section 1.4 that an n-globular set is a presheaf on the category $\mathbb{G}_n$ and that there is a forgetful functor from **Str-n-Cat** to $[\mathbb{G}_n^{op}, \textbf{Set}]$. Theorem F.2.1 implies that this induces a cartesian monad $T_{(n)}$ on $[\mathbb{G}_n^{op}, \textbf{Set}]$. A $T_{(n)}$-graph whose object-of-objects is 1 will be called an n-**collection**, a $T_{(n)}$-operad will be called an n-**globular operad** or simply an n-**operad**, and the category of n-operads will be written n-**Operad**.

The subtlety concerns operations in the top dimension. In a bicategory, for example, there are various ways of composing 2-cell diagrams of shape

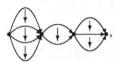

but each such way is uniquely determined once one has chosen how to compose the string of three 1-cells along the top and the string of three 1-cells along the bottom. This is essentially the 'all diagrams commute' coherence theorem. In creating the theory of weak ω-categories we never declared any two operations to be equal: the contraction on the operad L deferred the relation to the next dimension. In the finite-dimensional case we cannot always defer to the next dimension, and we need a notion of contraction sensitive to that.

Definition 9.3.1 Let $n \in \mathbb{N}$. A map $q : X \longrightarrow Y$ of n-globular sets is **tame** if any two parallel n-cells α^-, α^+ of X satisfying $q(\alpha^-) = q(\alpha^+)$ are equal.

Definition 9.3.2 Let $n \in \mathbb{N}$. A **precontraction** on a map $q : X \longrightarrow Y$ of n-globular sets is a family of functions

$$\left(\text{Par}_q(\phi) \xrightarrow{\kappa_\phi} X(m) \right)_{1 \leq m \leq n, \phi \in Y(m)}$$

satisfying the same conditions as those in Definition 9.1.2. A **contraction** is a precontraction on a tame map.

Tameness means injectivity in the top dimension (relative to sources and targets), so contractibility continues to mean something like 'injectivity on homotopy', as in the infinite-dimensional case.

Definition 9.3.3 An n-collection $(P \xrightarrow{d} T_{(n)}1)$ is **tame** if the map d is tame, and a **(pre)contraction** on the n-collection is a (pre)contraction on the map d.

Tameness of an n-collection P means that for all $\pi \in \mathbf{pd}(n)$, parallel elements of $P(\pi)$ are equal, or equivalently the function

$$(s, t) : P(\pi) \longrightarrow \mathrm{Par}_P(\pi)$$

is injective. If P is precontractible then (s, t) is surjective, with one-sided inverse κ_π. Hence a precontraction is a contraction if and only if (s, t) is bijective for all $\pi \in \mathbf{pd}(n)$. This implies that a contractible n-operad is entirely determined by its lower $(n-1)$-dimensional part; if $\pi \in \mathbf{pd}(n)$ then $P(\pi)$ must be the set of parallel pairs of elements of $P(\partial\pi)$.

Example 9.3.4 A **sesquicategory** is a category X together with a functor $\mathrm{HOM} : X^{\mathrm{op}} \times X \longrightarrow \mathbf{Cat}$ such that

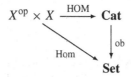

commutes. The objects of X are called the **0-cells** of the sesquicategory, and if a and b are 0-cells then the objects and arrows of the category $\mathrm{HOM}(a, b)$ are called **1-cells** and **2-cells** respectively. Any strict 2-category has an underlying sesquicategory.

Concretely, a sesquicategory consists of a 2-globular set together with structure and axioms saying that any diagram of shape

has a unique composite 1-cell and any diagram of shape

has a unique composite 2-cell. (The sequences of cells shown can have any length, including zero.) See Leinster (2001a, III.1) or Street (1996, §2) for a more detailed presentation. What a sesquicategory does not have is a canonical horizontal composition of 2-cells: given a diagram

$$\tag{9.9}$$

the only resulting 2-cells of the form

are the derived composites $(\alpha' g) \circ (f' \alpha)$ and $(g' \alpha) \circ (\alpha' f)$, which in general are not equal.

There is a 2-operad **Ssq** whose algebras are sesquicategories. If π is a 0- or 1-pasting diagram then $\mathbf{Ssq}(\pi) = 1$. A 2-pasting diagram is a finite sequence $(k_1, \ldots, k_n)$ of natural numbers, as observed in Section 8.1; pictorially, n is the width of the diagram and k_i the height of the ith column. Writing $\mathbf{m} = \{1, \ldots, m\}$, an element of $\mathbf{Ssq}(k_1, \ldots, k_n)$ is a total order on the disjoint union $\mathbf{k_1} + \cdots + \mathbf{k_n}$ that restricts to the standard order on each $\mathbf{k_i}$; hence

$$|\mathbf{Ssq}(k_1, \ldots, k_n)| = \frac{(k_1 + \cdots + k_n)!}{k_1! k_2! \cdots k_n!}.$$

A precontraction on **Ssq** amounts to a choice of element of $\mathbf{Ssq}(k_1, \ldots, k_n)$ for each $n, k_1, \ldots, k_n \in \mathbb{N}$, and since this set always has at least one element, **Ssq** is precontractible. But it is not contractible, or equivalently not tame: for since all elements of $\mathbf{Ssq}(k_1, \ldots, k_n)$ are parallel, this would say that $|\mathbf{Ssq}(k_1, \ldots, k_n)| = 1$ for all $n, k_1, \ldots, k_n$, which is false.

The categories n-**OP** of n-operads-with-precontraction and n-**OC** of n-operads-with-contraction are defined analogously to **OC** (9.2.1).

Proposition 9.3.5 *For each $n \in \mathbb{N}$, the category n-**OC** has an initial object.*

We write (L_n, λ_n) for the initial object. The proof comes later (9.3.10).

Definition 9.3.6 Let $n \in \mathbb{N}$. A **weak n-category** is an L_n-algebra.

We write **Wk-n-Cat** for $\mathbf{Alg}(L_n)$, the category of weak n-categories and strict n-functors. This contains **Str-n-Cat** as a full subcategory, just as in the infinite-dimensional case (9.2.5).

We now embark on a comparison of the theories of n-categories for varying values of n, including $n = \omega$. Among other things we prove Proposition 9.3.5, the strategy for which is as follows. Imagine L_n being built from the bottom dimension upwards, just as we did for L: then it is clear that L_n and L should be the same up to and including dimension $n - 1$. They are different in dimension n, as discussed above in the case $n = 2$: if π is an n-pasting diagram then an element of $L_n(\pi)$ is a parallel pair of elements of $L_n(\partial\pi)$. So we define an n-operad from L by discarding all the operations of dimension n and higher, then adding in parallel pairs as the operations in dimension n, and we prove that this is the initial n-operad-with-contraction.

Let $m, M \in \mathbb{N} \cup \{\omega\}$ with $m \leq M$. We compare m- and M-dimensional structures, using the following natural conventions for the ω-dimensional case:

$$\mathbb{G}_\omega = \mathbb{G}, \quad T_{(\omega)} = T, \quad \omega\text{-}\mathbf{Operad} = T\text{-}\mathbf{Operad}, \quad \omega\text{-}\mathbf{OP} = \omega\text{-}\mathbf{OC} = \mathbf{OC}.$$

We start with globular sets, and the adjunction

$$[\mathbb{G}_M^{\mathrm{op}}, \mathbf{Set}]$$
$$R_m^M \Big\downarrow \dashv I_m^M$$
$$[\mathbb{G}_m^{\mathrm{op}}, \mathbf{Set}].$$

Here R_M^m is **restriction**: if X is an M-globular set then $(R_m^M X)(k) = X(k)$ for all $k \leq m$. Its right adjoint I_m^M forms the **indiscrete** M-globular set on an m-globular set Y,

$$\cdots \overset{1}{\underset{1}{\rightrightarrows}} S \overset{1}{\underset{1}{\rightrightarrows}} S \overset{\mathrm{pr}_1}{\underset{\mathrm{pr}_2}{\rightrightarrows}} Y(m) \overset{s}{\underset{t}{\rightrightarrows}} Y(m-1) \overset{s}{\underset{t}{\rightrightarrows}} \cdots \overset{s}{\underset{t}{\rightrightarrows}} Y(0)$$

where S is the set of parallel pairs of elements of $Y(m)$. This adjunction is familiar in the case $M = 1$, $m = 0$: R_0^1 assigns to a directed graph its set of objects (vertices), and I_0^1 forms the indiscrete graph on a set (9.2.6). Formally, R_m^M is composition with, and I_m^M right Kan extension along, the obvious inclusion $\mathbb{G}_m \hookrightarrow \mathbb{G}_M$.

Next we move to the level of operads. The functor R_m^M is naturally a weak map of monads

$$([\mathbb{G}_M^{\mathrm{op}}, \mathbf{Set}], T_{(M)}) \longrightarrow ([\mathbb{G}_m^{\mathrm{op}}, \mathbf{Set}], T_{(m)})$$

(as shown in Appendix F), and preserves limits. Hence R_m^M is naturally a cartesian colax map of monads, and by Example 6.7.3, there is an induced adjunction

$$M\text{-}\mathbf{Operad}$$

$$(R_m^M)_* \dashv (I_m^M)_*$$

$$m\text{-}\mathbf{Operad}.$$

The functor $(R_m^M)_*$ forgets the top $(M - m)$ dimensions of an M-operad, and $(I_m^M)_*$ is defined on an m-operad Q at a k-pasting diagram π by

$$\left((I_m^M)_* Q\right)(\pi) = \begin{cases} \mathrm{Par}_Q(\partial^{k-m-1}\pi) & \text{if } k > m \\ Q(\pi) & \text{if } k \leq m. \end{cases}$$

Now we bring in precontractions. A precontraction on an M-operad P restricts to a precontraction on the m-operad $(R_m^M)_* P$, and similarly for $(I_m^M)_*$, so the previous adjunction lifts to an adjunction

$$M\text{-}\mathbf{OP}$$

$$(R_m^M)_* \dashv (I_m^M)_*$$

$$m\text{-}\mathbf{OP}.$$

So far, we have proved:

Proposition 9.3.7 *For any $m, M \in \mathbb{N} \cup \{\omega\}$ with $m \leq M$, there are adjunctions*

$[\mathbb{G}_M^{\mathrm{op}}, \mathbf{Set}]$	$M\text{-}\mathbf{Operad}$	$M\text{-}\mathbf{OP}$
$R_m^M \dashv I_m^M$	$(R_m^M)_* \dashv (I_m^M)_*$	$(R_m^M)_* \dashv (I_m^M)_*$
$[\mathbb{G}_m^{\mathrm{op}}, \mathbf{Set}]$	$m\text{-}\mathbf{Operad}$	$m\text{-}\mathbf{OP}$

defined by restriction R_m^M and indiscrete extension I_m^M. □

Finally, we consider contractions themselves. We have seen that an n-operad-with-contraction is entirely determined by its underlying $(n-1)$-operad-with-precontraction. In fact, the two types of structure are the same:

Proposition 9.3.8 *For any positive integer n, the third adjunction of Proposition 9.3.7 in the case $M = n$, $m = n - 1$ restricts to an equivalence*

$$n\text{-}\mathbf{OC}$$

$$(R_{n-1}^n)_* \simeq (I_{n-1}^n)_*$$

$$(n-1)\text{-}\mathbf{OP}.$$

Proof Any adjunction $\mathcal{C} \underset{\longleftarrow}{\overset{\longrightarrow}{\perp}} \mathcal{D}$ restricts to an equivalence between the full subcategory of $\mathcal{C}$ consisting of those objects at which the unit of the adjunction is an isomorphism and the full subcategory of $\mathcal{D}$ defined dually. In the case at hand we have $(R^n_{n-1})_* \circ (I^n_{n-1})_* = 1$, and the counit is the identity transformation. Given an n-operad P with precontraction κ, the unit map

$$P \longrightarrow (I^n_{n-1})_*(R^n_{n-1})_* P$$

is the identity in dimensions less than n, and in dimension n is made up of the functions

$$(s, t) : P(\pi) \longrightarrow \mathrm{Par}_P(\pi)$$

($\pi \in \mathbf{pd}(n)$). So P belongs to the relevant subcategory of n-**OP** if and only if this map is a bijection for all $\pi \in \mathbf{pd}(n)$, that is, κ is a contraction. $\qquad\square$

Example 9.3.9 Let **Ssq** be the 2-operad for sesquicategories (9.3.4). The resulting 3-operad $\mathbf{Gy} = (I^3_2)_*(\mathbf{Ssq})$ is given on k-pasting diagrams π by

$$\mathbf{Gy}(\pi) = \begin{cases} \mathbf{Ssq}(\partial\pi) \times \mathbf{Ssq}(\partial\pi) & \text{if } k = 3, \\ \mathbf{Ssq}(\pi) & \text{if } k \leq 2. \end{cases}$$

A **Gy**-algebra is what we will call a **Gray-category**. It consists of a 3-globular set X with a sesquicategory structure on its 0-, 1- and 2-cells and further structure in the top dimension: given a 3-pasting diagram labelled by cells of X, the ways of composing it to yield a single 3-cell correspond one-to-one with the ways of composing both the 2-dimensional diagram at its source and the 2-dimensional diagram at its target.

Let us explore the 3-dimensional operations. By considering 3-pasting diagrams π for which $\partial^2\pi$ is a single 1-cell, we see that for any two 0-cells a and b, the 2-globular set $X(a, b)$ has the structure of a strict 2-category. More subtly, take the 2-pasting diagram

$$\rho = \langle\!\!\!\overset{\downarrow}{\times}\!\!\!\overset{\downarrow}{\rangle},$$

and recall that $\partial(1_\rho) = \rho$ (p. 266). We saw in 9.3.4 that $\mathbf{Gy}(\rho) = \mathbf{Ssq}(\rho)$ has exactly two elements θ_1, θ_2, sending the data in (9.9) to the derived composites

$$\gamma_1 = (\alpha'g) \circ (f'\alpha), \qquad \gamma_2 = (g'\alpha) \circ (\alpha' f)$$

respectively. So $\mathbf{Gy}(1_\rho) = \{\theta_1, \theta_2\}^2$. The element (θ_1, θ_2) of $\mathbf{Gy}(1_\rho)$ sends a cell diagram (9.9) in a Gray-category to a 3-cell of the form

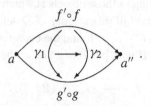

Similarly, the data of (9.9) is sent by (θ_2, θ_1) to a 3-cell $\gamma_2 \longrightarrow \gamma_1$, and by (θ_i, θ_i) to the identity 3-cell on γ_i. Since these are the only four elements of $\mathbf{Gy}(1_\rho)$, the two 2-cells $\gamma_2 \rightleftarrows \gamma_1$ are mutually inverse. In other words, in a Gray-category there is a specified isomorphism between the two derived horizontal compositions of 2-cells.

Here Gray-categories are treated as structures in their own right, but they were introduced by Gordon, Power and Street (1995) as a special kind of tri-category (their notion of weak 3-category). The embedding of **Gy**-algebras into the class of tricategories is non-canonical, amounting to the choice of one of the two derived horizontal compositions of 2-cells. 'Left-handed Gray-categories' and 'right-handed Gray-categories' therefore form (different) subclasses of the class of all tricategories. Gordon, Power and Street made the arbitrary decision to consider the left-handed version (let us say), and proved the important re-sult that every tricategory is equivalent to some left-handed Gray-category. By duality, the same result also holds for right-handed Gray-categories.

We do not set up a notion of weak equivalence of our weak n-categories, so cannot attempt an analogous coherence theorem, but we can show how to real-ize Gray-categories as weak 3-categories. Choose a precontraction on **Ssq**. By the last proposition, this induces a contraction on $(I_2^3)_*(\mathbf{Ssq}) = \mathbf{Gy}$, hence a map $L_3 \longrightarrow \mathbf{Gy}$ of 3-operads, hence a functor from Gray-categories to weak 3-categories (full and faithful, in fact). The final functor depends on the pre-contraction chosen. A precontraction is (9.3.4) a choice for each $n, k_1, \ldots, k_n$ of a total order on the set $\mathbf{k_1} + \cdots + \mathbf{k_n}$ restricting to the standard order on each $\mathbf{k_i}$, and there are two particularly obvious ones: order the set $\{1, \ldots, n\}$ either backwards or forwards, then order $\mathbf{k_1} + \cdots + \mathbf{k_n}$ lexicographically. The two embeddings induced are the analogues in our unbiased world of the left-and right-handed embeddings in the biased world of tricategories.

We can now read off results relating the theories of n- and ω-categories.

Corollary 9.3.10 $(I_{n-1}^n)_*(R_{n-1}^\omega)_*(L, \lambda)$ *is an initial object of* n-**OC**, *for any positive integer* n.

Proof By 9.3.7 and 9.3.8, we have a diagram

$$\omega\text{-}\mathbf{OC}$$

$$(R_{n-1}^\omega)_* \left\downarrow \dashv \right\uparrow (I_{n-1}^\omega)_*$$

$$(n-1)\text{-}\mathbf{OP} \xleftarrow[\ (I_{n-1}^n)_*\ ]{\overset{(R_{n-1}^n)_*}{\underset{\simeq}{\longleftarrow}}} n\text{-}\mathbf{OC},$$

and left adjoints and equivalences preserve initial objects. □

This proves Proposition 9.3.5, the existence of an initial n-operad-with-contraction, for $n \geq 1$, and tells us that

$$(L_n, \lambda_n) \cong (I_{n-1}^n)_*(R_{n-1}^\omega)_*(L, \lambda).$$

(The case $n = 0$ is done explicitly in Section 9.4.) So L_n is constructed from L by first forgetting all of L above dimension $n - 1$, then adjoining the only possible family of n-dimensional operations that will make the resulting n-operad contractible. In particular, L_n and L agree up to and including dimension $n - 1$, as do their associated contractions:

Corollary 9.3.11 *For any positive integer n, there is an isomorphism*

$$(R_{n-1}^n)_*(L_n, \lambda_n) \cong (R_{n-1}^\omega)_*(L, \lambda)$$

of $(n - 1)$-operads-with-precontraction. □

The ideas we have discussed suggest two alternative definitions of weak n-category, which we now formulate and prove equivalent to the main one.

The first starts from the thought that the cells of dimension at most n in a weak ω-category do not usually form a weak n-category, but they should do if the composition of n-cells is strict enough.

Definition 9.3.12 Let $n \in \mathbb{N}$ and let P be an n-operad. A P-algebra X is **tame** if

$$\overline{\theta^-} = \overline{\theta^+} : (TX)(\pi) \longrightarrow X(n)$$

for any $\pi \in \mathbf{pd}(n)$ and parallel $\theta^-, \theta^+ \in P(\pi)$.

Any algebra for a tame n-operad, and in particular any algebra for a contractible n-operad, is tame. A tame $((R_n^\omega)_*L)$-algebra ought to be the same thing as a weak n-category.

To state this precisely, note that the unit of the adjunction $(R_{n-1}^n)_* \dashv (I_{n-1}^n)_*$ gives a map

$$\alpha : (R_n^\omega)_* L \longrightarrow (I_{n-1}^n)_*(R_{n-1}^n)_*(R_n^\omega)_* L \cong L_n \qquad (9.10)$$

of operads, which induces a functor

$$\textbf{Wk-}n\textbf{-Cat} \longrightarrow \textbf{Alg}((R_n^\omega)_* L). \qquad (9.11)$$

Proposition 9.3.13 *The functor (9.11) restricts to an equivalence between* **Wk-**n**-Cat** *and the full subcategory of* **Alg**$((R_n^\omega)_* L)$ *consisting of the tame algebras.*

The proof uses a rather technical lemma, whose own proof is straightforward:

Lemma 9.3.14 *Let S and S' be monads on a category $\mathcal{C}$ and let $\psi : S \longrightarrow S'$ be a natural transformation commuting with the monad structures. Write $\mathcal{D}$ for the full subcategory of $\mathcal{C}^S$ consisting of the S-algebras $(SX \xrightarrow{h} X)$ for which h factors through ψ_X. If each component of ψ is split epi (has a right inverse) then the induced functor $\mathcal{C}^{S'} \longrightarrow \mathcal{C}^S$ restricts to an equivalence $\mathcal{C}^{S'} \longrightarrow \mathcal{D}$.*

□

Proof of Proposition 9.3.13 The n-operads $(R_n^\omega)_* L$ and L_n induce respective monads $(T_{(n)})_{(R_n^\omega)_* L}$ and $(T_{(n)})_{L_n}$ on $[\mathbb{G}_n^{\text{op}}, \textbf{Set}]$, and the map α of (9.10) induces a transformation ψ from the first monad to the second, commuting with the monad structures. If X is any n-globular set then the map ψ_X of n-globular sets is split epi: in dimension k it is the function

$$\coprod_{\pi \in \textbf{pd}(k)} L(\pi) \times (T_{(n)}X)(\pi) \xrightarrow{\coprod \alpha_\pi \times 1} \coprod_{\pi \in \textbf{pd}(k)} L_n(\pi) \times (T_{(n)}X)(\pi),$$

and α_π is bijective when $k < n$ (Corollary 9.3.11), so it is enough to show that α_π is surjective when $k = n$; this in turn is true because

$$\alpha_\pi = (s, t) : L(\pi) \longrightarrow \text{Par}_L(\pi)$$

and L is precontractible.

The lemma now applies, and we have only to check that $\mathcal{D}$ is the subcategory consisting of the tame $((R_n^\omega)_* L)$-algebras. Since ψ is the identity in dimensions less than n, an $((R_n^\omega)_* L)$-algebra X is in $\mathcal{D}$ if and only if there is a factorization

$$L(\pi) \times (T_{(n)}X)(\pi) \xrightarrow{\alpha_\pi \times 1} L_n(\pi) \times (T_{(n)}X)(\pi)$$

action   X(n)

in the category of sets for each $\pi \in \mathbf{pd}(n)$. We have already seen that α_π identifies two elements of $L(\pi)$ just when they are parallel, so this is indeed tameness. □

Corollary 9.3.15 *Let $n \in \mathbb{N}$ and $N \in \mathbb{N} \cup \{\omega\}$, with $n \leq N$. If X is a weak N-category with the property that for all $\pi \in \mathbf{pd}(n)$ and parallel $\theta^-, \theta^+ \in L_N(\pi)$,*

$$\overline{\theta^-} = \overline{\theta^+} : (T_{(N)}X)(\pi) \longrightarrow X(n),$$

then its n-dimensional restriction $R_n^N X$ inherits the structure of a weak n-category.

Proof The L_N-algebra structure on X induces an $((R_n^N)_* L_N)$-algebra structure on $R_n^N X$ (see Section 6.7). If $n = N$ the result is trivial; otherwise $(R_n^N)_* L_N \cong (R_n^\omega)_* L$ by 9.3.11, and the result follows from 9.3.13. □

The second alternative definition says that a weak n-category is a weak ω-category with only identity cells in dimensions higher than n. We show that this is equivalent to the main definition of weak n-category. More generally, we show that if $n \leq N$ then a weak N-category trivial above dimension n is the same thing as a weak n-category. The most simple case is that a discrete category is the same thing as a set.

Definition 9.3.16 Let $n \in \mathbb{N}$ and $N \in \mathbb{N} \cup \{\omega\}$, with $n \leq N$. An N-globular set X is n-**dimensional** if for all $n \leq k < N$, the maps

$$X(k+1) \overset{s}{\underset{t}{\rightrightarrows}} X(k)$$

are equal and bijective.

Any n-dimensional N-globular set is isomorphic to a 'strictly n-dimensional' N-globular set, that is, one of the form

$$\cdots \overset{1}{\underset{1}{\rightrightarrows}} X(n) \overset{1}{\underset{1}{\rightrightarrows}} X(n) \overset{s}{\underset{t}{\rightrightarrows}} X(n-1) \overset{s}{\underset{t}{\rightrightarrows}} \cdots \overset{s}{\underset{t}{\rightrightarrows}} X(0).$$

$$(9.12)$$

Theorem 9.3.17 *Let $n \in \mathbb{N}$ and $N \in \mathbb{N} \cup \{\omega\}$, with $n \leq N$. There is an equivalence of categories*

$$\mathbf{Wk}\text{-}n\text{-}\mathbf{Cat} \simeq (n\text{-}dimensional\ weak\ N\text{-}categories)$$

where the right-hand side is a full subcategory of $\mathbf{Wk}\text{-}N\text{-}\mathbf{Cat}$.

To prove this we construct the discrete weak N-category on a weak n-category, then show that the N-categories so arising are exactly the n-dimensional ones.

We start with the discrete construction in the setting of *strict* higher categories, and derive from it the weak version. Almost all of the steps involved are thought-free applications of previously-established theory. Let

$$D_n^N : [\mathbb{G}_n^{\mathrm{op}}, \mathbf{Set}] \longrightarrow [\mathbb{G}_N^{\mathrm{op}}, \mathbf{Set}]$$

be the functor sending an n-globular set X to the N-globular set of (9.12). This lifts to a functor

$$\mathbf{Str}\text{-}n\text{-}\mathbf{Cat} \lhook\joinrel\longrightarrow \mathbf{Str}\text{-}N\text{-}\mathbf{Cat},$$

and by Lemma 6.1.1, there is a corresponding lax map of monads

$$(D_n^N, \psi) : ([\mathbb{G}_n^{\mathrm{op}}, \mathbf{Set}], T_{(n)}) \longrightarrow ([\mathbb{G}_N^{\mathrm{op}}, \mathbf{Set}], T_{(N)}),$$

which, since D_n^N preserves finite limits, induces in turn a functor

$$(D_n^N)_* : n\text{-}\mathbf{Operad} \longrightarrow N\text{-}\mathbf{Operad}.$$

Explicitly, if Q is an n-operad and π a k-pasting diagram then

$$((D_n^N)_* Q)(\pi) = \begin{cases} Q(\partial^{k-n}\pi) & \text{if } k \geq n \\ Q(\pi) & \text{if } k \leq n, \end{cases}$$

from which it follows that $(D_n^N)_*$ lifts naturally to a functor

$$(D_n^N)_* : n\text{-}\mathbf{OC} \longrightarrow N\text{-}\mathbf{OC}.$$

Since L_N is initial in N-**OC**, there is a canonical map of operads

$$L_N \longrightarrow (D_n^N)_* L_n,$$

inducing a functor

$$\mathbf{Alg}((D_n^N)_* L_n) \longrightarrow \mathbf{Alg}(L_N) = \mathbf{Wk}\text{-}N\text{-}\mathbf{Cat}.$$

But we also have from Section 6.7 a functor

$$\mathbf{Wk}\text{-}n\text{-}\mathbf{Cat} = \mathbf{Alg}(L_n) \longrightarrow \mathbf{Alg}((D_n^N)_* L_n),$$

and so obtain a composite functor

$$D_n^N : \mathbf{Wk}\text{-}n\text{-}\mathbf{Cat} \longrightarrow \mathbf{Wk}\text{-}N\text{-}\mathbf{Cat}, \qquad (9.13)$$

as required. A weak N-category isomorphic to $D_n^N Y$ for some weak n-category Y will be called a **discrete** weak N-category on a weak n-category.

On underlying globular sets, the discrete functor (9.13) is merely the original functor D_n^N, so (9.13) restricts to a functor

$$D_n^N : \textbf{Wk-}n\textbf{-Cat} \longrightarrow (n\text{-dimensional weak } N\text{-categories}). \qquad (9.14)$$

A sharper statement of Theorem 9.3.17 is that this is an equivalence of categories.

Proof of 9.3.17 We show that restriction R_n^N is inverse to the functor D_n^N of (9.14).

First, if X is an n-dimensional weak N-category then Corollary 9.3.15 applies to give $R_n^N X$ the structure of a weak n-category. For let $\pi \in \textbf{pd}(n)$ and let θ^-, θ^+ be parallel elements of $L_N(\pi)$. By contractibility of L_N, there exists $\theta \in L_N(1_\pi)$ satisfying $s(\theta) = \theta^-$ and $t(\theta) = \theta^+$. Since any $(n + 1)$-cell of X has the same source and target, we have

$$\overline{\theta^-} = \overline{s(\theta)} = s \circ \overline{\theta} = t \circ \overline{\theta} = \overline{t(\theta)} = \overline{\theta^+}.$$

So by 9.3.15, R_n^N induces a functor

$$R_n^N : (n\text{-dimensional weak } N\text{-categories}) \longrightarrow \textbf{Wk-}n\textbf{-Cat}.$$

The composite functor $R_n^N \circ D_n^N$ on $\textbf{Wk-}n\textbf{-Cat}$ is the identity, ultimately because the same is true for strict n-categories. Conversely, let X be an n-dimensional weak N-category. We may assume that X is strictly n-dimensional (9.12), which means that there is an equality $D_n^N R_n^N X = X$ of globular sets. It remains only to check that the two L_N-algebra structures on X agree; but certainly they agree in dimensions n and lower, and n-dimensionality of X guarantees that they agree in dimensions higher than n too. $\qquad \square$

We finish with a method for turning higher-dimensional categories into lower-dimensional ones. It is an analogue of the path space construction in topology (or with slightly different analogies, the loop space construction or desuspension): given an n-category, we forget the 0-cells and decrease all the dimensions by 1 to produce an $(n - 1)$-category.

Again, we start by doing it in the strict setting. A strict n-category is a category enriched in $\textbf{Str-}n\textbf{-Cat}$, and a finite-limit-preserving functor $V \longrightarrow W$ between categories with finite limits induces a finite-limit-preserving functor $V\textbf{-Cat} \longrightarrow W\textbf{-Cat}$, so the functor $\textbf{Cat} \longrightarrow \textbf{Set}$ sending a category to its set of arrows induces a finite-limit-preserving functor

$$\textbf{Str-}n\textbf{-Cat} \longrightarrow \textbf{Str-}(n - 1)\textbf{-Cat}$$

for each positive integer n. This is a lift of the functor

$$J_n : \quad [\mathbb{G}_n^{op}, \textbf{Set}] \quad \longrightarrow \quad [\mathbb{G}_{n-1}^{op}, \textbf{Set}],$$

$$X \quad \longmapsto \quad \left(X(n) \underset{t}{\overset{s}{\rightrightarrows}} \cdots \underset{t}{\overset{s}{\rightrightarrows}} X(1) \right).$$

By 6.1.1, J_n has the structure of a lax map of monads

$$\left([\mathbb{G}_n^{op}, \textbf{Set}], T_{(n)} \right) \quad \longrightarrow \quad \left([\mathbb{G}_{n-1}^{op}, \textbf{Set}], T_{(n-1)} \right),$$

which, since J_n preserves finite limits, induces a functor

$$(J_n)_* : n\text{-}\textbf{Operad} \longrightarrow (n-1)\text{-}\textbf{Operad}.$$

To describe $(J_n)_*$ explicitly we use the **suspension** operator Σ : $\textbf{pd}(k) \longrightarrow \textbf{pd}(k+1)$, defined by $\Sigma\pi = (\pi)$. Here $\textbf{pd}(k+1)$ is regarded as the free monoid on $\textbf{pd}(k)$ and (π) is a sequence of length one. An example explains the name:

Now, if P is an n-operad, $0 \leq k \leq n-1$, and $\pi \in \textbf{pd}(k)$, we have

$$((J_n)_* P)(\pi) = P(\Sigma\pi),$$

and using the equation $\partial \Sigma\pi = \Sigma\partial\pi$, we find that $(J_n)_*$ lifts naturally to a functor

$$(J_n)_* : n\text{-}\textbf{OC} \longrightarrow (n-1)\text{-}\textbf{OC}.$$

Just as for the discrete construction, this induces a functor

$$J_n : \textbf{Wk-}n\text{-}\textbf{Cat} \longrightarrow \textbf{Wk-}(n-1)\text{-}\textbf{Cat}$$

whose effect on the underlying globular sets is the original 'shift' functor J_n.

The $(n-1)$-category $J_n X$ arising from an n-category X is called the **localization** of X. Its underlying $(n-1)$-globular set is the disjoint union over all $a, b \in X(0)$ of the $(n-1)$-globular sets $X(a, b)$ defined in the proof of 1.4.9. Since the functor $T_{(n-1)}$ preserves coproducts (F.2.1), it follows from the lemma below that the $(n-1)$-category structure on $J_n X$ amounts to an

$(n-1)$-category structure on each $X(a, b)$, in both the strict and the weak settings. So localization defines functors

$$\textbf{Str-}n\textbf{-Cat} \longrightarrow (\textbf{Str-}(n-1)\textbf{-Cat})\textbf{-Gph},$$
$$\textbf{Wk-}n\textbf{-Cat} \longrightarrow (\textbf{Wk-}(n-1)\textbf{-Cat})\textbf{-Gph}.$$

In the strict case, an n-category is a graph of $(n-1)$-categories together with composition functors obeying simple laws; this is just ordinary enrichment. In the weak case it is much more difficult to say what extra structure is needed.

Lemma 9.3.18 *Let S be a cartesian monad on a presheaf category $\mathcal{E}$, such that the functor part of S preserves coproducts. Let P be an S-operad and let $(X_i)_{i \in I}$ be a family of objects of $\mathcal{E}$. Then a P-algebra structure on $\coprod X_i$ amounts to a P-algebra structure on each X_i.*

Proof It is enough to show that the functor S_P preserves coproducts. Since S_P is defined using S and pullback, and pullbacks interact well with coproducts in presheaf categories, this follows from the same property of S. $\qquad\square$

Finally, localization works just as well for ω-categories. The localization functors $\textbf{Str-}n\textbf{-Cat} \longrightarrow \textbf{Str-}(n-1)\textbf{-Cat}$ induce in the limit an endofunctor of $\textbf{Str-}\omega\textbf{-Cat}$, which on underlying globular sets is the endofunctor J of $[\mathbb{G}, \textbf{Set}]$ forgetting 0-cells. So exactly as in the finite-dimensional case, a weak ω-category X gives rise to a family $(X(a, b))_{a, b \in X(0)}$ of weak ω-categories.

9.4 Weak 2-categories

A polite person proposing a definition of weak n-category should explain what happens when $n = 2$. With our definition, $\textbf{Wk-2-Cat}$ turns out to be equivalent to $\textbf{UBicat}_{\text{str}}$, the category of small unbiased bicategories and unbiased strict functors.

Observe that since the maps in $\textbf{Wk-2-Cat}$ are *strict* functors, we obtain an equivalence with $\textbf{UBicat}_{\text{str}}$, not $\textbf{UBicat}_{\text{wk}}$ or $\textbf{UBicat}_{\text{lax}}$; and unlike its weak and lax siblings, $\textbf{UBicat}_{\text{str}}$ is not equivalent to the analogous category of classical bicategories (or at least, the obvious functor is not an equivalence). So we cannot conclude that $\textbf{Wk-2-Cat}$ is equivalent to $\textbf{Bicat}_{\text{str}}$. Nevertheless, the results of Section 3.4 mean that it is fair to regard classical bicategories as essentially the same as unbiased bicategories, and therefore, by the results below, essentially the same as weak 2-categories. One would expect that if the definition of weak functor between n-categories were in place, $\textbf{Bicat}_{\text{wk}}$ would be equivalent to the category of weak 2-categories and weak functors.

Theorem 9.4.1 *There are equivalences of categories*

$$\textbf{Wk-0-Cat} \simeq \textbf{Set},$$
$$\textbf{Wk-1-Cat} \simeq \textbf{Cat},$$
$$\textbf{Wk-2-Cat} \simeq \textbf{UBicat}_{\text{str}}.$$

So far we have ignored weak 0-categories; indeed, we have not even proved that there is an initial 0-operad-with-contraction. A 0-globular set is a set and the monad $T_{(0)}$ is the identity, so a 0-operad is a monoid and an algebra for a 0-operad is a set acted on by the corresponding monoid. There is a unique precontraction on every 0-operad, which is a contraction just when the corresponding monoid has cardinality 1. So 0-**OC** is the category of one-element monoids, any object L_0 of which is initial, and

$$\textbf{Wk-0-Cat} \simeq \textbf{Set}.$$

A weak 1-category is a 1-dimensional weak 2-category (Theorem 9.3.17), so the middle equivalence of Theorem 9.4.1 will follow from the last. It is, however, easy enough to prove directly. We have just seen that a 0-operad-with-precontraction is a monoid, so the initial such is also the terminal such. The equivalence 1-**OC** $\simeq$ 0-**OP** of 9.3.8 then tells us that the initial 1-operad-with-contraction is the terminal 1-operad. But algebras for the terminal $T_{(1)}$-operad are just $T_{(1)}$-algebras, so

$$\textbf{Wk-1-Cat} \simeq \textbf{Cat}.$$

It is not prohibitively difficult to prove the 2-dimensional equivalence result explicitly, as in Leinster (2000b, 4.8); 2-operads are just about manageable. Here we use an abstract method instead, taking advantage of some earlier calculations.

Notation: we write

- W for both the free monoid monad on **Set** and the free strict monoidal category monad on **Cat**
- $\mathcal{V}$-**Gph** for the category of graphs enriched in a given finite product category $\mathcal{V}$ (1.3.1), and **Gph** for **Set-Gph**
- $\Sigma : \mathcal{V} \longrightarrow \mathcal{V}$-**Gph** for the functor sending an object V of $\mathcal{V}$ to the one-object $\mathcal{V}$-graph whose single hom-set is V
- $\text{fc}_{\mathcal{V}}$ for the free $\mathcal{V}$-enriched category monad on $\mathcal{V}$-**Gph** (when it exists).

A 1-globular set is a directed graph and $T_{(1)}$ is the free category monad **fc** of Chapter 5, so a 1-operad is an **fc**-operad, which is an **fc**-multicategory with only one 0-cell and one horizontal 1-cell. A precontraction κ on an **fc**-operad assigns to each $r \in \mathbb{N}$ and pair (f, f') of vertical 1-cells a 2-cell

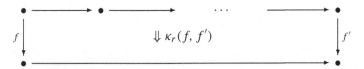

with r 1-cells along the top. Recall from 5.1.7 that there is an embedding

$$\textbf{Operad} \hookrightarrow \textbf{fc-Operad}$$

identifying plain operads with **fc**-operads having only one 0-cell, one vertical 1-cell and one horizontal 1-cell. There we called the embedding Σ; here we call it Σ_*, because it is induced by the weak map of monads $\Sigma :$ $(\textbf{Set}, W) \longrightarrow (\textbf{Gph}, \textbf{fc})$. If P is a plain operad then a precontraction on $\Sigma_* P$ consists of an element $\kappa_r \in P(r)$ for each $r \in \mathbb{N}$. Take the plain operad **tr** of trees and the r-leafed corolla $\nu_r \in \textbf{tr}(r)$ for each $r \in \mathbb{N}$ (2.3.3): then using the fact that **tr** is the free operad containing an operation of each arity, it is easy to show that the corresponding 1-operad-with-precontraction $\Sigma_*\textbf{tr}$ is initial. So $L_2 = (I_1^2)_* \Sigma_*\textbf{tr}$, and we have

$$\textbf{Wk-2-Cat} \cong \textbf{Alg}((I_1^2)_* \Sigma_*\textbf{tr}). \tag{9.15}$$

On the other hand, we saw earlier that the theory of unbiased bicategories is also generated by the operad of trees. Specifically, we showed in Section 3.4 that

$$\textbf{UBicat}_{\text{str}} \cong \textbf{1-Bicat}_{\text{str}} = \textbf{CatAlg}_{\text{str}} I_* \textbf{tr}.$$

The functors I_* and $\textbf{CatAlg}_{\text{str}}$ can be described as follows. We have maps of monads

$$(\textbf{Set}, W) \xrightarrow{\;I\;} (\textbf{Cat}, W) \xrightarrow{\;\Sigma\;} (\textbf{Cat-Gph}, \textbf{fc}_{\textbf{Cat}})$$

where I is the indiscrete category functor (p. 24); I is lax and Σ is weak. Recalling from 4.2.17 that a $(\textbf{Cat}, W)$-operad is a **Cat**-operad, we obtain induced functors

$$\textbf{Operad} \xrightarrow{\;I_*\;} \textbf{Cat-Operad} \xrightarrow{\;\Sigma_*\;} \textbf{fc}_{\textbf{Cat}}\textbf{-Operad} \xrightarrow{\;\textbf{Alg}\;} \textbf{CAT}^{\text{op}}.$$

The functor I_* is the same as the one used in Chapter 3, and the composite of the last two functors is $\textbf{CatAlg}_{\text{str}}$, so

$$\textbf{UBicat}_{\text{str}} \cong \textbf{Alg}(\Sigma_* I_* \textbf{tr}).$$

Comparing with (9.15), it is enough to prove

Lemma 9.4.2 *For any plain operad P, there is an isomorphism of categories*

$$\mathbf{Alg}((I_1^2)_* \Sigma_* P) \cong \mathbf{Alg}(\Sigma_* I_* P).$$

Proof We have three monads on the category **Gph-Gph** of 2-globular sets: first, **fc-Gph**, the result of applying the 2-functor ( )-**Gph** : **CAT** $\longrightarrow$ **CAT** to the monad **fc** on **Gph**; second, **fc$_{\mathbf{Gph}}$**, the free **Gph**-enriched category monad; third, $T_{(2)}$. We show in the proof of F.1.1 that $T_{(2)}$ is the result of gluing **fc-Gph** to **fc$_{\mathbf{Gph}}$** by a distributive law

$$\lambda : (\mathbf{fc\text{-}Gph}) \circ \mathbf{fc_{Gph}} \longrightarrow \mathbf{fc_{Gph}} \circ (\mathbf{fc\text{-}Gph}).$$

We also saw in Lemma 6.1.4 that a distributive law gives rise to a monad '$\widetilde{S}$', which in this case is the monad **fc$_{\mathbf{Cat}}$** on **Cat-Gph**, and in Lemma 6.1.5 that it gives rise to a lax map of monads, which in this case is of the form

$$(U, \psi) : (\mathbf{Cat\text{-}Gph}, \mathbf{fc_{Cat}}) \longrightarrow (\mathbf{Gph\text{-}Gph}, T_{(2)})$$

where U is the forgetful functor.

It is straightforward to check that there is an equality of natural transformations

$$
\begin{array}{ccc}
\mathbf{Set} \xrightarrow{W} \mathbf{Set} & & \mathbf{Set} \xrightarrow{W} \mathbf{Set} \\
\Sigma \downarrow \quad /\!/ \quad \downarrow \Sigma & & I \downarrow \quad \nearrow \quad \downarrow I \\
\mathbf{Gph} \xrightarrow{T_{(1)}} \mathbf{Gph} & = & \mathbf{Cat} \xrightarrow{W} \mathbf{Cat} \\
I_1^2 \downarrow \quad \nearrow \quad \downarrow I_1^2 & & \Sigma \downarrow \quad /\!/ \quad \downarrow \Sigma \\
\mathbf{Gph\text{-}Gph} \xrightarrow[T_{(2)}]{} \mathbf{Gph\text{-}Gph} & & \mathbf{Cat\text{-}Gph} \xrightarrow{\mathbf{fc_{Cat}}} \mathbf{Cat\text{-}Gph} \\
& & U \downarrow \quad \nearrow \psi \quad \downarrow U \\
& & \mathbf{Gph\text{-}Gph} \xrightarrow[T_{(2)}]{} \mathbf{Gph\text{-}Gph}
\end{array}
$$

where the unmarked transformations are the ones referred to above. So

$$(I_1^2)_* \Sigma_* = U_* \Sigma_* I_* : \mathbf{Operad} \longrightarrow T_{(2)}\text{-}\mathbf{Operad},$$

and it is enough to prove that for any **fc$_{\mathbf{Cat}}$**-operad Q,

$$\mathbf{Alg}(Q) \cong \mathbf{Alg}(U_* Q).$$

This follows immediately from Proposition 6.7.5. $\square$

That completes the proof of Theorem 9.4.1.

In Section 9.1 we said what we wanted the operad L to look like in low dimensions: if • is the unique 0-pasting diagram then $L(\bullet)$ should be a one-element set, and if χ_k is the 1-pasting diagram made up of k arrows then $L(\chi_k)$ should be $\mathbf{tr}(k)$. We now know that our wishes were met: for by 9.3.7, the 1-dimensional restriction $(R_1^\omega)_*L$ is the initial 1-operad-with-precontraction, and we have shown this to be $\Sigma_*\mathbf{tr}$.

What about 3-categories? It should be possible to write down an explicit definition of unbiased tricategory (similar to that of unbiased bicategory, 3.4.1) and to prove that the category of unbiased tricategories and strict maps is equivalent to **Wk-3-Cat**. This would be a lot of work, and it is not clear that the result would have any advantage over the abstract definition of weak 3-category.

We could also try to compare our weak 3-categories with the tricategories of Gordon, Power and Street (1995). The analogous comparison one dimension down, between weak 2-categories and classical bicategories, is already difficult because we do not have a notion of weak functor between n-categories (see the beginning of this section). A further difficulty is that Gordon, Power and Street's definition is not quite algebraic; put another way, the forgetful functor

$$(\text{tricategories} \; + \; \text{strict maps}) \longrightarrow [\mathbb{G}^{\mathrm{op}}, \mathbf{Set}]$$

seems highly unlikely to be monadic. If true, this means that there can be no 3-operad whose algebras are precisely tricategories (in apparent contradiction to Batanin, 1998a, p. 94).

The reason why the theory of tricategories is not quite algebraic is as follows. Most of the definition of tricategory consists of some data subject to some equations, but a small part does not: in items (TD5) and (TD6), it is stipulated that certain transformations of bicategories are equivalences. This is not an algebraic axiom, as there are many different choices of weak inverses and none has been specified. Compare the fact that the forgetful functor from non-empty sets to **Set** is not monadic (indeed, has no left adjoint), in contrast to the forgetful functor from pointed sets to **Set**. To make the definition algebraic we would have to add in as data a weak inverse for each of these equivalences, together with two invertible modifications witnessing that it is a weak inverse, and then add more coherence axioms (saying, among other things, that this data forms an *adjoint* equivalence). The result would be an even more complicated, but conceptually pure, notion of tricategory.

Notes

Contractions were introduced in my (2001a). When I wrote and made public the first version of that paper I believed that I was explaining Batanin's notion of contraction, but in fact I was inventing a new one; see Section 10.1 for an explanation of the difference.

Some of the results here on n- and 2-categories appeared in my (2000b).

See Crans (1999, 2.3) for a completely elementary definition of Gray-category (9.3.9). I learned that the operad for Gray-categories can be defined from the operad for sesquicategories from Batanin (1998a, p. 94). Constructions of the fundamental ω-groupoid of a space (Example 9.2.7) can also be found in Batanin (1998a, §9) and Berger (2002, 4.20(4)).

Chapter 10

Other definitions of weak n-category

Zounds! I was never so bethump'd with words!
Shakespeare (1596)

The definition of weak n-category studied in the previous chapter is, of course, just one of a host of proposed definitions. Ten of them were described in my (2001b) survey, all except one in formal, precise terms. However, the format of that paper did not allow for serious discussion of the interrelationships, and one might get the impression from it that the ten definitions embodied eight or so completely different approaches to the subject.

I hope to correct that impression here. Fundamentally, there seem to be only two approaches.

In the first, a weak n-category is regarded as a presheaf with *structure*. Usually 'presheaf' means n-globular set, and 'structure' means S-algebra structure for some monad S, often coming from a globular operad. The definition studied in the previous chapter is of this type.

In the second approach, a weak n-category is regarded as a presheaf with *properties*. There is no hope that a weak n-category could be defined as an n-globular set with properties, so the category $\mathbb{C}_n$ on which we are taking presheaves must be larger than $\mathbb{G}_n$; presheaves on $\mathbb{C}_n$ must somehow have composition built in. The case $n = 1$ makes this clear. In the first approach, a category is defined as a directed graph (presheaf on $\mathbb{G}_1$) with structure. In the second, a category cannot be defined a category as a presheaf-with-properties on $\mathbb{G}_1$, but it can be defined as a presheaf-with-properties on the larger category Δ: this is the standard characterization of a category by its nerve (p. 116).

There are other descriptions of the difference between the two approaches. In the sense of the introduction to Chapter 3, the first is algebraic (the various types of composition in a weak n-category are *bona fide* operations) and the second is non-algebraic (composites are not determined uniquely, only up to

309

equivalence). Or, the first approach can be summarized as 'take the theory of strict n-categories, weaken it, then take models for the weakened theory', and the second as 'take weak models for the theory of strict n-categories'.

The definitions following the two approaches are discussed in Sections 10.1 and 10.2 respectively. What I have chosen to say about each definition (and which definitions I have chosen to say anything about at all) is dictated by how much I feel capable of saying in a simple and not too technical way; the emphasis is therefore rather uneven. In particular, there is more on the definitions close to that of the previous chapter than on those further away.

10.1 Algebraic definitions

Here we discuss the definitions proposed by Batanin, Penon, Trimble, and May. We continue to use the notation of the previous chapter: T is the free strict ω-category monad, $\mathbf{pd} = T1$, and so on.

Batanin's definition

The definition of weak ω-category studied in the previous chapter is a simplification of Batanin's (1998a) definition. There are two main differences. The less significant is bias: where our definition treats composition of all shapes equally, Batanin's gives special status to binary and nullary compositions. For instance, our weak 2-categories are unbiased bicategories, but his are classical, biased, bicategories. The more significant difference is conceptual. We integrated composition and coherence into the single notion of contraction; Batanin keeps the two separate. This makes his definition more complicated to state, but more obvious from the traditional point of view.

Composition is handled as follows. The unit map $\eta_1 : 1 \longrightarrow T1 = \mathbf{pd}$ picks out, for each $m \in \mathbb{N}$, the m-pasting diagram ι_m looking like a single m-cell. Define a collection $\mathbf{binpd} \hookrightarrow \mathbf{pd}$ by

$$\mathbf{binpd}(m) = \{\iota_m \circ_0 \iota_m, \ \iota_m \circ_1 \iota_m, \ \ldots, \ \iota_m \circ_m \iota_m\} \subseteq \mathbf{pd}(m)$$

where $\circ_p$ is composition in the strict ω-category $\mathbf{pd}$ and $\iota_m \circ_m \iota_m$ means ι_m; then $\mathbf{binpd}$ consists of binary and unary diagrams such as

Definition 10.1.1 Let P be a globular operad. A **system of (binary) compositions** in P is a map of collections $\textbf{binpd} \longrightarrow P$, written

$$\left(\iota_m \circ_p \iota_m \in \textbf{binpd}(m) \right) \longmapsto \left(\delta_p^m \in P(\iota_m \circ_p \iota_m) \right),$$

such that δ_m^m is the identity operation $1_m \in P(\iota_m)$ for each $m \in \mathbb{N}$.

So, for instance, $s(\delta_0^2) = t(\delta_0^2) = \delta_0^1$ and $s(\delta_1^2) = t(\delta_1^2) = 1_1$. (If we wanted to do an unbiased version of Batanin's definition, we could replace **binpd** by **pd**.)

Example 10.1.2 A contraction κ on an operad P canonically determines a system $\delta_\bullet^\bullet$ of compositions, defined inductively by

$$\delta_p^m = \begin{cases} \kappa_{\iota_m \circ_p \iota_m}(\delta_p^{m-1}, \delta_p^{m-1}) & \text{if } p < m, \\ 1_m & \text{if } p = m. \end{cases}$$

To describe coherence we use the notion of a **reflexive globular set**, that is, a globular set Y together with functions

$$\cdots \xleftarrow{\ i\ } Y(k+1) \xleftarrow{\ i\ } Y(k) \xleftarrow{\ i\ } \cdots \xleftarrow{\ i\ } Y(0),$$

written $i(\psi) = 1_\psi$, such that $s(1_\psi) = t(1_\psi) = \psi$ for each $k \geq 0$ and $\psi \in Y(k)$. Reflexive globular sets form a presheaf category $[\mathbb{R}^{\mathrm{op}}, \textbf{Set}]$. The underlying globular set of a strict ω-category is canonically reflexive, taking the identity cells 1_ψ. This applies in particular to $T1 = \textbf{pd}$; if $\pi \in \textbf{pd}(k)$ then $1_\pi \in \textbf{pd}(k+1)$ is the degenerate $(k+1)$-pasting diagram represented by the same picture as π.

Definition 10.1.3 Let X be a globular set, Y a reflexive globular set, and $q : X \longrightarrow Y$ a map of globular sets. For $k \geq 0$ and $\psi \in Y(k)$, write (Fig. 10-A)

$$\mathrm{Par}_q'(k) = \{(\theta^-, \theta^+) \in X(k) \times X(k) \mid \theta^- \text{ and } \theta^+ \text{ are parallel,} \\ q(\theta^-) = q(\theta^+)\}.$$

A **coherence** ζ on q is a family of functions

$$\left(\mathrm{Par}_q'(k) \xrightarrow{\ \zeta_k\ } X(k+1) \right)_{k \geq 0}$$

such that for all $k \geq 0$ and $(\theta^-, \theta^+) \in \mathrm{Par}_q'(k)$, writing ζ_k as ζ,

$$s(\zeta(\theta^-, \theta^+)) = \theta^-, \quad t(\zeta(\theta^-, \theta^+)) = \theta^+, \\ q(\zeta(\theta^-, \theta^+)) = 1_{q(\theta^-)}(= 1_{q(\theta^+)}).$$

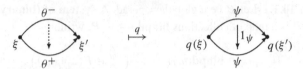

Fig. 10-A. Effect of a coherence ζ, shown for $k = 1$. The dotted arrow is $\zeta(\theta^-, \theta^+)$, and $\psi = q(\theta^-) = q(\theta^+)$

Example 10.1.4 Any contraction on a map canonically determines a coherence, as is clear from a comparison of Figs. 9-A (p. 281) and 10-A. Formally, $\text{Par}_q'(k) = \coprod_{\psi \in Y(k)} \text{Par}_q(1_\psi)$ and a contraction κ determines the coherence ζ given by $\zeta(\theta^-, \theta^+) = \kappa_{1_\psi}(\theta^-, \theta^+)$ where $\psi = q(\theta^-) = q(\theta^+)$. The class of contractible maps is closed under composition, but the class of maps admitting a coherence is not.

Definition 10.1.5 A **coherence** on a collection $(P \xrightarrow{d} T1)$ is a coherence on the map d, and a **coherence** on an operad is a coherence on its underlying collection. A map, collection, or operad is **coherent** if it admits a coherence.

Explicitly, a coherence on an operad P assigns to each $\pi \in \mathbf{pd}(k)$ and parallel pair $\theta^-, \theta^+ \in P(\pi)$ an element $\zeta(\theta^-, \theta^+)$ of $P(1_\pi)$ with source θ^- and target θ^+. Let X be a P-algebra; then since $\widehat{\pi} = \widehat{1_\pi}$ (p. 270), we have $(TX)(\pi) = (TX)(1_\pi)$, so if $\mathbf{x} \in (TX)(\pi)$ then there is a $(k+1)$-cell

$$\overline{\zeta(\theta^-, \theta^+)}(\mathbf{x}) : \overline{\theta^-}(\mathbf{x}) \longrightarrow \overline{\theta^+}(\mathbf{x})$$

in X connecting the two 'composites' $\overline{\theta^-}(\mathbf{x})$ and $\overline{\theta^+}(\mathbf{x})$ of $\mathbf{x}$. Taking $\pi = \iota_k$ and $\theta^- = \theta^+ = 1_k$, this provides in particular a reflexive structure on the underlying globular set of X.

Example 10.1.6 Any contractible operad is coherent (by the previous example), but not conversely. For instance, there is an operad R whose algebras are reflexive globular sets; it is uniquely determined by

$$R(\pi) = \begin{cases} 1 & \text{if } \pi \in \{\iota_k, 1_{\iota_{k-1}}, 1_{1_{\iota_{k-2}}}, \ldots\} \\ \varnothing & \text{otherwise} \end{cases}$$

($\pi \in \mathbf{pd}(k)$). This operad is coherent (trivially) but not contractible (since some of the sets $R(\pi)$ are empty). So a given globular set X is an algebra for some coherent operad if and only if it admits a reflexive structure; on the other hand, by 9.2.4, it is an algebra for some contractible operad if and only if it admits a weak ω-category structure.

A coherence is what Batanin calls a contraction. As we have seen, our contractions are more powerful, providing both a coherence and a system of compositions.

Rather confusingly, at several points in Batanin (1998a) the word 'contractible' is used as an abbreviation for 'contractible [coherent] and admitting a system of compositions'. In particular, weak ω- and n-categories are often referred to as algebras for a 'universal contractible operad'. This is not meant literally: 'universal' means weakly initial (in other words, there is at least one map from it to any other contractible operad), and the operad R of 10.1.6 is in Batanin's terminology contractible, so any genuine 'universal contractible operad' P satisfies $P(\pi) = \emptyset$ for almost all pasting diagrams π. A P-algebra is then nothing like an ω-category. Indeed, R is in Batanin's terminology the initial operad equipped with a contraction, and its algebras are mere reflexive globular sets. The system of compositions is a vital ingredient; left out, the situation degenerates almost entirely.

I believe it is the case that given a system of compositions and a coherence on an operad, a contraction can be built. This is the non-canonical converse to the canonical process in the other direction; the situation is like that of biased vs. unbiased bicategories. So it appears that despite an abuse of terminology and two different definitions of contractibility, the term 'contractible operad' means exactly the same in Batanin's work as here.

The two ingredients – composition and coherence – can be combined to make a definition of weak ω-category in several possible ways:

- Imitate the definition of the previous chapter. In other words, take the category of globular operads equipped with both a system of compositions and a coherence, prove that it has an initial object $(B, \delta_{\bullet}^{\bullet}, \kappa)$, and define a weak ω-category as a B-algebra. This is Definition **B1** in my (2001b) survey.
- Use Batanin's operad K, constructed in his Theorem 8.1. He proves that K is weakly initial in the full subcategory of T-**Operad** consisting of the coherent operads admitting a system of compositions. Weak initiality does not characterize K up to isomorphism, so one needs some further information about K in order to use this definition. It seems to be claimed that K is initial in the category of operads equipped with a system of compositions and a coherence (Remark 2 after the proof of Theorem 8.1), but it does not seem obvious that this claim is true, essentially because of the set-theoretic complement taken in the proof of Lemma 8.1.
- Define a weak ω-category as a pair (P, X) where P is a coherent operad admitting a system of compositions and X is a P-algebra. Given such a pair (P, X), we can choose a system of compositions and a coherence on P,

and this turns X into a B-algebra – that is, a weak ω-category in the sense of the first method. The present method has some variants: we might insist that $P(\cdot) = 1$, where $\cdot$ is the unique 0-pasting diagram, or we might drop the condition that P admits a system of compositions and replace it with the more relaxed requirement that $P(\pi) \neq \emptyset$ for all pasting diagrams π (giving Batanin's Definition 8.6 of 'weak ω-categorical object').

Batanin's weak ω-categories can be compared with the weak ω-categories of the previous chapter. We have already shown that a contraction on a globular operad gives rise canonically to a system of compositions and a coherence (10.1.2, 10.1.6). This is true in particular of the operad L, so there is a canonical map $B \longrightarrow L$ of operads, inducing in turn a canonical functor from L-algebras to B-algebras. Conversely, B is non-canonically contractible, so there is a non-canonical functor in the other direction. In the case $n = 2$, this is the comparison of biased and unbiased bicategories.

Penon's definition

The definition of weak ω-category proposed by Penon (1999) does not use the language of operads, but is nevertheless close in spirit to the definition of Batanin. It can be stated very quickly.

Definition 10.1.7 Let $q : X \longrightarrow Y$ be a map of reflexive globular sets. A coherence ζ on q is **normal** if $\zeta(\theta, \theta) = 1_\theta$ for all $k \geq 0$ and $\theta \in X(k)$.

Penon calls a normal coherence an étirement, or stretching. This might seem to conflict with the contraction terminology, but it is only a matter of viewpoint: X is being shrunk, Y stretched.

Definition 10.1.8 An ω-**magma** is a reflexive globular set X equipped with a binary composition function $\circ_p : X(m) \times_p X(m) \longrightarrow X(m)$ for each $m > p \geq 0$, satisfying the source and target axioms of 1.4.8(a).

An ω-magma is a very wild structure, and a strict ω-category very tame; weak ω-categories are somewhere in between.

Let $\mathcal{Q}$ be the category whose objects are quadruples (X, Y, q, ζ) with X an ω-magma, Y a strict ω-category, $q : X \longrightarrow Y$ a map of ω-magmas, and ζ a normal coherence on q. Maps

$$(X, Y, q, \zeta) \longrightarrow (X', Y', q', \zeta')$$

in $\mathcal{Q}$ are pairs $(X \longrightarrow X', Y \longrightarrow Y')$ of maps commuting with all the structure present. There is a forgetful functor U from $\mathcal{Q}$ to the category $[\mathbb{R}^{op}, \mathbf{Set}]$

of reflexive globular sets, sending (X, Y, q, ζ) to the underlying reflexive glob-ular set of X. This has a left adjoint F, and a weak ω-category is defined as an algebra for the induced monad $U \circ F$ on $[\mathbb{R}^{op}, \mathbf{Set}]$.

Observe that U takes the underlying reflexive globular set of X, not of Y. The object (X, Y, q, ζ) of $\mathcal{Q}$ should therefore be regarded as X (not Y) equipped with extra structure, making it perhaps more apt to think of an object of $\mathcal{Q}$ as a shrinking rather than a stretching.

Recent work of Cheng relates Penon's definition to the definition of the previous chapter through a series of intermediate definitions; older work of Batanin (2002a) relates Penon's definition to his own. One point can be ex-plained immediately. Take the operad B equipped with its system of composi-tions $\delta_\bullet^\bullet$ and its coherence κ. There is a unique reflexive structure on the under-lying globular set of B for which κ is normal, namely $1_\theta = \kappa(\theta, \theta)$. Also, the system of compositions in the operad B makes its underlying globular set into an ω-magma: given $0 \leq p < m$ and $(\theta_1, \theta_2) \in B(m) \times_{B(p)} B(m)$, put

$$\theta_1 \circ_p \theta_2 = \delta_p^m \circ (\theta_1, \theta_2)$$

where the $\circ$ on the right-hand side is operadic composition. This gives $B \longrightarrow T1$ the structure of an object of $\mathcal{Q}$. Now, if X is any B-algebra, we have a pullback square

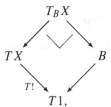

and the $\mathcal{Q}$-object structure on $B \longrightarrow T1$ induces a $\mathcal{Q}$-object structure on $T_B X \longrightarrow TX$. So any B-algebra gives rise to an object of $\mathcal{Q}$ and hence, applying the comparison functor for the monad $U \circ F$, a Penon weak ω-category.

Trimble's and May's definitions

Trimble has also proposed a simple definition of n-category, unpublished but written up as Definition **Tr** in my (2001b) survey. He was not quite so ambi-tious as to attempt a fully weak notion of n-category; rather, he sought a no-tion just general enough to capture the fundamental n-groupoids of topological spaces. He called his structures 'flabby n-categories'.

Trimble's definition uses simple operad language. Let $\mathcal{A}$ be a category with finite products and let P be a (non-symmetric) operad in $\mathcal{A}$. Extending slightly the terminology of p. 166, a **categorical P-algebra** is an $\mathcal{A}$-graph X together with a map

$$P(k) \times X(x_0, x_1) \times \cdots \times X(x_{k-1}, x_k) \longrightarrow X(x_0, x_k) \qquad (10.1)$$

for each $k \in \mathbb{N}$ and $x_0, \ldots, x_k \in X_0$, satisfying the evident axioms. The category of categorical P-algebras is written **CatAlg**(P), and itself has finite products. If $F : \mathcal{A} \longrightarrow \mathcal{A}'$ is a finite-product-preserving functor then there is an induced operad F_*P in $\mathcal{A}'$ and a finite-product-preserving functor

$$\widehat{F} : \mathbf{CatAlg}(P) \longrightarrow \mathbf{CatAlg}(F_*P).$$

The topological content consists of two observations: first (Example 2.2.18), that if E is the operad of path re-parametrizations then there is a canonical functor

$$\Xi : \mathbf{Top} \longrightarrow \mathbf{CatAlg}(E),$$

and second, that taking path-components defines a finite-product-preserving functor $\Pi_0 : \mathbf{Top} \longrightarrow \mathbf{Set}$.

Applying the operadic constructions recursively to the topological data, we define for each $n \in \mathbb{N}$ a category **Flabby-n-Cat** with finite products and a functor

$$\Pi_n : \mathbf{Top} \longrightarrow \mathbf{Flabby\text{-}}n\text{-}\mathbf{Cat}$$

preserving finite products: **Flabby-0-Cat** $=$ **Set**, and Π_{n+1} is the composite functor

$$\mathbf{Top} \xrightarrow{\;\Xi\;} \mathbf{CatAlg}(E) \xrightarrow{\;\widehat{\Pi_n}\;} \mathbf{CatAlg}((\Pi_n)_*E) = \mathbf{Flabby\text{-}}(n+1)\text{-}\mathbf{Cat}.$$

That completes the definition.

Operads in **Top** can be regarded as T-operads, where T is the free topological monoid monad (4.1.14, 4.2.16). The operadic techniques used in Trimble's definition are then expressible in the language of generalized operads, and I make the following

Claim 10.1.9 *For each $n \in \mathbb{N}$, there is a contractible globular n-operad whose category of algebras is equivalent to* **Flabby-n-Cat**.

I hope to prove this elsewhere. It implies that every flabby n-category is a weak n-category in the sense of the previous chapter. The n-operad concerned is something like an n-dimensional version of the operad P of Example 9.2.7 (fundamental weak ω-groupoids), but a little smaller.

Alternatively, we can generalize Trimble's definition by considering operads in an arbitrary symmetric monoidal category. This leads us to the definition of enriched n-category proposed by May (2001) (although that is not what led May there).

Let $\mathcal{B}$ and $\mathcal{G}$ be symmetric monoidal categories and suppose that $\mathcal{B}$ acts on $\mathcal{G}$ in a way compatible with both monoidal structures. (Trivial example: $\mathcal{B} = \mathcal{G}$ with tensor as action.) Let P be an operad in $\mathcal{B}$. Then we can define a **categorical P-algebra in** $\mathcal{G}$ as a $\mathcal{G}$-graph X together with maps as in (10.1), with the first $\times$ replaced by the action $\odot$ and the others $\times$s by $\otimes$s, subject to the inevitable axioms. For example, if $\mathcal{B} = \textbf{Top}$ and $\mathcal{G} = \textbf{Flabby-}n\textbf{-Cat}$ with monoidal structures given by products then the functor $\Pi_n : \mathcal{B} \longrightarrow \mathcal{G}$ induces an action $\odot$ of $\mathcal{B}$ on $\mathcal{G}$ by $B \odot G = \Pi_n(B) \times G$, and a categorical P-algebra in $\mathcal{G}$ is what we previously called a categorical $((\Pi_n)_* P)$-algebra.

The idea now is that if we can find some substitute for the functor Ξ : $\textbf{Top} \longrightarrow \textbf{CatAlg}(E)$ then we can imitate Trimble's recursive definition in this more general setting. So, May starts with a 'base' symmetric monoidal category $\mathcal{B}$ and an operad P in $\mathcal{B}$, each carrying certain extra structure and satisfying certain extra properties, the details of which need not concern us here. He then considers symmetric monoidal categories $\mathcal{G}$, restricting his attention to just those that are 'good' in the sense that they too have certain extra structure and properties, including that they are acted on by $\mathcal{B}$. Then the point is that

- the trivial example $\mathcal{G} = \mathcal{B}$ is good
- if $\mathcal{G}$ is good then so is the category of categorical P-algebras in $\mathcal{G}$.

So given a category $\mathcal{B}$ and operad P in $\mathcal{B}$ as above, we can define for each $n \in \mathbb{N}$ the category $\mathcal{B}(n; P)$ of n-P-**categories enriched in** $\mathcal{B}$ as follows:

- $\mathcal{B}(0; P) = \mathcal{B}$
- $\mathcal{B}(n + 1; P) =$ (categorical P-algebras in $\mathcal{B}(n; P)$).

An operad P in $\mathcal{B}$ is called an A_∞-**operad** if for each $k \in \mathbb{N}$, the object $P(k)$ of $\mathcal{B}$ is weakly equivalent to the unit object; here 'weakly equivalent' refers to a Quillen model category structure on $\mathcal{B}$, which is part of the assumed structure. May proposes that when P is an A_∞-operad, n-P-categories enriched in $\mathcal{B}$ should be called weak n-categories enriched in $\mathcal{B}$. His definition, like Trimble's, aims for a slightly different target from most of the other definitions. It has enrichment built in; he writes

> In all of the earlier approaches, 0-categories are understood to be sets, whereas we prefer a context in which 0-categories come with their own homotopy theory.

So, for instance, we might take the category $\mathcal{B}$ of 0-categories to be a convenient category of topological spaces, or the category of simplicial sets, or a category of chain complexes.

10.2 Non-algebraic definitions

Most of the proposed non-algebraic definitions of weak n-category can be expressed neatly using a generalization of the standard nerve construction, which describes a category as a simplicial set with properties. We discuss nerves in general, then the definitions of Joyal, Tamsamani, Simpson, Baez and Dolan (and others), Street, and Leinster (Definition $\mathbf{L'}$ of (2001b)). We finish by looking at some structures approximating to the idea of a weak ω-category in which all cells of dimension 2 and higher are invertible: A_∞-categories, Segal categories, and quasi-categories. These have been found especially useful in geometry.

Nerves

The nerve idea allows us to define species of mathematical structures by saying on the one hand what the structures look like locally, and on the other how the local pieces are allowed to be fitted together. We consider it in some generality.

The setting is a category $\mathcal{C}$ of 'mathematical structures' with a small subcategory $\mathbb{C}$ of 'local pieces' or 'building blocks'. The condition that every object of $\mathcal{C}$ is put together from objects of $\mathbb{C}$ is called density, defined in a moment.

Proposition 10.2.1 *Let $\mathbb{D}$ be a small category and $\mathcal{D}$ a category with small colimits. The following conditions on a functor $I : \mathbb{D} \longrightarrow \mathcal{D}$ are equivalent:*

a. for each $Y \in \mathcal{D}$, the canonical map

$$\int^{D \in \mathbb{D}} \mathcal{D}(ID, Y) \times ID \longrightarrow Y$$

    is an isomorphism

b. the functor

$$\begin{aligned} \mathcal{D} &\longrightarrow [\mathbb{D}^{\mathrm{op}}, \mathbf{Set}], \\ Y &\longmapsto \mathcal{D}(I-, Y) \end{aligned}$$

    is full and faithful.

Proof See Mac Lane (1971, X.6). $\square$

A functor $I : \mathbb{D} \longrightarrow \mathcal{D}$ is **dense** if it satisfies the equivalent conditions of the Proposition, and a subcategory $\mathbb{C}$ of a category $\mathcal{C}$ is **dense** if the inclusion functor $\mathbb{C} \hookrightarrow \mathcal{C}$ is dense. Condition (a) formalizes the idea that objects of $\mathcal{C}$ are pasted-together objects of $\mathbb{C}$. Condition (b) is what we use in examples to prove density.

Example 10.2.2 Let $n \in \mathbb{N}$, let n-**Mfd** be the category of smooth n-manifolds and smooth maps, and let $\mathbb{U}_n$ be the subcategory whose objects are all open subsets of $\mathbb{R}^n$ and whose maps $f : U \longrightarrow U'$ are diffeomorphisms from U to an open subset of U'. Then $\mathbb{U}_n$ is dense in n-**Mfd**: every manifold is a pasting of Euclidean open sets.

Example 10.2.3 Let k be a field, let **Vect**$_k$ be the category of all vector spaces over k, and let **Mat**$_k$ be the category whose objects are the natural numbers and whose maps $m \longrightarrow n$ are $n \times m$ matrices over k. There is a natural inclusion **Mat**$_k \hookrightarrow$ **Vect**$_k$ (sending n to k^n), and **Mat**$_k$ is then dense in **Vect**$_k$. This reflects the fact that any vector space is the colimit (pasting-together) of its finite-dimensional subspaces, which in turn is true because the theory of vector spaces is finitary (its operations take only a finite number of arguments).

Example 10.2.4 The inclusion $\Delta \hookrightarrow$ **Cat** (p. 116) is also dense. This says informally that a category is built out of objects, arrows, commutative triangles, commutative tetrahedra, ..., and source, target and identity functions between them. We would not expect the commutative tetrahedra and higher-dimensional simplices to be necessary, and indeed, if Δ_2 denotes the full subcategory of Δ consisting of the objects [0], [1] and [2] then the inclusion $\Delta_2 \hookrightarrow$ **Cat** is also dense.

Generalizing the terminology of this example, if $\mathbb{C}$ is a dense subcategory of $\mathcal{C}$ and X is an object of $\mathcal{C}$ then the **nerve** (over $\mathbb{C}$) of X is the presheaf $\mathcal{C}(-, X)$ on $\mathbb{C}$.

For us, the crucial point about density is that it allows the mathematical structures (objects of $\mathcal{C}$) to be viewed as presheaves-with-properties on the category $\mathbb{C}$ of local pieces. That is, if $\mathbb{C}$ is dense in $\mathcal{C}$ then $\mathcal{C}$ is equivalent to the full subcategory of $[\mathbb{C}^{\text{op}}, \textbf{Set}]$ consisting of those presheaves isomorphic to the nerve of some object of $\mathcal{C}$. Sometimes the presheaves arising as nerves can be characterized intrinsically, yielding an alternative definition of $\mathcal{C}$ by presheaves on $\mathbb{C}$.

Example 10.2.5 A vector space over a field k can be *defined* as a presheaf $V : \textbf{Mat}_k^{\text{op}} \longrightarrow \textbf{Set}$ preserving finite limits.

Example 10.2.6 A smooth n-manifold can be *defined* as a presheaf on $\mathbb{U}_n$ with certain properties.

Example 10.2.7 A category can be *defined* as a presheaf on Δ (or indeed on Δ_2) with certain properties. There are various ways to express those properties: for instance, categories are functors $\Delta^{\text{op}} \longrightarrow \textbf{Set}$ preserving finite limits, or preserving certain pullbacks, or they are simplicial sets in which every inner

horn has a unique filler. (A horn is **inner** if the missing face is not the first or
the last one.)

The structures we want to define are weak n-categories, and the strategy is:

- find a small dense subcategory $\mathbb{C}$ of the category of *strict n-categories*
- find conditions on a presheaf on $\mathbb{C}$ equivalent to it being a nerve of a strict
  n-category
- relax those conditions to obtain a definition of *weak n-category*.

Most of the definitions of weak n-category described below can be regarded
as implementations of this strategy. A different way to put it is that we seek an
intrinsic characterization of presheaves on $\mathbb{C}$ of the form

$$C \longmapsto \{\text{weak functors } C \longrightarrow Y\}$$

for some weak n-category Y. Of course, we start from a position of not knowing what a weak n-category or functor is, but we choose the conditions on
presheaves to fit the usual intuitions.

Joyal's definition

Perhaps the most obvious implementation is to take the local pieces to be
all globular pasting diagrams. So, let Δ_ω be the category with object-set
$\coprod_{m \in \mathbb{N}} \mathbf{pd}(m)$ and hom-sets

$$\Delta_\omega(\sigma, \pi) = \mathbf{Str}\text{-}\omega\text{-}\mathbf{Cat}(F\hat{\sigma}, F\hat{\pi}) \cong [\mathbb{G}^{\mathrm{op}}, \mathbf{Set}](\hat{\sigma}, T\hat{\pi})$$

where $F : [\mathbb{G}^{\mathrm{op}}, \mathbf{Set}] \longrightarrow \mathbf{Str}\text{-}\omega\text{-}\mathbf{Cat}$ is the free strict ω-category functor and,
as in Chapter 9, T is the corresponding monad. There is an inclusion functor
$\Delta_\omega \hookrightarrow \mathbf{Str}\text{-}\omega\text{-}\mathbf{Cat}$, and a typical object of the corresponding subcategory of
$\mathbf{Str}\text{-}\omega\text{-}\mathbf{Cat}$ is the ω-category naturally depicted as

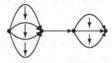

– that is, freely generated by the 0-, 1- and 2-cells shown, and with only identity
cells in dimensions 3 and above.

The subcategory Δ_ω of $\mathbf{Str}\text{-}\omega\text{-}\mathbf{Cat}$ is dense. That the induced functor
$\mathbf{Str}\text{-}\omega\text{-}\mathbf{Cat} \longrightarrow [\Delta_\omega^{\mathrm{op}}, \mathbf{Set}]$ is faithful follows from Δ_ω containing the trivial pasting diagrams

representing single m-cells. That it is full follows from Δ_ω containing maps
such as

$$\text{(10.2)}$$

which, for appropriately chosen f and g, induce 2-cell identities and vertical
2-cell composition respectively.

A strict ω-category can therefore be defined as a presheaf on Δ_ω with properties. Joyal (1997) proposed, effectively, a way of describing and then relaxing
those properties: a strict ω-category is a presheaf on Δ_ω for which 'every inner
horn has a unique filler', and a weak ω-category is defined by simply dropping
the uniqueness.

We can do the same with ω replaced by any finite n (taking care in the top
dimension). Recall from p. 270 that if 1_π denotes the $(m+1)$-pasting diagram
resembling an m-pasting diagram π then $\widehat{\widehat{\pi}} \cong \widehat{1_\pi}$: so Δ_n is equivalent to its
full subcategory consisting of just the n-pasting diagrams. For instance, Δ_1 is
equivalent to the usual category Δ of 1-pasting diagrams

and we recover the standard nerve construction for categories.

Joyal also noted a duality. The category Δ is equivalent to the opposite
of the category $\mathbb{I}$ of **finite strict intervals,** that is, finite totally ordered sets
with distinct least and greatest elements (to be preserved by the maps). Generalizing this, he defined a category $\mathbb{I}_\omega$ of 'finite disks', which turns out to be
equivalent to the opposite of Δ_ω; see the Notes for references. So his weak
ω-categories were defined as functors $\mathbb{I}_\omega \longrightarrow$ **Set** satisfying horn-filling
conditions.

Tamsamani's and Simpson's definitions

A very similar story can be told for the definitions proposed by Tamsamani (1995) and Simpson (1997). Observe that in the proof of the density
of Δ_ω in **Str-ω-Cat**, we did not use many of the pasting diagrams, so we
can replace Δ_ω by a smaller category. Tamsamani and Simpson consider just
'cuboidal' pasting diagrams such as

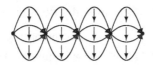

and similarly 'cuboidal' maps between the strict ω-categories that they generate.

Let us restrict ourselves to the n-dimensional case, since that is a little easier. There is a functor

$$I : \Delta^n \longrightarrow \textbf{Str-}n\textbf{-Cat}$$

which, for instance, when $n = 2$, sends ([4], [3]) to the free strict 2-category on the diagram above. In general, each $(r_1, \ldots, r_n) \in \mathbb{N}^n$ determines an n-pasting diagram $\pi_{r_1,\ldots,r_n}$, given inductively by

$$\pi_{r_1,\ldots,r_n} = \left(\pi_{r_2,\ldots,r_n}, \ldots, \pi_{r_2,\ldots,r_n} \right)$$

with r_1 terms on the right-hand side, and then

$$I([r_1], \ldots, [r_n]) = F\widehat{\pi_{r_1,\ldots,r_n}}.$$

To describe I on maps, take, for instance, $n = 2$ and the map (id, δ) : $([1], [1]) \longrightarrow ([1], [2])$ in which δ is the injection omitting $1 \in [2]$ from its image; then $I(\mathrm{id}, \delta)$ is the map g of (10.2).

By exactly the same argument as for Joyal's definition, the functor $\Delta^n \longrightarrow \textbf{Str-}n\textbf{-Cat}$ is dense. A strict n-category is therefore the same thing as a presheaf on Δ^n (a 'multisimplicial set') with properties, and relaxing those properties gives a definition of weak n-category.

Nerves of strict n-categories are characterized among functors $(\Delta^n)^{\mathrm{op}} \longrightarrow \textbf{Set}$ by the properties that the functor is degenerate in certain ways (to give us n-categories rather than n-tuple categories) and, more significantly, that certain pullbacks are preserved. Tamsamani sets up a notion of equivalence, and defines weak n-category by asking only that the pullbacks are preserved up to equivalence. Simpson does the same, but with a more stringent notion of equivalence that he calls 'easy equivalence'. It is indeed easier, and is nearly the same as the notion of contractibility of a map of globular sets: see my survey (2001b) for details. In the special case of one-object 2-categories, Tamsamani's definition gives the homotopy monoidal categories of Section 3.3, and Simpson's gives the same but with the extra condition that the functors $\xi^{(k)}$ of Proposition 3.3.6, which for homotopy monoidal categories are required to be equivalences, are *genuinely* surjective on objects.

Opetopic definitions

We have already (Section 7.5) looked at the opetopic definitions of weak n-category: that of Baez and Dolan and subsequent variants. They are all of the form 'a weak n-category is an opetopic set with properties', for varying meanings of 'opetopic set' and varying lists of properties. Here we see how this fits in with the nerve idea. The situation is, as we shall see, slightly different from that in the definitions of Joyal, Tamsamani, and Simpson.

We start with the category $\mathbb{O}$ of opetopes. As mentioned on p. 247, there is an embedding $\mathbb{O} \hookrightarrow \mathbf{Str}\text{-}\omega\text{-}\mathbf{Cat}$, and this induces a functor

$$U : \mathbf{Str}\text{-}\omega\text{-}\mathbf{Cat} \longrightarrow [\mathbb{O}^{\mathrm{op}}, \mathbf{Set}]$$

sending a strict ω-category to its underlying opetopic set.

This functor is faithful, because there is for each $m \in \mathbb{N}$ an m-opetope resembling a single globular m-cell. It is not, however, full. To see this, note that if $F : A \longrightarrow B$ is a strict map of strict ω-categories then the induced map $U(F) : U(A) \longrightarrow U(B)$ preserves universality of cells: for instance, if f and g are abutting 1-cells in A then $U(F)$ sends the canonical 2-cell

$$
\begin{array}{c}
f \nearrow \underset{\Downarrow}{} \nwarrow g \\[-4pt]
\xrightarrow{\hspace{1.2cm}} \\[-6pt]
g \circ f
\end{array}
$$

in $U(A)$ to the canonical 2-cell

$$
\begin{array}{c}
Ff \nearrow \underset{\Downarrow}{} \nwarrow Fg \\[-4pt]
\xrightarrow{\hspace{1.4cm}} \\[-6pt]
(Fg) \circ (Ff)
\end{array}
$$

in $U(B)$. But not every map $U(A) \longrightarrow U(B)$ of opetopic sets preserves universality; indeed, any *lax* map $A \longrightarrow B$ of strict ω-categories ought (in principle, at least) to induce a map $U(A) \longrightarrow U(B)$, and this will preserve universality if and only if the lax map is weak. Compare the relationship between monoidal categories and plain multicategories (2.1.10, 3.3).

So $\mathbb{O}$ is not dense in $\mathbf{Str}\text{-}\omega\text{-}\mathbf{Cat}$, and correspondingly U does not define an equivalence between $\mathbf{Str}\text{-}\omega\text{-}\mathbf{Cat}$ and a full subcategory of $[\mathbb{O}^{\mathrm{op}}, \mathbf{Set}]$. But with a slight modification, the nerve idea can still be made to work. For the above arguments suggest that U defines an equivalence between

(strict ω-categories + weak maps)

and a full subcategory of

(opetopic sets + universality-preserving maps),

and it is then, as usual, a matter of identifying the characteristic properties of those opetopic sets arising from strict ω-categories, then relaxing the properties to obtain a definition of weak ω-category. So a weak ω-category is defined as an opetopic set with properties, and a weak ω-functor as a map of opetopic sets preserving universality.

Street's definition

The definition of weak ω-category proposed by Street has the distinctions of being the first and probably the most tentatively phrased; it hides in the last paragraph of his paper of (1987). It was part of the inspiration for Baez and Dolan's definition, and has much in common with it, but uses simplicial rather than opetopic sets.

Street follows the nerve idea explicitly. He first constructs an embedding

$$I : \Delta \hookrightarrow \textbf{Str-}\omega\textbf{-Cat},$$

where $I(m)$ is the mth 'oriental', the free strict ω-category on an m-simplex. For example, $I(3)$ is the strict ω-category generated freely by 0-, 1-, 2- and 3-cells

$$
\begin{array}{ccc}
a_1 \xrightarrow{f_{12}} a_2 & & a_1 \xrightarrow{f_{12}} a_2 \\
\llap{f_{01}}\Big/ \; {}^{\alpha_{012}} \;\; \Big\backslash\rlap{f_{23}} & \xRightarrow{\;\Gamma\;} & \llap{f_{01}}\Big/ \; {}^{\alpha_{123}} \;\; \Big\backslash\rlap{f_{23}} \\
a_0 \underset{f_{02}}{} \Uparrow^{\alpha_{023}} a_3 & & a_0 \;\; {}_{\alpha_{013}}\Uparrow \;\; f_{13} \; a_3. \\
a_0 \xrightarrow[f_{03}]{} a_3 & & a_0 \xrightarrow[f_{03}]{} a_3.
\end{array}
$$

(Orientation needs care.) This induces a functor

$$U : \textbf{Str-}\omega\textbf{-Cat} \longrightarrow [\Delta^{\text{op}}, \textbf{Set}]$$

and, roughly speaking, a weak ω-category is defined as a simplicial set with horn-filling properties.

This is, however, a slightly inaccurate account. For similar reasons to those in the opetopic case, Δ is not dense in **Str-ω-Cat**; the functor U is again faithful but not full. Street's original solution was to replace $[\Delta^{\text{op}}, \textbf{Set}]$ by the category **Sss** of **stratified simplicial sets**, that is, simplicial sets equipped with a class of distinguished cells in each dimension (to be thought of as 'universal', 'hollow', or 'thin'). The underlying simplicial set of a strict ω-category has a canonical stratification, so U lifts to a functor

$$U' : \textbf{Str-}\omega\textbf{-Cat} \longrightarrow \textbf{Sss},$$

and U' *is* full and faithful. Detailed work by Street (1987, 1988) and Verity (unpublished) gives precise conditions for an object of **Sss** to be in the image of U'. So a strict ω-category is the same thing as a simplicial set equipped with a class of distinguished cells satisfying some conditions. One of the conditions is that certain horns have unique fillers, and dropping the uniqueness gives Street's proposed definition of weak ω-category.

The most vexing aspect of this proposal is that extra structure is required on the simplicial set. It would seem more satisfactory if, as in the opetopic approach, the universal cells could be recognized intrinsically. A recent paper of Street (2003) aims to repair this apparent defect, proposing a similar definition in which a weak ω-category is genuinely a simplicial set with properties.

Contractible multicategories

The last definition of weak n-category that we consider was introduced as Definition **L**$'$ in my (2001b) survey. Like the other definitions in this section, it is non-algebraic and can be described in terms of nerves. The nerve description seems, however, to be rather complicated (the shapes involved being a combination of globular and opetopic) and not especially helpful, so we approach it from another angle instead.

The idea is that a weak ω-category is meant to be a 'weak algebra' for the free strict ω-category monad on globular sets. We saw in 4.2.22 that for any cartesian monad T, a *strict T-algebra* is the same thing as a T-multicategory whose domain map is the identity – in other words, with underlying graph of the form

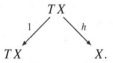

To define 'weak T-algebra' we relax the condition that the domain map is the identity, asking only that it be an equivalence in some sense. Contractibility together with surjectivity on 0-cells is a reasonable notion of equivalence: contractible means something like 'injective on homotopy', and any contractible map surjective on 0-cells is surjective on m-cells for all $m \in \mathbb{N}$. For 1-dimensional structures, it means full, faithful and surjective on objects. We also ask that the domain map is injective on 0-cells, expressing the thought that 0-cells in an ω-category should not be composable.

Definition **L**$'$ says, then, that a weak ω-category is a globular multicategory C whose domain map $C_1 \longrightarrow TC_0$ is bijective on 0-cells and contractible.

Any weak ω-category in the sense of the previous chapter gives rise canonically to one in the sense of **L**$'$. This is the 'multicategory of elements' construction of Section 6.3: if $(T_L X \xrightarrow{h} X)$ is an L-algebra then there is a

commutative diagram

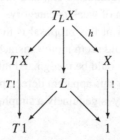

the left-hand half of which is a pullback square and the top part of which forms
a multicategory C^X with $C_0^X = X$ and $C_1^X = T_L X$. The map $L \longrightarrow T1$ is
contractible (by definition of L) and bijective on 0-cells (because, as we saw
on p. 307, if $\cdot$ denotes the 0-pasting diagram then $L(\cdot) = 1$). Contractibility
and bijectivity on 0-cells are stable under pullback, so the multicategory C^X is
a weak ω-category in the sense of **L**$'$.

Unpicking this construction explains further the idea behind **L**$'$. An m-cell
of $T_L X$ is a pair

$$(\theta, \mathbf{x}) \in \coprod_{\pi \in \mathbf{pd}(m)} L(\pi) \times [\mathbb{G}^{\mathrm{op}}, \mathbf{Set}](\widehat{\pi}, X),$$

which lies over $\mathbf{x} \in (TX)(m)$ and $\overline{\theta}(\mathbf{x}) \in X(m)$. It is usefully regarded as a
'way of composing' the labelled pasting diagram $\mathbf{x}$. (Contractibility guarantees
that there are plenty of ways of composing.) Among all weak ω-categories in
the sense of **L**$'$, those of the form C^X have the special feature that the set of
ways of composing a labelled pasting diagram $\mathbf{x} \in (TX)(\pi)$ depends only on
the pasting diagram π, not on the labels: it is just $L(\pi)$.

So in the definition of the previous chapter, the ways of composition available in a weak ω-category are prescribed once and for all; in the present definition, they are allowed to vary from ω-category to ω-category. This is precisely
analogous to the difference between the loop space machinery of Boardman–
Vogt and May (with fixed parameter spaces forming an operad) and that of
Segal (with a variable, flabby, structure): see Adams (1978, p. 60).

Locally groupoidal structures

Many of the weak ω-categories of interest in geometry have the property that
all cells of dimension 2 and higher are equivalences (weakly invertible). Some
examples were given in 'Motivation for topologists'. There are several structures in use that, roughly speaking, aim to formalize the idea of such a weak
ω-category. I will describe three of them here.

A weak ω-category in which all cells (of dimension 1 and higher) are equivalences is called a **weak ω-groupoid**. This is, of course, subject to precise definitions of weak ω-category and equivalence. From the topological viewpoint one of the main purposes of ω-groupoids is to model homotopy types of spaces (see Grothendieck's letter of (1983), for instance), so it is reasonable to replace ω-groupoids by spaces, or perhaps simplicial sets or chain complexes. The structures we seek are, therefore, graphs $(X(x, x'))_{x,x' \in X_0}$ of spaces, simplicial sets, or chain complexes, together with extra data determining some kind of weak composition.

We have already seen one version of this: an A_∞-category (p. 236) is a graph X of chain complexes together with various composition maps

$$X(x_{k-1}, x_k) \otimes \cdots \otimes X(x_0, x_1) \longrightarrow X(x_0, x_k)$$

parametrized by elements of the operad A_∞. There is a similar notion with spaces in place of complexes. These are algebraic definitions (so properly belong in the previous section).

A similar but non-algebraic notion is that of a Segal category (sometimes called by other names: see the Notes below). Take a bisimplicial set, expressed as a functor

$$X : \Delta^{\mathrm{op}} \longrightarrow [\Delta^{\mathrm{op}}, \mathbf{Set}],$$

and suppose that the simplicial set $X[0]$ is discrete (that is, the functor $X[0]$ is constant). Write X_0 for the set of points (constant value) of $X[0]$. Then for each $k \in \mathbb{N}$, the simplicial set $X[k]$ decomposes naturally as a coproduct

$$X[k] \cong \coprod_{x_0,\ldots,x_k \in X_0} X(x_0, \ldots, x_k),$$

and for each $k \in \mathbb{N}$ and $x_0, \ldots, x_k \in X_0$, there is a natural map

$$X(x_0, \ldots, x_k) \longrightarrow X(x_{k-1}, x_k) \times \cdots \times X(x_0, x_1).$$

A bisimplicial set X is called a **Segal category** if $X[0]$ is discrete and each of these canonical maps is a weak equivalence of simplicial sets, in the homotopy-theoretic sense. This definition is very closely related to the definitions of n-category proposed by Simpson and Tamsamani, and in particular to the definition of homotopy bicategory in Chapter 3.

Finally, there are the quasi-categories of Joyal, Boardman, and Vogt. We have a diagram

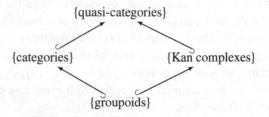

of classes of simplicial sets, in which

- categories are identified with simplicial sets in which every inner horn has a unique filler (10.2.7)
- **Kan complexes** are simplicial sets in which every horn has at least one filler (the principal example being the underlying simplicial set of a space)
- groupoids are simplicial sets in which every horn has a unique filler
- **quasi-categories** are simplicial sets in which every inner horn has at least one filler.

Large amounts of the theory of ordinary categories can be reproduced for quasi-categories, although requiring much longer proofs. Given a simplicial set X and elements $x, x' \in X[0]$, there is a simplicial set $X(x, x')$, the analogue of the space of paths from x to x' in topology, and it can be shown that if X is a quasi-category then each $X(x, x')$ is a Kan complex. So quasi-categories do indeed approximate the idea of a weak ω-category in which all cells of dimension at least 2 are invertible.

Notes

An extensive bibliography and historical discussion of proposed definitions of n-category is in my survey paper (2001b).

Various people have confused the contractions of Chapter 9 with the contractions of Batanin, principally me (see the Notes to Chapter 9) but also Berger (2002, 1.20). I apologize to Batanin for stealing his word, using it for something else, then renaming his original concept (10.1.5).

Berger (2002) has investigated nerves of ω-categories in detail, making connections to various proposed definitions of weak ω-category. Other work on higher-dimensional nerves has been done by Street (1996, §10, and references therein) and Duskin (2002, 2001). Berger's Proposition 2.2 provides most of a proof that the category $\mathbb{I}_\omega$ of finite discs used in Joyal's definition is the opposite of the category Δ_ω of globular pasting diagrams. A different proof was given by Makkai and Zawadowski (2001), and Joyal's definition has also been illuminated by Batanin and Street (2000).

The observation that most geometrically interesting ω-categories are locally groupoidal was made to me by Bertrand Toën. His (2002) paper with Vezzosi gives an introduction to Segal categories, as well as further references. In particular, they point to Dwyer, Kan and Smith (1989), where Segal categories were called special bisimplicial sets, and Schwänzl and Vogt (1992), where they were called Δ-categories.

Joyal's quasi-categories are a renaming of Boardman and Vogt's 'restricted Kan complexes' (1973). His (2002) paper contains the basic results, but much of his work remains unpublished.

I thank Sjoerd Crans for useful conversations on the details of Batanin's definition, Jacques Penon for the observation that coherent maps do not compose, and Michael Batanin and Ross Street for useful comments on their respective definitions.

Appendices

Appendix A

Symmetric structures

For my birthday I got a humidifier and a de-humidifier ... I put them in
the same room and let them fight it out.

Steven Wright

Here we meet an alternative definition of symmetric multicategory and prove
it equivalent to the usual one. This has two purposes. First, the alternative definition is in some ways nicer and more natural than the usual one, avoiding as it
does the delicate matter of formulating the symmetry axioms (2.2.20, 2.2.21).
Second, it will be used in Appendix C to show that every symmetric multicategory gives rise to a T-multicategory for each T belonging to a certain large
class of cartesian monads.

Actually, we start (Section A.1) with an alternative definition of commutative monoid. Although this could hardly be shorter than the standard definition, it acts as a warm-up to the alternative definition of symmetric multicategory (Section A.2).

A.1 Commutative monoids

Here we define 'fat commutative monoids' and prove that they are essentially
the same as the ordinary kind. The idea is to have some direct way of summing
arbitrary finite families $(a_x)_{x \in X}$ of elements of a commutative monoid A, not
just ordered sequences $a_1, \ldots, a_n$.

Definition A.1.1 A **fat commutative monoid** is a set A equipped with a function $\sum_X : A^X \longrightarrow A$ for each finite set X, satisfying the axioms below. We
write elements of A^X as families $(a_x)_{x \in X}$ (where $a_x \in A$) and $\sum_X (a_x)_{x \in X}$ as

333

$\sum_{x \in X} a_x$. The axioms are:

- for any map $s : X \longrightarrow Y$ of finite sets and any family $(a_x)_{x \in X}$ of elements of A,

$$\sum_{y \in Y} \sum_{x \in s^{-1}\{y\}} a_x = \sum_{x \in X} a_x$$

- for any one-element set X and any $a \in A$,

$$a = \sum_{x \in X} a.$$

A **map** $A \longrightarrow A'$ of fat commutative monoids is a function $f : A \longrightarrow A'$ such that for all finite sets X, the square

$$
\begin{array}{ccc}
A^X & \xrightarrow{f^X} & A'^X \\
\Sigma_X \big\downarrow & & \big\downarrow \Sigma_X \\
A & \xrightarrow{\;\;f\;\;} & A'
\end{array}
$$

commutes.

Observe the following crucial property immediately:

Lemma A.1.2 *Let A be a fat commutative monoid, $s : X \longrightarrow Y$ a bijection between finite sets, and $(b_y)_{y \in Y}$ an indexed family of elements of A. Then*

$$\sum_{x \in X} b_{s(x)} = \sum_{y \in Y} b_y.$$

Proof Define a family $(a_x)_{x \in X}$ by $a_x = b_{s(x)}$. Then

$$\sum_{x \in X} a_x = \sum_{y \in Y} \sum_{x \in s^{-1}\{y\}} a_x = \sum_{y \in Y} b_y,$$

by the first and second axioms respectively. $\qquad\square$

This allows us to take the expected liberties with notation. If, for example, we have a family of elements $a_{v,w,x}$ indexed over finite sets V, W and X, then we may write $\sum_{v \in V, w \in W, x \in X} a_{v,w,x}$ without ambiguity; the sum could 'officially' be interpreted as either of

$$\sum_{((v,w),x) \in (V \times W) \times X} a_{v,w,x} \quad \text{or} \quad \sum_{(v,(w,x)) \in V \times (W \times X)} a_{v,w,x}$$

(or some further possibility), but these expressions are equal.

Those concerned with foundations might feel uneasy about the idea of specifying a function $\sum_X : A^X \longrightarrow A$ 'for each finite set X'. The remedy is to choose a small full subcategory $\mathbb{F}$ of the category of finite sets and functions, such that $\mathbb{F}$ contains at least one object of each finite cardinality, and interpret 'finite set' as 'object of $\mathbb{F}$'; everything works just as well. In particular, you might choose to replace the category of finite sets with its skeleton whose objects are the natural numbers $\mathbf{n} = \{1, \ldots, n\}$, and this might seem like a simplifying move, but it can actually make fat commutative monoids confusing to work with: for instance, whereas bijections

$$X \xrightarrow{\ s\ } Y \xrightarrow{\ t\ } Z$$

can be composed in only one possible order, permutations $s, t \in S_n$ can be composed in two. I will stick with 'all finite sets'.

Write **FatCommMon** and **CommMon** for the categories of fat and ordinary commutative monoids, respectively.

Theorem A.1.3 *There is an isomorphism of categories*

$$\textbf{FatCommMon} \cong \textbf{CommMon}.$$

Proof We show that both sides are isomorphic to the category **UCommMon** of unbiased commutative monoids. By definition, an **unbiased commutative monoid** is a set A equipped with an n-ary addition operation

$$\begin{array}{ccc} A^n & \longrightarrow & A \\ (a_1, \ldots, a_n) & \longmapsto & (a_1 + \cdots + a_n) \end{array}$$

for each $n \in \mathbb{N}$, satisfying the three axioms displayed in Example 2.2.2 (written there with $\cdot$ instead of $+$ and xs instead of as). We have **UCommMon** $\cong$ **CommMon**, easily.

Given a fat commutative monoid $(A, \sum)$, define an unbiased commutative monoid structure $+$ on A by

$$(a_1 + \cdots + a_n) = \sum_{x \in \{1, \ldots, n\}} a_x.$$

The first two axioms for an unbiased commutative monoid follow from the two axioms for a fat commutative monoid, and the third follows from Lemma A.1.2.

Conversely, take an unbiased commutative monoid $(A, +)$ and define a fat commutative monoid structure $\sum$ on A as follows. For any finite set X, let $n_X \in \mathbb{N}$ be the cardinality of X and choose a bijection $t_X :$

$\{1, \ldots, n_X\} \xrightarrow{\sim} X$; then define $\sum_X : A^X \longrightarrow A$ by

$$\sum_{x \in X} a_x = (a_{t_X(1)} + \cdots + a_{t_X(n_X)}).$$

By commutativity, this definition is independent of the choice of t_X. Clearly the axioms for a fat commutative monoid are satisfied.

It is straightforward to check that these two processes are mutually inverse and extend to an isomorphism of categories.   □

A.2 Symmetric multicategories

As in the previous section, we reformulate a notion of symmetric structure by moving from finite sequences to finite families: so in a 'fat symmetric multicategory', maps look like

$$(a_x)_{x \in X} \xrightarrow{\theta} b$$

rather than

$$a_1, \ldots, a_n \xrightarrow{\theta} b.$$

Definition A.2.1 A **fat symmetric multicategory** A consists of

- a set A_0, whose elements are called the **objects** of A
- for each finite set X, family $(a_x)_{x \in X}$ of objects, and object b, a set $C((a_x)_{x \in X}; b)$, whose elements θ are called **maps** in A and written

$$(a_x)_{x \in X} \xrightarrow{\theta} b$$

- for each function $s : X \longrightarrow Y$ between finite sets, family $(a_x)_{x \in X}$ of objects, family $(b_y)_{y \in Y}$ of objects, and object c, a function

$$A((b_y)_{y \in Y}; c) \times \prod_{y \in Y} A((a_x)_{x \in s^{-1}\{y\}}; b_y) \longrightarrow A((a_x)_{x \in X}; c),$$

  called **composition** and written

$$(\phi, (\theta_y)_{y \in Y}) \longmapsto \phi \circ (\theta_y)_{y \in Y}$$

- for each one-element set X and object a, an **identity** map

$$1_a^X \in A((a)_{x \in X}; a),$$

satisfying

- associativity: if $X \xrightarrow{s} Y \xrightarrow{t} Z$ are functions between finite sets and

$$(a_x)_{x \in s^{-1}\{y\}} \xrightarrow{\theta_y} b_y, \qquad (b_y)_{y \in t^{-1}\{z\}} \xrightarrow{\phi_z} c_z, \qquad (c_z)_{z \in Z} \xrightarrow{\psi} d$$

are maps in A (for $y \in Y, z \in Z$), then

$$(\psi \circ (\phi_z)_{z \in Z}) \circ (\theta_y)_{y \in Y} = \psi \circ (\phi_z \circ (\theta_y)_{y \in t^{-1}\{z\}})_{z \in Z}$$

- left identity axiom: if X is a finite set, Y a one-element set, and $\theta :$ $(a_x)_{x \in X} \longrightarrow b$ a map in A, then

$$1_b^Y \circ (\theta)_{y \in Y} = \theta$$

- right identity axiom: if X is a finite set and $\theta : (a_x)_{x \in X} \longrightarrow b$ a map in A, then

$$\theta \circ (1_{a_x}^{\{x\}})_{x \in X} = \theta.$$

A **map** $f : A \longrightarrow A'$ of fat symmetric multicategories consists of

- a function $f : A_0 \longrightarrow A_0'$
- for each finite set X, family $(a_x)_{x \in X}$ of objects of A, and object b of A, a function

$$A((a_x)_{x \in X}; b) \longrightarrow A'((fa_x)_{x \in X}; fb),$$

also written as f,

such that

- $f(\phi \circ (\theta_y)_{y \in Y}) = f(\phi) \circ (f(\theta_y))_{y \in Y}$ whenever these composites make sense
- $f(1_a^X) = 1_{fa}^X$ whenever a is an object of A and X is a one-element set.

This defines a category **FatSymMulticat**.

As promised, this definition avoids the delicate symmetry axioms present in the traditional version. The following lemma, analogous to Lemma A.1.2, shows that the symmetric group actions really are hiding in there.

Lemma A.2.2 *Let A be a fat symmetric multicategory. Then any bijection $s : X \longrightarrow Y$ between finite sets and map $\phi : (b_y)_{y \in Y} \longrightarrow c$ in A give rise to a map $\phi \cdot s : (b_{s(x)})_{x \in X} \longrightarrow c$ in A. This construction satisfies*

$$\phi \cdot (s \circ r) = (\phi \cdot s) \cdot r, \qquad \theta \cdot 1_Y = \theta,$$

where $W \xrightarrow{r} X \xrightarrow{s} Y$ in the first equation. Moreover, if f is a map of fat symmetric multicategories then $f(\phi \cdot s) = f(\phi) \cdot s$ whenever these expressions make sense.

Proof Take s and ϕ as in the statement. Define a family $(a_x)_{x \in X}$ by $a_x = b_{s(x)}$. For each $y \in Y$ we have the map

$$1_{b_y}^{s^{-1}\{y\}} : (b_y)_{x \in s^{-1}\{y\}} \longrightarrow b_y;$$

but $b_y = a_{s^{-1}(y)}$, so the domain of this map is $(a_x)_{x \in s^{-1}\{y\}}$. We may therefore define

$$\phi \cdot s = \phi \circ \left(1_{b_y}^{s^{-1}\{y\}}\right)_{y \in Y} : (a_x)_{x \in X} \longrightarrow c.$$

The two equations follow from the associativity and identity axioms for a fat symmetric multicategory, respectively. That maps of fat symmetric multicategories preserve $\cdot$ is immediate from the definitions. $\qquad\square$

This lemma is very useful in the proof below that fat and ordinary symmetric multicategories are essentially the same. Like Lemma A.1.2, it also allows us to take notational liberties. If, for example, we have a family of objects $a_{v,w,x}$ indexed over finite sets V, W and X, then we may safely speak of 'maps

$$(a_{v,w,x})_{v \in V, w \in W, x \in X} \longrightarrow b$$

in A'; it does not matter whether the indexing set in the domain is meant to be $(V \times W) \times X$ or $V \times (W \times X)$ (or some other 3-fold product), as the canonical isomorphism between them induces a canonical isomorphism of homsets,

$$A((a_{v,w,x})_{((v,w),x) \in (V \times W) \times X}; b) \xrightarrow{\sim} A((a_{v,w,x})_{(v,(w,x)) \in V \times (W \times X)}; b).$$

Example A.2.3 A **fat symmetric operad** P is, of course, a fat symmetric multicategory with only one object; it consists of a set $P(X)$ for each finite set X, together with composition and identity operations.

Many well-known examples of symmetric operads are naturally regarded as fat symmetric operads. For instance, there is a fat symmetric operad $\mathbf{D}$ where an element of $\mathbf{D}(X)$ is an X-indexed family $(\alpha_x)_{x \in X}$ of disjoint little discs inside the unit disc (compare 2.2.16). Or, for any set S there is a fat symmetric operad $\mathbf{End}(S)$ defined by $(\mathbf{End}(S))(X) = \mathbf{Set}(S^X, S)$ (compare 2.2.8). Or, there is a fat symmetric operad O in which $O(X)$ is the set of total orders on X, with composition done lexicographically; under the equivalence we are about to establish, it corresponds to the ordinary operad $\mathbf{S}$ of symmetries. (This last example appeared in Beilinson and Drinfeld ($c.$1997, 1.1.4) and 2.2.20 above.)

Theorem A.2.4 *There is a canonical equivalence of categories*

$$\overline{(\ )} : \mathbf{FatSymMulticat} \xrightarrow{\sim} \mathbf{SymMulticat}.$$

Proof We define the functor $\overline{(\ )}$ and show that it is full, faithful and (genuinely) surjective on objects. Details are omitted.

To define $\overline{(\ )}$ on objects, take a fat symmetric multicategory A. The symmetric multicategory $\overline{A}$ has the same objects as A and hom-sets

$$\overline{A}(a_1, \ldots, a_n; b) = A((a_x)_{x \in [1,n]}; b),$$

where for $m, n \in \mathbb{N}$ we write

$$[m, n] = \{l \in \mathbb{N} \mid m \leq l \leq n\}.$$

For composition, take maps

$$a_1^1, \ldots, a_1^{k_1} \xrightarrow{\theta_1} b_1, \quad \ldots, \quad a_n^1, \ldots, a_n^{k_n} \xrightarrow{\theta_n} b_n,$$
$$b_1, \ldots, b_n \xrightarrow{\phi} c$$

in $\overline{A}$. Define objects $a_1, \ldots, a_{k_1 + \cdots + k_n}$ by the equation of formal sequences

$$(a_1, \ldots, a_{k_1 + \cdots + k_n}) = (a_1^1, \ldots, a_1^{k_1}, \ldots, a_n^1, \ldots, a_n^{k_n}).$$

For each $x \in [1, n]$ there is an obvious bijection

$$t_x : [k_1 + \cdots + k_{x-1} + 1, k_1 + \cdots + k_{x-1} + k_x] \xrightarrow{\sim} [1, k_x]$$

defined by subtraction: so the map

$$(a_{k_1 + \cdots + k_{x-1} + y})_{y \in [1, k_x]} = (a_x^y)_{y \in [1, k_n]} \xrightarrow{\theta_x} b_x$$

in A gives rise to a map

$$(a_z)_{z \in [k_1 + \cdots + k_{x-1} + 1, k_1 + \cdots + k_{x-1} + k_x]} \xrightarrow{\theta_x \cdot t_x} b_x$$

in A, by Lemma A.2.2. It now makes sense to define composition in $\overline{A}$ by

$$\phi \circ (\theta_1, \ldots, \theta_n) = \phi \circ (\theta_x \cdot t_x)_{x \in [1,n]},$$

since the domain of this map is $(a_z)_{z \in [1, k_1 + \cdots + k_n]}$. Identities in $\overline{A}$ are easier: for $a \in A$, put $1_a = 1_a^{[1,1]}$. The structure $\overline{A}$ just defined really is a symmetric multicategory, as is straightforward to prove with the aid of Lemma A.2.2.

The definition of the functor $\overline{(\ )}$ on morphisms and the proof of functoriality are also straightforward.

Now we show that $\overline{(\ )}$ is surjective on objects. For each finite set X, choose a bijection $s_X : [1, n_X] \longrightarrow X$, where $n_X = \mathrm{card}(X)$. In the case that $X = [m + 1, m + n]$ for some $m, n \in \mathbb{N}$, choose s_X to be the obvious bijection (add m). Let C be a symmetric multicategory; our task is to define a fat symmetric multicategory A such that $\overline{A} = C$. (This is the uphill direction and is bound to require more work.) We define the objects of A to be the objects of C. If $(a_x)_{x \in X}$ is a finite family of objects then we define

$$A((a_x)_{x \in X}; b) = C(a_{s_X(1)}, \ldots, a_{s_X(n_X)}; b).$$

If X is a one-element set and a an object then we put

$$1_a^X = 1_a \in C(a; a) = A((a)_{x \in X}; a).$$

The definition of composition is clear in principle but fiddly in practice, hence omitted. The idea is that a composite in A can almost be defined as a composite in C, but because the bijections s_X were chosen at random, we have to apply a symmetry after composing in C – the unique symmetry that makes the domain come out right. The axioms for the fat symmetric multicategory A then follow, with some effort, from the ordinary symmetric multicategory axioms on C.

We also have to show that $\overline{A} = C$. Certainly their object-sets are equal, and their hom-sets are equal because

$$\overline{A}(a_1, \ldots, a_n; b) = A((a_x)_{x \in [1,n]}; b) = C(a_{s_{[1,n]}(1)}, \ldots, a_{s_{[1,n]}(n)}; b)$$

and we chose $s_{[1,n]}$ to be the identity. Composition in $\overline{A}$ was defined using the obvious bijections

$$[m+1, m+k] \xrightarrow{\sim} [1, k]$$

for certain values of m and k, and to show that it coincides with composition in A we use the fact that $s_{[m+1,m+k]}$ was also chosen to be the obvious bijection.

Next, $\overline{(\ )}$ is full. Let A and A' be fat symmetric multicategories and $h : \overline{A} \longrightarrow \overline{A'}$ a map of ordinary symmetric multicategories; we define a map $f : A \longrightarrow A'$ such that $\overline{f} = h$. On objects, $f(a) = h(a)$. Given a map $\theta : (a_x)_{x \in X} \longrightarrow b$ in A, choose a bijection $s : [1, n] \longrightarrow X$, where $n = \mathrm{card}(X)$. Then we have the map

$$\theta \cdot s : (a_{s(y)})_{y \in [1,n]} \longrightarrow b$$

in A, that is, we have

$$\theta \cdot s : a_{s(1)}, \ldots, a_{s(n)} \longrightarrow b$$

in $\overline{A}$; so we obtain the map

$$h(\theta \cdot s) : f a_{s(1)}, \ldots, f a_{s(n)} \longrightarrow f b$$

in $\overline{A'}$, that is,

$$h(\theta \cdot s) : (f a_{s(y)})_{y \in [1,n]} \longrightarrow f b$$

in A'. It therefore makes sense to define

$$f(\theta) = h(\theta \cdot s) \cdot s^{-1} : (fa_x)_{x \in X} \longrightarrow fb.$$

This definition is independent of the choice of s: note that any other choice is of the form $s \circ \sigma$ for some $\sigma \in S_n$, then use Lemma A.2.2 and the fact that h preserves symmetric group actions. It is straightforward to check that f really is a map of fat symmetric multicategories and that $\overline{f} = h$.

Finally, $\overline{(\ )}$ is faithful. Let $A \overset{f}{\underset{g}{\rightrightarrows}} A'$ be a pair of maps of fat symmetric multicategories satisfying $\overline{f} = \overline{g}$. Certainly f and g agree on objects. Given $\theta : (a_x)_{x \in X} \longrightarrow b$ in A, choose a bijection $s : [1, n] \longrightarrow X$, where $n = \mathrm{card}(X)$; then we have a map

$$\theta \cdot s : a_{s(1)}, \ldots, a_{s(n)} \longrightarrow b$$

in $\overline{A}$. Using Lemma A.2.2,

$$f(\theta) = f(\theta \cdot s \cdot s^{-1}) = f(\theta \cdot s) \cdot s^{-1} = \overline{f}(\theta \cdot s) \cdot s^{-1},$$

and similarly $g(\theta) = \overline{g}(\theta \cdot s) \cdot s^{-1}$, so $f(\theta) = g(\theta)$. □

Notes

Something very close to the notion of fat symmetric multicategories appeared in Beilinson and Drinfeld (c.1997, §1.1), under the name of 'pseudo-tensor categories'. The idea has probably appeared elsewhere too. Beilinson and Drinfeld insisted that the domain $(a_x)_{x \in X}$ of a map should be a *non-empty* finite family of objects, and correspondingly that the functions called $s : X \longrightarrow Y$ in Definition A.2.1 should be surjective. This amounts to excluding the possibility of nullary maps, which we have no reason to do. They also handled one-member families slightly differently.

Appendix B
Coherence for monoidal categories

Here we prove the 'descriptive' coherence theorems for unbiased and classical monoidal categories and functors (3.2.2, 3.2.3).

Unbiased and classical monoidal categories were defined concretely (1.2.5, 3.1.1), whereas Σ-monoidal categories were defined abstractly (3.2.1). To carry out the comparisons, we put the definition of Σ-monoidal category into concrete terms. Let $\Sigma \in \mathbf{Set}^{\mathbb{N}}$. Unwinding the abstract definition, we find that a Σ-monoidal category is a triple $A = (A, \otimes, \delta)$ consisting of

- a small category A
- a functor $\otimes_\tau : A^n \longrightarrow A$ for each $n \in \mathbb{N}$, $\tau \in (F\Sigma)(n)$
- an isomorphism

$$(\delta_{\tau,\tau'})_{a_1,\ldots,a_n} : \otimes_\tau(a_1, \ldots, a_n) \overset{\sim}{\longrightarrow} \otimes_{\tau'}(a_1, \ldots, a_n)$$

in A for each $n \in \mathbb{N}$, $\tau, \tau' \in (F\Sigma)(n)$, $a_i \in A$ (usually just written $\delta_{\tau,\tau'}$)

satisfying

MC1 $(\delta_{\tau,\tau'})_{a_1,\ldots,a_n}$ is natural in $a_1, \ldots, a_n \in A$, for each $n \in \mathbb{N}$, $\tau, \tau' \in (F\Sigma)(n)$

MC2 $\delta_{\tau',\tau''} \circ \delta_{\tau,\tau'} = \delta_{\tau,\tau''}$ and $1 = \delta_{\tau,\tau}$, for all $n \in \mathbb{N}$, $\tau, \tau', \tau'' \in (F\Sigma)(n)$

MC3 $\otimes_{\tau \circ (\tau_1, \ldots, \tau_n)} = (A^{k_1 + \cdots + k_n} \xrightarrow{\otimes_{\tau_1} \times \cdots \times \otimes_{\tau_n}} A^n \xrightarrow{\otimes_\tau} A)$ for all $n, k_i \in \mathbb{N}$, $\tau \in (F\Sigma)(n)$, $\tau_i \in (F\Sigma)(k_i)$; and $1_A = \otimes_|$

MC4 the diagram

$$
\begin{array}{ll}
\otimes_\tau(\otimes_{\tau_1}(a_1^1,\ldots,a_1^{k_1}),\ldots, & \overline{}\ \otimes_{\tau\circ(\tau_1,\ldots,\tau_n)}(a_1^1,\ldots,a_n^{k_n}) \\
\quad \otimes_{\tau_n}(a_n^1,\ldots,a_n^{k_n})) &
\end{array}
$$

$$
{\scriptstyle \otimes_\tau(\delta_{\tau_1,\tau_1'},\ldots,\delta_{\tau_n,\tau_n'})}\Big\downarrow
$$

$$
\begin{array}{l}
\otimes_\tau(\otimes_{\tau_1'}(a_1^1,\ldots,a_1^{k_1}),\ldots, \\
\quad \otimes_{\tau_n'}(a_n^1,\ldots,a_n^{k_n}))
\end{array}
\qquad\qquad {\scriptstyle \delta_{\tau\circ(\tau_1,\ldots,\tau_n),\,\tau'\circ(\tau_1',\ldots,\tau_n')}}
$$

$$
{\scriptstyle \delta_{\tau,\tau'}}\Big\downarrow
$$

$$
\begin{array}{ll}
\otimes_{\tau'}(\otimes_{\tau_1'}(a_1^1,\ldots,a_1^{k_1}),\ldots, & \\
\quad \otimes_{\tau_n'}(a_n^1,\ldots,a_n^{k_n})) & \overline{}\ \otimes_{\tau'\circ(\tau_1',\ldots,\tau_n')}(a_1^1,\ldots,a_n^{k_n})
\end{array}
$$

commutes, for all $n, k_i \in \mathbb{N}$, $\tau, \tau' \in (F\Sigma)(n)$, $\tau_i, \tau_i' \in (F\Sigma)(k_i)$, $a_i^j \in A$.

A lax monoidal functor $(P, \pi) : (A, \otimes, \delta) \longrightarrow (A', \otimes, \delta)$ (where, in an abuse of notation, we write $\otimes$ for the tensor and δ for the coherence maps in both A and A') consists of

- a functor $P : A \longrightarrow A'$
- a map

$$
(\pi_\tau)_{a_1,\ldots,a_n} : \otimes_\tau(Pa_1,\ldots,Pa_n) \longrightarrow P\otimes_\tau(a_1,\ldots,a_n)
$$

(usually just written π_τ) for each $n \in \mathbb{N}$, $\tau \in (F\Sigma)(n)$, $a_i \in A$

satisfying

MF1 $(\pi_\tau)_{a_1,\ldots,a_n}$ is natural in $a_1,\ldots,a_n \in A$
MF2 the diagram

$$
\begin{array}{ccc}
\otimes_\tau(Pa_1,\ldots,Pa_n) & \xrightarrow{\ \delta_{\tau,\tau'}\ } & \otimes_{\tau'}(Pa_1,\ldots,Pa_n) \\
{\scriptstyle \pi_\tau}\Big\downarrow & & \Big\downarrow{\scriptstyle \pi_{\tau'}} \\
P\otimes_\tau(a_1,\ldots,a_n) & \xrightarrow[\ P\delta_{\tau,\tau'}\ ]{} & P\otimes_{\tau'}(a_1,\ldots,a_n)
\end{array}
$$

commutes, for all $n \in \mathbb{N}$, $\tau, \tau' \in (F\Sigma)(n)$, $a_i \in A$

MF3 the diagram

$$\otimes_\tau(\otimes_{\tau_1}(Pa_1^1,\ldots,Pa_1^{k_1}),\ldots, \quad = \otimes_{\tau\circ(\tau_1,\ldots,\tau_n)}(Pa_1^1,\ldots,Pa_n^{k_n})$$
$$\otimes_{\tau_n}(Pa_n^1,\ldots,Pa_n^{k_n}))$$

$$\downarrow{\scriptstyle\otimes_\tau(\pi_{\tau_1},\ldots,\pi_{\tau_n})}$$

$$\otimes_\tau(P\otimes_{\tau_1}(a_1^1,\ldots,a_1^{k_1}),\ldots,$$
$$P\otimes_{\tau_n}(a_n^1,\ldots,a_n^{k_n})) \qquad\qquad\Big\downarrow{\scriptstyle\pi_{\tau\circ(\tau_1,\ldots,\tau_n)}}$$

$$\downarrow{\scriptstyle\pi_\tau}$$

$$P\otimes_\tau(\otimes_{\tau_1}(a_1^1,\ldots,a_1^{k_1}),\ldots,$$
$$\otimes_{\tau_n}(a_n^1,\ldots,a_n^{k_n})) \qquad = P\otimes_{\tau\circ(\tau_1,\ldots,\tau_n)}(a_1^1,\ldots,a_n^{k_n})$$

commutes for all $n, k_i \in \mathbb{N}$, $\tau \in (F\Sigma)(n)$, $\tau_i \in (F\Sigma)(k_i)$, $a_i^j \in A$, and the diagram

$$
\begin{array}{ccc}
Pa & == & \otimes_1 Pa \\
{\scriptstyle 1}\downarrow & & \downarrow{\scriptstyle \pi_1} \\
Pa & == & P\otimes_1 a
\end{array}
$$

commutes for all $a \in A$ (where in the expression 'π_1', the 1 is the unit of the operad $F\Sigma$).

A weak (respectively, strict) monoidal functor is a lax monoidal functor (P, π) in which all the π_τs are isomorphisms (respectively, identities).

In the cases at hand, the object Σ of $\mathbf{Set}^{\mathbb{N}}$ is either the terminal object 1 (for unbiased monoidal categories) or the object Σ_c defined in Theorem 3.2.3 (for classical monoidal categories). The $\mathbf{Set}$-operad $F\Sigma$ is then, respectively, either the operad $\mathbf{tr}$ of all trees (2.3.3) or the operad $\mathbf{ctr}$ of classical trees (2.3.4).

Sections B.1 and B.2 prove coherence for unbiased and classical monoidal categories, respectively. The unbiased case contains some scary-looking expressions but is completely straightforward: one only needs to be awake, not clever. By contrast, the classical case requires guile, cunning and trickery – attributes not displayed here, but vital to the proofs of the coherence theorems upon which we rely.

All of the results that we prove continue to hold in the more general setting of Σ-bicategories (Section 3.4). The proofs need only superficial changes of the kind described in Section 3.4; for instance, the category A becomes a $\mathbf{Cat}$-graph B and tensor becomes composition.

B.1 Unbiased monoidal categories

The plan is to define a functor

$$J : \text{1-MonCat}_{\text{lax}} \longrightarrow \text{UMonCat}_{\text{lax}},$$

to prove that it is an isomorphism (by showing in turn that it is injective on objects, surjective on objects, faithful, and full), and then to prove that it restricts to isomorphisms

$$\text{1-MonCat}_{\text{wk}} \xrightarrow{\;\sim\;} \text{UMonCat}_{\text{wk}}, \qquad \text{1-MonCat}_{\text{str}} \xrightarrow{\;\sim\;} \text{UMonCat}_{\text{str}}$$

at the weak and strict levels.

Recall from 2.3.3 that $|\ \in \textbf{tr}(1)$ denotes the 'null' or identity tree, and that $\nu_n \in \textbf{tr}(n)$ denotes the simplest n-leafed tree $\quad$. (Recall also that $\nu_1 = $

$\quad \neq\ |\ $.) We will be composing trees – that is, composing in the operad $\textbf{tr}$ – and in particular we will often use the composite tree

$$\nu_n \circ (\nu_{k_1}, \ldots, \nu_{k_n}) = \qquad .$$

Our first task is to define a functor $J : \text{1-MonCat}_{\text{lax}} \longrightarrow \text{UMonCat}_{\text{lax}}$.

On objects Let $(A, \otimes, \delta)$ be an object of $\text{1-MonCat}_{\text{lax}}$. The unbiased monoidal category $J(A, \otimes, \delta)$ is given by taking the underlying category to be A, the n-fold tensor $\otimes_n : A^n \longrightarrow A$ to be $\otimes_{\nu_n}$, and the coherence maps to be

$$\gamma_{((a_1^1,\ldots,a_1^{k_1}),\ldots,(a_n^1,\ldots,a_n^{k_n}))} = (\delta_{\nu_n \circ (\nu_{k_1},\ldots,\nu_{k_n}),\, \nu_{k_1+\cdots+k_n}})_{a_1^1,\ldots,a_n^{k_n}} :$$
$$((a_1^1 \otimes \cdots \otimes a_1^{k_1}) \otimes \cdots \otimes (a_n^1 \otimes \cdots \otimes a_n^{k_n})) \longrightarrow (a_1^1 \otimes \cdots \otimes a_n^{k_n})$$

and

$$\iota_a = (\delta_{|,\nu_1})_a : a \longrightarrow (a).$$

On maps Let $(P, \pi) : A \longrightarrow A'$ be a map in $\text{1-MonCat}_{\text{lax}}$. The lax monoidal functor $J(P, \pi) : J(A) \longrightarrow J(A')$ is given by taking the same underlying functor P, and by taking the coherence map

$$\pi_{a_1,\ldots,a_n} : (Pa_1 \otimes \cdots \otimes Pa_n) \longrightarrow P(a_1 \otimes \cdots \otimes a_n)$$

(re-using the letter π, in another slight abuse) to be $(\pi_{\nu_n})_{a_1,\ldots,a_n}$.

Lemma B.1.1 *This defines a functor* $J : $ 1-**MonCat**$_{\text{lax}} \longrightarrow$ **UMonCat**$_{\text{lax}}$.

Proof We have to check three things:

- $J(A, \otimes, \delta)$ as defined above really is an unbiased monoidal category – in other words, the axioms in Definition 3.1.1 hold. Naturality and invertibility of γ and ι follow from the same properties for δ. The associativity axiom holds because both routes around the square are

$$\left(\delta_{v_n \circ (v_{m_1}, \ldots, v_{m_n}) \circ (v_{k_1^1}, \ldots, v_{k_n^{mn}}), v_{k_1^1 + \cdots + k_n^{mn}}}\right) a_{1,1,\ldots,a_{n,m_n,k_n^{mn}}},$$

as can be shown from axioms MC2 and MC4. The identity axioms hold by similar reasoning.

- $J(P, \pi)$ as defined above really is a lax monoidal functor between unbiased monoidal categories (Definition 3.1.3). Naturality of $\pi_{a_1,\ldots,a_n}$ in the a_is follows from the same naturality for π_{v_n}. The coherence axioms can be deduced from axioms MF1–MF3.

- J preserves composition and identities. This is trivial.   □

Lemma B.1.2 *The functor* $J : $ 1-**MonCat**$_{\text{lax}} \longrightarrow$ **UMonCat**$_{\text{lax}}$ *is injective on objects.*

Proof Suppose that $(A, \otimes, \delta)$ and $(A', \otimes', \delta')$ are 1-monoidal categories with $J(A, \otimes, \delta) = J(A', \otimes', \delta')$. (Just for once, the tensor of A' is written $\otimes'$ rather than $\otimes$, and similarly δ' rather than δ.) Then:

- $A = A'$ immediately.

- $\otimes_\tau = \otimes'_\tau$ for all $n \in \mathbb{N}$ and $\tau \in \mathbf{tr}(n)$. This is proved by induction on τ, using the description of $\mathbf{tr}$ in 2.3.3. Since there will be many inductions on trees in this appendix, I will write this one out in full and leave the others to the virtuous reader. First, $\otimes_| = 1_A = \otimes'_|$ by MC3. Second, take $\tau_1 \in \mathbf{tr}(k_1), \ldots, \tau_n \in \mathbf{tr}(k_n)$. Then

$$\otimes_{(\tau_1,\ldots,\tau_n)} = \otimes_{v_n \circ (\tau_1,\ldots,\tau_n)}$$
$$= \otimes_{v_n} \circ (\otimes_{\tau_1} \times \cdots \times \otimes_{\tau_n}),$$

again by MC3. But $\otimes_{v_n} = \otimes'_{v_n}$ since $J(A, \otimes, \delta) = J(A', \otimes', \delta')$, and $\otimes_{\tau_i} = \otimes'_{\tau_i}$ by inductive hypothesis, so $\otimes_{(\tau_1,\ldots,\tau_n)} = \otimes'_{(\tau_1,\ldots,\tau_n)}$, completing the induction.

- $\delta_{\tau,\tau'} = \delta'_{\tau,\tau'}$ for all $\tau, \tau' \in \mathbf{tr}(n)$. Since $\delta_{\tau,\tau'} = \delta^{-1}_{\tau',v_n} \circ \delta_{\tau,v_n}$, it is enough to prove this in the case $\tau' = v_n$; and that is done by another short induction on τ.   □

Lemma B.1.3 *The functor* J : 1-**MonCat**$_{\text{lax}}$ $\longrightarrow$ **UMonCat**$_{\text{lax}}$ *is surjective on objects.*

Proof Take an unbiased monoidal category $(A, \otimes, \gamma, \iota)$. Attempt to define a 1-monoidal category $(A, \otimes, \delta)$ as follows:

- The underlying category A is the same.
- The tensor $\otimes_\tau : A^n \longrightarrow A$, for $\tau \in \mathbf{tr}(n)$, is defined inductively on τ by $\otimes_| = 1_A$ and

$$\otimes_{(\tau_1,\ldots,\tau_n)} = (A^{k_1+\cdots+k_n} \xrightarrow{\otimes_{\tau_1} \times \cdots \times \otimes_{\tau_n}} A^n \xrightarrow{\otimes_n} A).$$

- The coherence isomorphisms are defined by $\delta_{\tau,\tau'} = \delta_{\tau'}^{-1} \circ \delta_\tau$, where in turn

$$\delta_\tau : \otimes_\tau(a_1, \ldots, a_n) \xrightarrow{\sim} (a_1 \otimes \cdots \otimes a_n)$$

is defined by taking $\delta_| = \iota$ and taking $\delta_{(\tau_1,\ldots,\tau_n)}$ to be the composite

$$(\otimes_{\tau_1}(a_1^1, \ldots, a_1^{k_1}) \otimes \cdots \otimes \otimes_{\tau_n}(a_n^1, \ldots, a_n^{k_n}))$$
$$\xrightarrow{(\delta_{\tau_1} \otimes \cdots \otimes \delta_{\tau_n})} ((a_1^1 \otimes \cdots \otimes a_1^{k_1}) \otimes \cdots \otimes (a_n^1 \otimes \cdots \otimes a_n^{k_n}))$$
$$\xrightarrow{\gamma} (a_1^1 \otimes \cdots \otimes a_n^{k_n}).$$

This does indeed satisfy the axioms for a 1-monoidal category:

MC1, MC2 Immediate.

MC3 That $\otimes_| = 1_A$ is immediate. The other equation can be proved by induction on τ, or by using the fact that $\mathbf{tr}$ is the free operad on $1 \in \mathbf{Set}^{\mathbb{N}}$ and that $(\mathbf{Cat}(A^n, A))_{n \in \mathbb{N}}$ forms an operad.

MC4 It is enough to show that this axiom is satisfied when $\tau' = \nu_n$, $\tau_1' = \nu_{k_1}, \ldots, \tau_n' = \nu_{k_n}$: in other words, that

$$
\begin{array}{ccc}
\otimes_\tau(\otimes_{\tau_1}(a_1^1, \ldots, a_1^{k_1}), \ldots, & & \otimes_{\tau \circ (\tau_1,\ldots,\tau_n)}(a_1^1, \ldots, a_n^{k_n}) \\
\otimes_{\tau_n}(a_n^1, \ldots, a_n^{k_n})) & = & \\
{\scriptstyle \otimes_\tau(\delta_{\tau_1},\ldots,\delta_{\tau_n})} \downarrow & & \downarrow {\scriptstyle \delta_{\tau \circ (\tau_1,\ldots,\tau_n)}} \\
\otimes_\tau((a_1^1 \otimes \cdots \otimes a_1^{k_1}) \otimes \cdots & & \\
\otimes(a_n^1 \otimes \cdots \otimes a_n^{k_n})) & & (a_1^1 \otimes \cdots \otimes a_n^{k_n}) \\
{\scriptstyle \delta_\tau} \downarrow & & \uparrow {\scriptstyle \delta_{\nu_n \circ (\nu_{k_1},\ldots,\nu_{k_n})} = \gamma} \\
((a_1^1 \otimes \cdots \otimes a_1^{k_1}) \otimes \cdots & & ((a_1^1 \otimes \cdots \otimes a_1^{k_1}) \otimes \cdots \\
\otimes(a_n^1 \otimes \cdots \otimes a_n^{k_n})) & = & \otimes(a_n^1 \otimes \cdots \otimes a_n^{k_n}))
\end{array}
$$

commutes. This is done by induction on τ, using the associativity axiom for unbiased monoidal categories.

Finally, $J(A, \otimes, \delta) = (A, \otimes, \gamma, \iota)$:

- Clearly the underlying categories agree, both being A.
- We have

$$\otimes_{\nu_n} = \otimes_{(|,\dots,|)} = \otimes_n \circ (\otimes_| \times \cdots \times \otimes_|) = \otimes_n,$$

so the tensor products agree.

- Certainly $\delta_| = \iota$. Also

$$\delta_{\nu_k} = \gamma \circ (\delta_| \otimes \cdots \otimes \delta_|) = \gamma \circ (\iota \otimes \cdots \otimes \iota) = 1$$

for any k, so

$$\delta_{\nu_n \circ (\nu_{k_1}, \dots, \nu_{k_n})} = \gamma \circ (\delta_{\nu_{k_1}} \otimes \cdots \otimes \delta_{\nu_{k_n}}) = \gamma.$$

Hence the coherence maps ι and γ also agree. $\qquad \square$

Lemma B.1.4 *The functor* $J : 1\text{-}\mathbf{MonCat}_{\text{lax}} \longrightarrow \mathbf{UMonCat}_{\text{lax}}$ *is faithful.*

Proof Suppose that $A \overset{(P,\pi)}{\underset{(Q,\chi)}{\rightrightarrows}} A'$ in $1\text{-}\mathbf{MonCat}_{\text{lax}}$ with $J(P, \pi) = J(Q, \chi)$. Then:

- $P = Q$ immediately.
- $\pi_\tau = \chi_\tau$ for all $\tau \in \mathbf{tr}(n)$, by an induction on τ similar to the one written out in the proof of B.1.2. $\qquad \square$

Lemma B.1.5 *The functor* $J : 1\text{-}\mathbf{MonCat}_{\text{lax}} \longrightarrow \mathbf{UMonCat}_{\text{lax}}$ *is full.*

Proof Let $A, A' \in 1\text{-}\mathbf{MonCat}_{\text{lax}}$ and let $J(A) \xrightarrow{(P,\pi)} J(A')$ be a map in $\mathbf{UMonCat}_{\text{lax}}$. Attempt to define a map $A \xrightarrow{(P,\pi)} A'$ in $1\text{-}\mathbf{MonCat}_{\text{lax}}$ (where as usual we abuse notation by recycling the name (P, π)) as follows:

- The underlying functor P is the same.
- The coherence maps π_τ are defined by induction on τ. We take $\pi_| = 1$ and take

$$\otimes_{(\tau_1, \dots, \tau_n)} (Pa_1^1, \dots, Pa_n^{k_n}) \xrightarrow{\pi_{(\tau_1, \dots, \tau_n)}} P \otimes_{(\tau_1, \dots, \tau_n)} (a_1^1, \dots, a_n^{k_n})$$

to be the composite

$$(\otimes_{\tau_1} (Pa_1^1, \dots, Pa_1^{k_1}) \otimes \cdots \otimes \otimes_{\tau_n} (Pa_n^1, \dots, Pa_n^{k_n}))$$

$$\xrightarrow{(\pi_{\tau_1} \otimes \cdots \otimes \pi_{\tau_n})}$$

$$(P \otimes_{\tau_1} (a_1^1, \dots, a_1^{k_1}) \otimes \cdots \otimes P \otimes_{\tau_n} (a_n^1, \dots, a_n^{k_n}))$$

$$\xrightarrow{\pi_{\otimes_{\tau_1} (a_1^1, \dots, a_1^{k_1}), \dots, \otimes_{\tau_n} (a_n^1, \dots, a_n^{k_n})}} P(a_1^1 \otimes \cdots \otimes a_n^{k_n}).$$

This does satisfy the axioms above for a lax monoidal functor between 1-monoidal categories:

MF1 Immediate.
MF2 It is enough to prove this when $\tau' = \nu_n$; in other words, that

$$
\begin{array}{ccc}
\otimes_\tau (Pa_1, \ldots, Pa_n) & \xrightarrow{\delta_\tau} & (Pa_1 \otimes \cdots \otimes Pa_n) \\
\pi_\tau \downarrow & & \downarrow \pi_{a_1,\ldots,a_n} \\
P \otimes_\tau (a_1, \ldots, a_n) & \xrightarrow[P\delta_\tau]{} & P(a_1 \otimes \cdots \otimes a_n)
\end{array}
$$

commutes. This is done by induction on τ, using the coherence axioms for lax monoidal functors between unbiased monoidal categories.
MF3 This is, inevitably, another induction on τ.

Finally, $J(P, \pi) = (P, \pi)$:

- The underlying functors agree, both being P.
- We have

$$
\begin{aligned}
(\pi_{\nu_n})_{a_1,\ldots,a_n} &= \pi_{\otimes|(a_1),\ldots,\otimes|(a_n)} \circ ((\pi_|)_{a_1} \otimes \cdots \otimes (\pi_|)_{a_n}) \\
&= \pi_{a_1,\ldots,a_n} \circ (1 \otimes \cdots \otimes 1) \\
&= \pi_{a_1,\ldots,a_n},
\end{aligned}
$$

so the coherence maps also agree. □

Theorem 3.2.2 now follows in the lax case: **UMonCat**$_{\text{lax}} \cong$ **1-MonCat**$_{\text{lax}}$. The weak and strict cases follow from:

Lemma B.1.6 *The isomorphism* $J :$ **1-MonCat**$_{\text{lax}} \xrightarrow{\sim}$ **UMonCat**$_{\text{lax}}$ *restricts to isomorphisms*

$$\text{\bf 1-MonCat}_{\text{wk}} \xrightarrow{\sim} \text{\bf UMonCat}_{\text{wk}}, \qquad \text{\bf 1-MonCat}_{\text{str}} \xrightarrow{\sim} \text{\bf UMonCat}_{\text{str}}.$$

Proof Trivial. □

B.2 Classical monoidal categories

The strategy is the same as in the unbiased setting: we define a functor

$$J : \Sigma_c\text{-}\textbf{MonCat}_{\text{lax}} \longrightarrow \textbf{MonCat}_{\text{lax}},$$

prove it is an isomorphism, then prove that it restricts to isomorphisms at the levels of weak and strict maps.

We will use the operad **ctr** of classical or unitrivalent trees (2.3.4). In particular, we have trees

$$\nu_0 = \,\wr\; \in \mathbf{ctr}(0), \qquad \nu_2 = \,\mathsf{Y}\; \in \mathbf{ctr}(2),$$

the identity tree $|\; \in \mathbf{ctr}(1)$, and composite trees

$$\nu_2 \circ (\nu_2, |) = \;\mathsf{Y}\; \in \mathbf{ctr}(3), \qquad \nu_2 \circ (|, \nu_2) = \;\mathsf{Y}\; \in \mathbf{ctr}(3),$$

$$\nu_2 \circ (\nu_0, |) = \;\mathsf{Y}\; \in \mathbf{ctr}(1), \qquad \nu_2 \circ (|, \nu_0) = \;\mathsf{Y}\; \in \mathbf{ctr}(1).$$

To prove our result, we will certainly want to use the fact that 'all diagrams commute' in a classical monoidal category, and some similar statement for monoidal functors. In other words, the proof of our unified coherence theorem for classical monoidal categories and functors will depend on pre-established coherence theorems. A rather vague description of those theorems was given in Section 1.2. Using the language of trees we can now be more precise.

So, let $A = (A, \otimes, I, \alpha, \lambda, \rho)$ be a classical monoidal category. We define for each $n \in \mathbb{N}$ and $\tau \in \mathbf{ctr}(n)$ a functor $\otimes_\tau : A^n \longrightarrow A$. This definition is by induction on τ (using the description of **ctr** in 2.3.4), as follows:

- $\otimes_| : A^1 \longrightarrow A$ is the canonical isomorphism
- $\otimes_\wr : A^0 \longrightarrow A$ 'is' the unit object I of A
- if $\tau_1 \in \mathbf{ctr}(k_1)$ and $\tau_2 \in \mathbf{ctr}(k_2)$ then $\otimes_{(\tau_1, \tau_2)}$ is the composite

$$A^{k_1 + k_2} \xrightarrow{\otimes_{\tau_1} \times \otimes_{\tau_2}} A^2 \xrightarrow{\;\otimes\;} A.$$

The coherence results from Section 1.2 can now be stated as:

- Let A be a classical monoidal category. Then for each $n \in \mathbb{N}$ and $\tau, \tau' \in \mathbf{ctr}(n)$, there is a unique natural transformation $A^n \overset{\otimes_\tau}{\underset{\otimes_{\tau'}}{\Downarrow}} A$ built out of α, λ and ρ.

- Let $(P, \pi) : A \longrightarrow A'$ be a lax monoidal functor. Then for each $n \in \mathbb{N}$ and $\tau, \tau' \in \mathbf{ctr}(n)$, there is a unique natural transformation

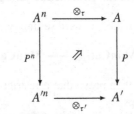

built out of the coherence isomorphisms α, λ and ρ of A and A' and the coherence maps $\pi_{-,-}$ and $\pi_.$.

I will refer to these as **informal coherence** for classical monoidal categories and functors. They could be made precise by defining 'built out of', but I hope the reader will be content to use them in their present imprecise form.

Our first task is to define a functor $J : \Sigma_c\text{-}\mathbf{MonCat}_{\text{lax}} \longrightarrow \mathbf{MonCat}_{\text{lax}}$.

On objects Let $(A, \otimes, \delta)$ be an object of $\Sigma_c\text{-}\mathbf{MonCat}_{\text{lax}}$, as described in the introduction to this appendix. Then the classical monoidal category $J(A, \otimes, \delta)$ is given by taking the underlying category to be A, the tensor to be $\otimes_{\curlyvee}$, the unit object to be $\otimes_{\uparrow}$, and the coherence isomorphisms

to be

$$\alpha = \delta_{\tau_1, \tau_1'}, \qquad \lambda = \delta_{\tau_2, |}, \qquad \rho = \delta_{\tau_3, |}$$

where

$$\tau_1 = \curlyvee\!\!\!\curlyvee, \qquad \tau_1' = \curlyvee^{\curlyvee}, \qquad \tau_2 = \curlyvee, \qquad \tau_3 = \curlyvee.$$

On maps Let $(P, \pi) : A \longrightarrow A'$ be a map in $\Sigma_c\text{-}\mathbf{MonCat}_{\text{lax}}$, also as described in the introductory section. Then the lax monoidal functor $J(P, \pi) : J(A) \longrightarrow J(A')$ is given by taking the same underlying category P and coherence maps

$$\pi_{-,-} = \pi_{\curlyvee}, \qquad \pi_. = \pi_{\uparrow}.$$

Lemma B.2.1 *This defines a functor* $J : \Sigma_c\text{-}\mathbf{MonCat}_{\text{lax}} \longrightarrow \mathbf{MonCat}_{\text{lax}}$.

Proof We have to check three things:

- $J(A, \otimes, \delta)$ as defined above really is a classical monoidal category. Naturality and invertibility of α, λ and ρ follow from the same properties of δ. The pentagon (1.2.5) commutes because each route around it is $\delta_{\tau, \tau'}$, where

$$\tau = \curlyvee\!\!\!\curlyvee, \qquad \tau' = \curlyvee^{\curlyvee}.$$

Similarly, the triangle commutes because both ways round it are $\delta_{\sigma, \sigma'}$, where

$$\sigma = \curlyvee\!\!\curlyvee, \qquad \sigma' = \curlyvee.$$

- (J, π) as defined above really is a lax monoidal functor between classical monoidal categories. The axioms can be deduced from MF1–MF3.
- J preserves composition and identities. This is trivial. $\qquad\qquad\square$

Lemma B.2.2 *The functor* $J : \Sigma_c\text{-}\mathbf{MonCat}_{lax} \longrightarrow \mathbf{MonCat}_{lax}$ *is injective on objects.*

Proof Suppose that $(A, \otimes, \delta)$ and $(A', \otimes', \delta')$ are Σ_c-monoidal categories with $J(A, \otimes, \delta) = J(A', \otimes', \delta')$. Then:

- $A = A'$ immediately.
- $\otimes_\tau = \otimes'_\tau$ for all $n \in \mathbb{N}$ and $\tau \in \mathbf{tr}(n)$. This is proved by induction on τ, using the definition of **ctr** in 2.3.4. As in the previous section, there will be many proofs by induction on the structure of a tree; and as a sample of the technique in the classical case, I will write this one out in full. First, $\otimes_| = 1_A = \otimes'_|$ by MC3. Second, $\otimes_\bullet = I = \otimes'_\bullet$ since $J(A, \otimes, \delta) = J(A', \otimes', \delta')$. Third, take $\tau_1 \in \mathbf{ctr}(k_1)$ and $\tau_2 \in \mathbf{ctr}(k_2)$. Then

$$\otimes_{(\tau_1, \tau_2)} = \otimes_Y \circ_{(\tau_1, \tau_2)} = \otimes_Y \circ (\otimes_{\tau_1} \times \otimes_{\tau_2}),$$

  again by MC3. But $\otimes_Y = \otimes'_Y$ since $J(A, \otimes, \delta) = J(A', \otimes', \delta')$, and $\otimes_{\tau_i} = \otimes'_{\tau_i}$ by inductive hypothesis, so $\otimes_{(\tau_1, \tau_2)} = \otimes'_{(\tau_1, \tau_2)}$, completing the induction.
- $\delta_{\tau, \tau'} = \delta'_{\tau, \tau'}$ for all $\tau, \tau' \in \mathbf{ctr}(n)$. This follows from informal coherence for classical monoidal categories – in particular, from the fact that there is *at least* one natural transformation $\otimes_\tau \longrightarrow \otimes_{\tau'}$ built up from α, λ and ρ. $\square$

Lemma B.2.3 *The functor* $J : \Sigma_c\text{-}\mathbf{MonCat}_{lax} \longrightarrow \mathbf{MonCat}_{lax}$ *is surjective on objects.*

Proof Take a classical monoidal category $(A, \otimes, I, \alpha, \lambda, \rho)$. Attempt to define a Σ_c-monoidal category $(A, \otimes, \delta)$ as follows:

- The underlying category A is the same.
- The tensor $\otimes_\tau : A^n \longrightarrow A$, for $\tau \in \mathbf{ctr}(n)$, is defined inductively on τ by $\otimes_| = 1_A$, $\otimes_\bullet = I$, and

$$\otimes_{(\tau_1, \tau_2)} = (A^{k_1+k_2} \xrightarrow{\otimes_{\tau_1} \times \otimes_{\tau_2}} A^2 \xrightarrow{\otimes} A).$$

- The coherence isomorphism $\delta_{\tau, \tau'} : \otimes_\tau \longrightarrow \otimes_{\tau'}$ is the canonical natural isomorphism in the statement of informal coherence.

This does indeed satisfy the axioms for a Σ_c-monoidal category, as listed in the introductory section:

MC1, MC2 Immediate.

MC3 That $\otimes_| = 1_A$ is immediate. The other equation can be proved by induction on τ, or by using the fact that **ctr** is the free operad on $\Sigma_c \in \mathbf{Set}^{\mathbb{N}}$ and that $(\mathbf{Cat}(A^n, A))_{n \in \mathbb{N}}$ forms an operad.

MC4 Follows immediately from informal coherence.

Finally, $J(A, \otimes, \delta) = (A, \otimes, I, \alpha, \lambda, \rho)$:

- Clearly the underlying categories agree, both being A.
- We have

$$\otimes_{\curlyvee} = \otimes_{(|,|)} = \otimes \circ (\otimes_| \times \otimes_|) = \otimes$$

and $\otimes_{\uparrow} = I$, so the tensor products and unit objects agree.

- To prove that the coherence isomorphisms agree, we have to check that

$$\delta_{\curlyvee,\curlyvee} = \alpha, \qquad \delta_{\curlyvee,|} = \lambda, \qquad \delta_{\curlyvee,|} = \rho.$$

This can be done either by a short calculation or by applying informal coherence. $\square$

Lemma B.2.4 *The functor* $J : \Sigma_c\text{-}\mathbf{MonCat}_{\mathrm{lax}} \longrightarrow \mathbf{MonCat}_{\mathrm{lax}}$ *is faithful.*

Proof Suppose that $A \underset{(Q,\chi)}{\overset{(P,\pi)}{\rightrightarrows}} A'$ in $\Sigma_c\text{-}\mathbf{MonCat}_{\mathrm{lax}}$ with $J(P, \pi) = J(Q, \chi)$. Then:

- $P = Q$ immediately.
- $\pi_\tau = \chi_\tau$ for all $\tau \in \mathbf{ctr}(n)$. This can be proved either by an induction on τ similar to the one written out in the proof of Lemma B.2.2, or by applying informal coherence for lax monoidal functors. $\square$

Lemma B.2.5 *The functor* $J : \Sigma_c\text{-}\mathbf{MonCat}_{\mathrm{lax}} \longrightarrow \mathbf{MonCat}_{\mathrm{lax}}$ *is full.*

Proof Let $A, A' \in \Sigma_c\text{-}\mathbf{MonCat}_{\mathrm{lax}}$ and let $J(A) \xrightarrow{(P,\pi)} J(A')$ be a map in $\mathbf{MonCat}_{\mathrm{lax}}$. Attempt to define a map $A \xrightarrow{(P,\pi)} A'$ in $\Sigma_c\text{-}\mathbf{MonCat}_{\mathrm{lax}}$ (where again we abuse notation by re-using the name (P, π)) as follows:

- The underlying functor P is the same.
- The coherence maps π_τ are defined by induction on τ. We take $\pi_| = 1$; we take

$$(\pi_{\uparrow} : \otimes_{\uparrow} \longrightarrow P\otimes_{\uparrow}) = (\pi. : I \longrightarrow PI);$$

and we take the component of $\pi_{(\tau_1, \tau_2)}$ at $a_1^1, \ldots, a_1^{k_1}, a_2^1, \ldots, a_2^{k_2}$ to be the composite

$$(\otimes_{\tau_1}(Pa_1^1, \ldots, Pa_1^{k_1}) \otimes \otimes_{\tau_2}(Pa_2^1, \ldots, Pa_2^{k_2}))$$

$$\xrightarrow{(\pi_{\tau_1} \otimes \pi_{\tau_2})} \quad (P \otimes_{\tau_1}(a_1^1, \ldots, a_1^{k_1}) \otimes P \otimes_{\tau_2}(a_2^1, \ldots, a_2^{k_2}))$$

$$\xrightarrow{\pi_{\otimes_{\tau_1}(a_1^1, \ldots, a_1^{k_1}), \otimes_{\tau_2}(a_2^1, \ldots, a_2^{k_2})}} \quad P(\otimes_{\tau_1}(a_1^1, \ldots, a_1^{k_1}) \otimes \otimes_{\tau_2}(a_2^1, \ldots, a_2^{k_2})).$$

This satisfies axioms MF1–MF3 for a lax monoidal functor between Σ_c-monoidal categories, by informal coherence. We then have $J(P, \pi) = (P, \pi)$:

- The underlying functors agree, both being P.
- We have

$$(\pi_{\curlyvee})_{a_1, a_2} = \pi_{\otimes|(a_1), \otimes|(a_2)} \circ ((\pi|)_{a_1} \otimes (\pi|)_{a_2})$$
$$= \pi_{a_1, a_2} \circ (1 \otimes 1)$$
$$= \pi_{a_1, a_2}$$

and $\pi_{\centerdot} = \pi_{\centerdot}$, so the coherence maps also agree. $\square$

Theorem 3.2.3 now follows in the lax case: $\mathbf{MonCat}_{\mathrm{lax}} \cong \Sigma_c\text{-}\mathbf{MonCat}_{\mathrm{lax}}$. The weak and strict cases follow from:

Lemma B.2.6 *The isomorphism* $J : \Sigma_c\text{-}\mathbf{MonCat}_{\mathrm{lax}} \xrightarrow{\sim} \mathbf{MonCat}_{\mathrm{lax}}$ *restricts to isomorphisms*

$$\Sigma_c\text{-}\mathbf{MonCat}_{\mathrm{wk}} \xrightarrow{\sim} \mathbf{MonCat}_{\mathrm{wk}}, \qquad \Sigma_c\text{-}\mathbf{MonCat}_{\mathrm{str}} \xrightarrow{\sim} \mathbf{MonCat}_{\mathrm{str}}.$$

Proof Trivial. $\square$

Notes

References can be found in the Notes to Chapter 3.

Appendix C

Special cartesian monads

Pictures can't say 'ain't'.
Worth (1975)

We have met many monads, most of them cartesian. Some had special properties beyond being cartesian – for instance, some were the monads arising from operads, and, as will be explained, some admitted a certain explicit representation. Here we look at these special kinds of cartesian monad and prove some results supporting the theory in the main text.

First (Section C.1) we look at the monads arising from plain operads. Monads are algebraic theories, so we can ask which algebraic theories come from operads. The answer turns out to be the strongly regular theories (2.2.5).

In Section 6.2 we saw that the monad arising from an operad is cartesian. It now follows that the monad corresponding to a strongly regular theory is cartesian, a fact we used in many of the examples in Chapter 4. More precisely, we saw in 6.2.4 that a monad on **Set** arises from a plain operad if and only if it is cartesian and 'augmented over the free monoid monad', meaning that there exists a cartesian natural transformation from it into the free monoid functor, commuting with the monad structures. On the other hand, not every cartesian monad on **Set** possesses such an augmentation, as we shall see.

One possible drawback of the generalized operad approach to higher-dimensional category theory is that it can involve monads that are rather hard to describe explicitly. For instance, the sequence T_n of 'opetopic' monads (Chapter 7) was generated recursively using nothing more than the existence of free operads, and to describe T_n explicitly beyond low values of n is difficult. The second and third sections of this appendix ease this difficulty.

The basic result is that if a **Set**-valued functor preserves infinitary or 'wide' pullbacks – which is not asking much more than it be cartesian – then it is the

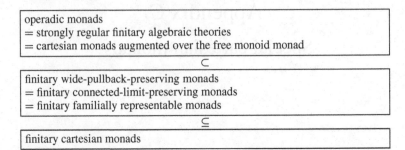

Fig. C-A. Classes of finitary cartesian monads on **Set**

coproduct of a family of representable functors. This goes some way towards providing the explicit form desired in the previous paragraph. In Section C.2 we look at monads on **Set** whose functor parts are 'familially representable' in this sense. (Finitary familially representable functors can also be viewed as a non-symmetric version of the analytic functors of Joyal, 1986.) In Section C.3 we examine monads (T, μ, η) on presheaf categories whose functor parts T satisfy an analogous condition. All the opetopic monads T_n are of this form, and this enables us to prove in Section C.4 that every symmetric multicategory gives rise naturally to a T_n-multicategory for each $n \in \mathbb{N}$.

The situation for the various types of finitary cartesian monad on **Set** is depicted in Fig. C-A. The terminology is defined and the proofs are given below. Example C.2.8 shows that the first inclusion is proper; I do not know whether the second is too.

C.1 Operads and algebraic theories

Some algebraic theories can be described by plain operads, and some cannot. In this section we will prove that the theories that can are precisely those that admit strongly regular presentations, in the sense of Example 2.2.5.

What exactly this means is as follows. Any plain operad P gives rise to a monad (T_P, μ^P, η^P) on **Set**, as we saw in Section 4.3. Say that a monad on **Set** is **operadic** if it is isomorphic to the monad (T_P, μ^P, η^P) arising from some plain operad P. On the other hand, any algebraic theory gives rise to a monad on **Set**, as we will discuss in a moment. Say that a monad on **Set** is **strongly regular** if it is isomorphic to the monad arising from some strongly regular finitary theory. Our result is:

Theorem C.1.1 *A monad on **Set** is operadic if and only if it is strongly regular.*

This means that if we have before us an algebraic theory presented by operations and strongly regular equations then we may deduce immediately that it can be described by an operad. Moreover, since any operadic monad is cartesian (Proposition 6.2.1), we obtain the useful corollary:

Corollary C.1.2 *Any strongly regular monad on* **Set** *is cartesian.* □

This is used throughout Section 4.1 to generate examples of cartesian monads on **Set**. It is also proved in Carboni and Johnstone (1995), Proposition 3.2 (part of which is wrong, as discussed on p. 370, but the part we are using here is right). The other half of Theorem C.1.1 says that any operad can be presented by a system of operations and strongly regular equations. This may not be so useful, but gives the story a tidy ending.

The definition of an algebraic theory and the way that one gives rise to a monad on **Set** are well known, and a detailed account can be found in, for instance, Manes (1976) or Borceux (1994b). Things are easier when the theory is strongly regular, and this case is all we will need, so I will describe it in full.

Let $\Sigma \in \mathbf{Set}^{\mathbb{N}}$. Write $F\Sigma$ for the free plain operad on Σ (or for the underlying object of $\mathbf{Set}^{\mathbb{N}}$), as constructed explicitly in Section 2.3. We think of Σ as a 'signature': $\Sigma(n)$ is the set of primitive n-ary operations, and $(F\Sigma)(n)$ is the set of n-ary operations derived from the primitive ones. A **strongly regular presentation of an algebraic theory** is a pair (Σ, E) where $\Sigma \in \mathbf{Set}^{\mathbb{N}}$ and E is a family $(E_n)_{n \in \mathbb{N}}$ with $E_n \subseteq (F\Sigma)(n) \times (F\Sigma)(n)$. We think of E as the system of equations. For example, the usual presentation of the theory of semigroups is

$$\Sigma(n) = \begin{cases} \{\sigma\} & \text{if } n = 2 \\ \emptyset & \text{otherwise} \end{cases}$$

$$E_n = \begin{cases} \{(\sigma \circ (\sigma, 1), \sigma \circ (1, \sigma))\} & \text{if } n = 3 \\ \emptyset & \text{otherwise.} \end{cases}$$

(Implicitly, we are using 'algebraic theory' to mean 'finitary, single-sorted algebraic theory'.)

The operad $F\Sigma$ induces a monad $(T_{F\Sigma}, \mu^{F\Sigma}, \eta^{F\Sigma})$ on **Set**. We have

$$T_{F\Sigma}A = \coprod_{n \in \mathbb{N}} (F\Sigma)(n) \times A^n$$

for any set A; this is usually called the set of 'Σ-terms in A'. Bringing in equations, if (Σ, E) is a strongly regular presentation of an algebraic theory then the induced monad $(T_{(\Sigma,E)}, \mu^{(\Sigma,E)}, \eta^{(\Sigma,E)})$ on **Set** is given as follows. For any set A,

$$T_{(\Sigma,E)}A = (T_{F\Sigma}A)/\sim_A$$

where $\sim_A$ is the smallest equivalence relation on $T_{F\Sigma}A$ such that

- $\sim_A$ is a congruence: if $\sigma \in \Sigma(n)$ and we have $\tau_i \in (F\Sigma)(k_i)$, $\hat{\tau}_i \in (F\Sigma)(\hat{k}_i)$, $a_i^j, \hat{a}_i^j \in A$ satisfying

$$(\tau_i, a_i^1, \ldots, a_i^{k_i}) \sim_A (\hat{\tau}_i, \hat{a}_i^1, \ldots, \hat{a}_i^{\hat{k}_i})$$

for each $i = 1, \ldots, n$, then

$$(\sigma \circ (\tau_1, \ldots, \tau_n), a_1^1, \ldots, a_n^{k_n}) \sim_A (\sigma \circ (\hat{\tau}_1, \ldots, \hat{\tau}_n), \hat{a}_1^1, \ldots, \hat{a}_n^{\hat{k}_n})$$

- the equations are satisfied: if $(\tau, \hat{\tau}) \in E_n$ and $\tau_1 \in (F\Sigma)(k_1), \ldots, \tau_n \in (F\Sigma)(k_n)$, $a_i^j \in A$, then

$$(\tau \circ (\tau_1, \ldots, \tau_n), a_1^1, \ldots, a_n^{k_n}) \sim_A (\hat{\tau} \circ (\tau_1, \ldots, \tau_n), a_1^1, \ldots, a_n^{k_n}).$$

This defines an endofunctor $T_{(\Sigma, E)}$ of **Set** and a natural transformation $\varepsilon :$ $T_{F\Sigma} \longrightarrow T_{(\Sigma, E)}$ (the quotient map). The multiplication $\mu_A^{(\Sigma, E)}$ and unit $\eta_A^{(\Sigma, E)}$ of the monad are the unique maps making

$$\begin{array}{ccccc}
T_{F\Sigma}^2 A & \xrightarrow{\mu_A^{F\Sigma}} & T_{F\Sigma}A & \xleftarrow{\eta_A^{F\Sigma}} & A \\
{\scriptstyle (\varepsilon * \varepsilon)_A} \downarrow & & \downarrow {\scriptstyle \varepsilon_A} & & \| \\
T_{(\Sigma, E)}^2 A & \xdashrightarrow[\mu_A^{(\Sigma, E)}]{} & T_{(\Sigma, E)}A & \xdashleftarrow[\eta_A^{(\Sigma, E)}]{} & A
\end{array} \qquad (C.1)$$

commute. By definition, a monad on **Set** is strongly regular if and only if it is isomorphic to a monad of the form $(T_{(\Sigma, E)}, \mu^{(\Sigma, E)}, \eta^{(\Sigma, E)})$.

We have now given sense to the terms in the theorem and can start to prove it. Fix a strongly regular presentation (Σ, E) of an algebraic theory. The first task is to simplify the clauses above defining $\sim_A$, which we do by reducing to the case $A = 1$. Note that $\sim_1$ is an equivalence relation on $\coprod_{n \in \mathbb{N}}(F\Sigma)(n)$.

Lemma C.1.3 *Let A be a set, $n, \hat{n} \in \mathbb{N}$, $\tau \in (F\Sigma)(n)$, $\hat{\tau} \in (F\Sigma)(\hat{n})$.*

a. *If $\tau \sim_1 \hat{\tau}$ then $n = \hat{n}$ and $(\tau, a_1, \ldots, a_n) \sim_A (\hat{\tau}, a_1, \ldots, a_n)$ for all $a_1, \ldots, a_n \in A$.*

b. *If $a_1, \ldots, a_n, \hat{a}_1, \ldots, \hat{a}_{\hat{n}} \in A$ and $(\tau, a_1, \ldots, a_n) \sim_A (\hat{\tau}, \hat{a}_1, \ldots, \hat{a}_{\hat{n}})$ then $n = \hat{n}$, $a_1 = \hat{a}_1, \ldots, a_n = \hat{a}_{\hat{n}}$, and $\tau \sim_1 \hat{\tau}$.*

Proof To prove (a), define a relation $\approx$ on $\coprod_{m \in \mathbb{N}}(F\Sigma)(m)$ as follows: for $\phi \in (F\Sigma)(m)$ and $\hat{\phi} \in (F\Sigma)(\hat{m})$,

$$\phi \approx \hat{\phi} \iff m = \hat{m} \text{ and } (\phi, a_1, \ldots, a_m) \sim_A (\hat{\phi}, a_1, \ldots, a_m)$$
$$\text{for all } a_1, \ldots, a_m \in A.$$

Then $\approx$ is an equivalence relation since $\sim_A$ is. If we can prove that $\approx$ also satisfies the two conditions for which $\sim_1$ is minimal then we will have $\sim_1 \subseteq \approx$, proving (a). So, for the first condition ($\approx$ is a congruence): suppose that $\sigma \in \Sigma(m)$ and that for each $i = 1, \ldots, m$ we have $\phi_i, \hat{\phi}_i \in (F\Sigma)(k_i)$ with $\phi_i \approx \hat{\phi}_i$. Then for all $a_1^1, \ldots, a_m^{k_m} \in A$, the fact that $\phi_i \approx \hat{\phi}_i$ implies that

$$(\phi_i, a_i^1, \ldots, a_i^{k_i}) \sim_A (\hat{\phi}_i, a_i^1, \ldots, a_i^{k_i}),$$

and so the fact that $\sim_A$ is a congruence implies that

$$(\sigma \circ (\phi_1, \ldots, \phi_m), a_1^1, \ldots, a_m^{k_m}) \sim_A (\sigma \circ (\hat{\phi}_1, \ldots, \hat{\phi}_m), a_1^1, \ldots, a_m^{k_m}).$$

Hence $\sigma \circ (\phi_1, \ldots, \phi_m) \approx \sigma \circ (\hat{\phi}_1, \ldots, \hat{\phi}_m)$, as required. For the second condition (the equations are satisfied): if $(\phi, \hat{\phi}) \in E_m$ then $\phi \approx \hat{\phi}$, by taking $\tau_1, \ldots, \tau_n$ all to be the identity $1 \in (F\Sigma)(1)$ in the second condition for $\sim_A$.

Several of the proofs here are of this type: we have an equivalence relation $\sim$ (in this case $\sim_1$) defined to be minimal such that certain conditions are satisfied, and we want to prove that if $x \sim y$ then some conclusion involving x and y holds. To do this we define a new equivalence relation $\approx$ by '$x \approx y$ if and only if the conclusion holds'; we then show that $\approx$ satisfies the conditions for which $\sim$ was minimal, and the result follows. These proofs, like diagram chases in homological algebra, are more illuminating to write than to read, so the remaining similar ones (including (b)) are omitted. None of them is significantly more complicated than the one just done. □

Corollary C.1.4 *For any set A, the quotient map $\varepsilon_A : T_{F\Sigma}A \longrightarrow T_{(\Sigma, E)}A$ induces a bijection*

$$\coprod_{n \in \mathbb{N}} ((F\Sigma)(n)/\sim_1) \times A^n \xrightarrow{\;\sim\;} T_{(\Sigma, E)}A.$$

□

Having expressed the equivalence relation $\sim_A$ for an arbitrary set A in terms of the single equivalence relation $\sim_1$, we now show that $\sim_1$ has a congruence property of an operadic kind.

Lemma C.1.5 *If $\tau, \hat{\tau} \in (F\Sigma)(n)$ with $\tau \sim_1 \hat{\tau}$, and $\tau_i, \hat{\tau}_i \in (F\Sigma)(k_i)$ with $\tau_i \sim_1 \hat{\tau}_i$ for each $i = 1, \ldots, n$, then*

$$\tau \circ (\tau_1, \ldots, \tau_n) \sim_1 \hat{\tau} \circ (\hat{\tau}_1, \ldots, \hat{\tau}_n).$$

Proof Define a relation $\approx$ on $\coprod_{m \in \mathbb{N}}(F\Sigma)(m)$ as follows: if $\phi \in (F\Sigma)(m)$ and $\hat{\phi} \in (F\Sigma)(\hat{m})$ then $\phi \approx \hat{\phi}$ if and only if $m = \hat{m}$ and

> for all $\quad \phi_i, \hat{\phi}_i \in (F\Sigma)(k_i)$ with $\phi_i \sim_1 \hat{\phi}_i$ ($i = 1, \ldots, n$),
> we have $\quad \phi \circ (\phi_1, \ldots, \phi_m) \sim_1 \hat{\phi} \circ (\hat{\phi}_1, \ldots, \hat{\phi}_m)$.

Then show that $\sim_1 \subseteq \approx$ in the usual way. $\qquad\square$

Define $P_{(\Sigma,E)}(n) = (F\Sigma)(n)/\sim_1$ for each n: then Lemma C.1.5 tells us that there is a unique operad structure on $P_{(\Sigma,E)}$ such that the quotient map $F\Sigma \longrightarrow P_{(\Sigma,E)}$ is a map of operads. So starting from a strongly regular presentation of a theory we have constructed an operad, and this operad induces the same monad as the theory:

Corollary C.1.6 *There is an isomorphism of monads* $T_{P_{(\Sigma,E)}} \cong T_{(\Sigma,E)}$. *Hence any strongly regular monad is operadic.*

Proof We have only to prove the first sentence; the second follows immediately. By Corollary C.1.4, there is an isomorphism of functors $T_{P_{(\Sigma,E)}} \cong T_{(\Sigma,E)}$ making the diagram

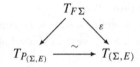

commute, where the left-hand arrow is the quotient map. The multiplication $\mu^{P_{(\Sigma,E)}}$ of the monad $T_{P_{(\Sigma,E)}}$ comes from composition in the operad $P_{(\Sigma,E)}$, and this in turn comes via the quotient map from composition in the operad $F\Sigma$, so it follows from diagram (C.1) (p. 358) that $\mu^{P_{(\Sigma,E)}}$ corresponds under the isomorphism to $\mu^{(\Sigma,E)}$. The same goes for units; hence the result. $\qquad\square$

This concludes the proof of the more 'useful' half of Theorem C.1.1. To prove the converse, we first establish one more fact about strongly regular presentations in general.

Lemma C.1.7 *Let* (Σ, E) *be a strongly regular presentation of an algebraic theory. Then the quotient map* $\varepsilon : F\Sigma \longrightarrow P_{(\Sigma,E)}$ *is the universal operad map out of* $F\Sigma$ *with the property that* $\varepsilon(\tau) = \varepsilon(\hat{\tau})$ *for all* $(\tau, \hat{\tau}) \in E_n$.

'Universal' means that if $\zeta : F\Sigma \longrightarrow Q$ is a map of operads satisfying $\zeta(\tau) = \zeta(\hat{\tau})$ for all $(\tau, \hat{\tau}) \in E_n$ then there is a unique operad map $\overline{\zeta} : P_{(\Sigma,E)} \longrightarrow Q$ such that $\overline{\zeta} \circ \varepsilon = \zeta$.

Proof Universality can be re-formulated as the following condition: $\sim_1$ is the smallest equivalence relation on $\coprod_{n \in \mathbb{N}}$ that is an operadic congruence (in other words, such that Lemma C.1.5 holds) and satisfies $\tau \sim_1 \hat{\tau}$ for all $(\tau, \hat{\tau}) \in E_n$. The proof is by the usual method. □

Now fix an operad P. Define $\Sigma_P \in \mathbf{Set}^{\mathbb{N}}$ by

$$\Sigma(n) = \{\sigma_\theta \mid \theta \in P(n)\}$$

where σ_θ is a formal symbol. Let E_P be the system of equations with elements

$$(\sigma_\theta \circ (\sigma_{\theta_1}, \ldots, \sigma_{\theta_n}), \sigma_{\theta \circ (\theta_1, \ldots, \theta_n)}) \in (E_P)_{k_1 + \cdots + k_n}$$

for each $\theta \in P(n)$, $\theta_i \in P(k_i)$, and

$$(1, \sigma_1) \in (E_P)_1.$$

Then (Σ_P, E_P) is a strongly regular presentation of an algebraic theory.

There is a map $\varepsilon : F\Sigma_P \longrightarrow P$ of operads, a component of the counit of the adjunction between **Operad** and $\mathbf{Set}^{\mathbb{N}}$. Concretely (Section 2.3), ε is described by the inductive clauses

$$\varepsilon(1) = 1,$$
$$\varepsilon(\sigma_\theta \circ (\tau_1, \ldots, \tau_n)) = \theta \circ (\varepsilon(\tau_1), \ldots, \varepsilon(\tau_n))$$

for $\theta \in P(n)$, $\tau_i \in (F\Sigma_P)(k_i)$.

Lemma C.1.8 $\varepsilon : F\Sigma_P \longrightarrow P$ *is the universal operad map out of $F\Sigma_P$ with the property that $\varepsilon(\tau) = \varepsilon(\hat{\tau})$ for all $(\tau, \hat{\tau}) \in (E_P)_n$.*

Proof That ε does have the property is easily verified. For universality, take a map $\zeta : F\Sigma \longrightarrow Q$ of operads such that $\zeta(\tau) = \zeta(\hat{\tau})$ for all $(\tau, \hat{\tau}) \in (E_P)_n$; we want there to be a unique operad map $\overline{\zeta} : P \longrightarrow Q$ such that $\overline{\zeta} \circ \varepsilon = \zeta$. Since $\varepsilon(\sigma_\theta) = \theta$ for all $\theta \in P(n)$, the only possibility is $\overline{\zeta}(\theta) = \zeta(\sigma_\theta)$, and it is easy to check that this map $\overline{\zeta}$ does satisfy the conditions required. □

Corollary C.1.9 *There is an isomorphism of monads $T_{(\Sigma_P, E_P)} \cong T_P$. Hence any operadic monad is strongly regular.*

Proof Lemmas C.1.7 and C.1.8 together imply that there is an isomorphism of operads $P_{(\Sigma_P, E_P)} \cong P$; so there is an isomorphism of monads $T_{P_{(\Sigma_P, E_P)}} \cong T_P$. Corollary C.1.6 implies that $T_{P_{(\Sigma_P, E_P)}} \cong T_{(\Sigma_P, E_P)}$. The result follows. □

The proof of Theorem C.1.1 is now complete.

Similar results can be envisaged for other kinds of operad. For example, any symmetric operad induces a monad on **Set** whose algebras are exactly the algebras for the operad; an obvious conjecture is that the monads arising in this way are those that can be presented by finitary operations and equations in which the same variables appear, without repetition, on each side (but not necessarily in the same order). Another possibility is to produce results of this kind for operads in a symmetric monoidal category; compare Example 2.2.6. We would like, for instance, a general principle telling us that the theory of Lie algebras can be described by a symmetric operad of vector spaces on the grounds that its governing equations

$$[x, y] + [y, x] = 0,$$
$$[x, [y, z]] + [y, [z, x]] + [z, [x, y]] = 0$$

are both 'good': the same variables are involved, without repetition, in each summand. This principle is well-known informally, but as far as I know has not been proved.

A more difficult generalization would be to algebraic theories on presheaf categories. For example, take the theory of strict ω-categories presented in the algebraic way (1.4.8). This is not a 'strongly regular' presentation in the simple-minded sense, because the interchange law

$$(\beta' \circ_p \beta) \circ_q (\alpha' \circ_p \alpha) = (\beta' \circ_q \alpha') \circ_p (\beta \circ_q \alpha)$$

does permute the order of the variables. Yet somehow this equation should be regarded as 'good', since when the appropriate picture is drawn – e.g.

for $p = 1$ and $q = 0$ – there is no movement in the positions of the cells. So we hope for some notion of a strongly regular theory on a presheaf category, and a result saying that the monad induced by such a theory is at least cartesian, and perhaps operadic in some sense. This would render unnecessary the *ad hoc* calculations of Appendix F showing that the free strict ω-category monad is cartesian. However, such a general result has yet to be found.

C.2 Familially representable monads on Set

Not every set-valued functor is representable, but every set-valued functor is a colimit of representables. It turns out that an intermediate condition is relevant to the theory of operads: that of being a *coproduct* of representables. Such a functor is said to be 'familially representable'.

Familial representability has been studied in, among other places, an important paper of Carboni and Johnstone (1995). This section is not much more than an account of some of their results. Our goal is to describe monads on **Set** whose functor parts are familially representable and whose natural transformation parts are cartesian; we lead up to this by considering, more generally, familially representable functors into **Set**.

But first, as a warm-up, consider ordinary representability. Here follow some very well-known facts, presented in a way that foreshadows the material on familial representability.

Let us say that a category $\mathcal{A}$ has the **adjoint functor property** if any limit-preserving functor from $\mathcal{A}$ to a locally small category has a left adjoint. ('Limit-preserving' really means 'small-limit-preserving'.) For example, the Special Adjoint Functor Theorem states that any complete, well-powered, locally small category with a cogenerating set has the adjoint functor property. In particular, **Set** has the adjoint functor property.

Proposition C.2.1 *Let $\mathcal{A}$ be a category with the adjoint functor property. The following conditions on a functor $T : \mathcal{A} \longrightarrow$ **Set** are equivalent:*

a. *T preserves limits*
b. *T has a left adjoint*
c. *T is representable.*

Proof

(a) $\Rightarrow$ **(b)** Adjoint functor property.
(b) $\Rightarrow$ **(c)** Take a left adjoint S to T: then T is represented by $S1$.
(c) $\Rightarrow$ **(a)** Standard.   □

Suppose we want to define the subcategory of $[\mathcal{A}, $ **Set**$]$ consisting of only those things that are representable. We know what it means for a functor $\mathcal{A} \longrightarrow$ **Set** to be representable. By rights, a natural transformation

$$\mathcal{A}(X, -) \longrightarrow \mathcal{A}(X', -) \tag{C.2}$$

should qualify as a map in our subcategory just when it is of the form $\mathcal{A}(f, -)$ for some map $f : X' \longrightarrow X$ in $\mathcal{A}$. But the Yoneda Lemma implies that *all* natural transformations (C.2) are of this form, so in fact the appropriate subcategory is just the full subcategory of representable functors.

Representable functors are certainly cartesian, so we might reasonably ask when a natural transformation between representables is cartesian. The answer is 'seldom': the natural transformation (C.2) induced by a map $f : X' \longrightarrow X$ is cartesian just when f is an isomorphism.

The basic result on familial representability is like Proposition C.2.1, but involves a weaker set of equivalent conditions. To express them we need some terminology.

Let $\mathcal{A}$ be a category, I a set, and $(X_i)_{i \in I}$ a family of objects of $\mathcal{A}$: then there is a functor

$$\coprod_{i \in I} \mathcal{A}(X_i, -) : \mathcal{A} \longrightarrow \mathbf{Set}.$$

Such a functor, or one isomorphic to it, is said to be **familially representable**; $(X_i)_{i \in I}$ is the **representing family**. Note that we can recover I as the value of the functor at the terminal object of $\mathcal{A}$, if it has one.

A category is **connected** if it is non-empty and any functor from it into a discrete category is constant. A **connected limit** is a limit of a functor whose domain is a connected category. The crucial fact about connected limits is:

Lemma C.2.2 *Connected limits commute with small coproducts in* **Set**.   □

Corollary C.2.3 *For any set I, the forgetful functor* $\mathbf{Set}/I \longrightarrow \mathbf{Set}$ *preserves and reflects connected limits.*

Proof $\mathbf{Set}/I$ is equivalent to $\mathbf{Set}^I$, and under this equivalence the forgetful functor becomes the functor $\Sigma : \mathbf{Set}^I \longrightarrow \mathbf{Set}$ defined by $\Sigma((X_i)_{i \in I}) = \coprod_{i \in I} X_i$. Lemma C.2.2 tells us that Σ preserves connected limits. Moreover, $\mathbf{Set}^I$ has all connected limits and Σ reflects isomorphisms, so preservation implies reflection.   □

One type of connected limit is of particular importance to us. A **wide pullback** is a limit of shape

where the top row indicates a set of any cardinality. Formally, for any set K let $\mathbb{P}_K$ be the (connected) category whose objects are the elements of K together with a further object ∞ and whose only non-identity maps are a single map $k \longrightarrow \infty$ for each $k \in K$; a wide pullback is a limit of a functor with domain $\mathbb{P}_K$ for some set K. If K has cardinality two then this is just an ordinary pullback.

If $T : \mathcal{A} \longrightarrow \mathcal{B}$ is a functor and $\mathcal{A}$ has a terminal object then we write $\widetilde{T} : \mathcal{A} \longrightarrow \mathcal{B}/T1$ for the functor

$$X \longmapsto \begin{pmatrix} TX \\ \downarrow T! \\ T1 \end{pmatrix}.$$

We recover T from $\widetilde{T}$ as the composite

$$T = (\mathcal{A} \xrightarrow{\ \widetilde{T}\ } \mathcal{B}/T1 \xrightarrow{\ \text{forgetful}\ } \mathcal{B}). \tag{C.3}$$

Proposition C.2.4 *Let $\mathcal{A}$ be a category with the adjoint functor property and a terminal object. The following conditions on a functor $T : \mathcal{A} \longrightarrow$ **Set** are equivalent:*

a. *T preserves connected limits*
b. *T preserves wide pullbacks*
c. *$\widetilde{T}$ preserves limits*
d. *$\widetilde{T}$ has a left adjoint*
e. *T is familially representable.*

Proof

(a) $\Rightarrow$ **(b)** A wide pullback is a connected limit.

(b) $\Rightarrow$ **(c)** The forgetful functor **Set**$/T1 \longrightarrow$ **Set** reflects wide pullbacks (Corollary C.2.3), and T preserves them, so by equation (C.3), $\widetilde{T}$ preserves them too. $\widetilde{T}$ also preserves terminal objects, trivially. So $\widetilde{T}$ preserves all limits.

(c) $\Rightarrow$ **(d)** Adjoint functor property.

(d) $\Rightarrow$ **(e)** Take a left adjoint S to $\widetilde{T}$. For each $i \in T1$ we have adjunctions

$$\mathcal{A} \underset{S}{\overset{\widetilde{T}}{\rightleftarrows}} \mathbf{Set}/T1 \underset{i_!}{\overset{(\ )_i}{\rightleftarrows}} \mathbf{Set},$$

where $(\ )_i$ takes the fibre over i and $i_! K$ is the function $K \longrightarrow T1$ constant at i. So

$$TX \cong \coprod_{i \in T1} (\widetilde{T}X)_i \cong \coprod_{i \in T1} \mathbf{Set}(1, (\widetilde{T}X)_i) \cong \coprod_{i \in T1} \mathcal{A}(Si_!1, X)$$

naturally in $X \in \mathcal{A}$.

(e) $\Rightarrow$ **(a)** Representables preserve limits (C.2.1) and coproducts commute with connected limits in **Set** (C.2.2), so any coproduct of representables preserves connected limits. $\qquad \square$

Let $(X_i)_{i \in I}$ and $(X'_{i'})_{i' \in I'}$ be families of objects in a category $\mathcal{A}$. A function $\phi : I \longrightarrow I'$ together with a family of maps $(X'_{\phi(i)} \xrightarrow{\ f_i\ } X_i)_{i \in I}$ induces a natural transformation

$$\coprod_{i \in I} \mathcal{A}(X_i, -) \longrightarrow \coprod_{i' \in I'} \mathcal{A}(X'_{i'}, -), \qquad (C.4)$$

and the Yoneda Lemma implies that, in fact, all natural transformations (C.4) arise uniquely in this way. Let $\mathbf{Fam}(\mathcal{A})$ be the category whose objects are families $(X_i)_{i \in I}$ of objects of $\mathcal{A}$ and whose maps $(X_i)_{i \in I} \longrightarrow (X'_{i'})_{i' \in I'}$ are pairs (ϕ, f) as above. (This is the free coproduct cocompletion of $\mathcal{A}^{\mathrm{op}}$.) There is a functor

$$\mathbf{y} : \quad \begin{array}{ccc} \mathbf{Fam}(\mathcal{A}) & \longrightarrow & [\mathcal{A}, \mathbf{Set}] \\ (X_i)_{i \in I} & \longmapsto & \coprod_{i \in I} \mathcal{A}(X_i, -), \end{array} \qquad (C.5)$$

and we have just seen that it is full and faithful. It therefore defines an equivalence between $\mathbf{Fam}(\mathcal{A})$ and the full subcategory of $[\mathcal{A}, \mathbf{Set}]$ formed by the familially representable functors.

A transformation (ϕ, f) as above is cartesian just when each of its pieces

$$\mathcal{A}(f_i, -) : \mathcal{A}(X_i, -) \longrightarrow \mathcal{A}(X'_{\phi(i)}, -)$$

$(i \in I)$ is cartesian, which, from our earlier result, happens just when each $f_i : X'_{\phi(i)} \longrightarrow X_i$ is an isomorphism.

We have already encountered many familially representable endofunctors of $\mathbf{Set}$:

Example C.2.5 For any plain operad P, the functor part of the induced monad T_P on $\mathbf{Set}$ is familially representable. Indeed,

$$T_P(X) \cong \coprod_{n \in \mathbb{N}} P(n) \times X^n \cong \coprod_{i \in I} [X_i, X]$$

where $I = \coprod_{n \in \mathbb{N}} P(n)$, X_i is an n-element set for $i \in P(n)$, and $[X_i, X]$ denotes the hom-set $\mathbf{Set}(X_i, X)$.

In this example all of the sets X_i are finite and, as we saw in Section C.1, the monad T_P corresponds to a finitary algebraic theory. Here is the general result on finiteness.

Proposition C.2.6 *The following are equivalent conditions on a family $(X_i)_{i \in I}$ of sets:*

a. the functor $\coprod_{i \in I} [X_i, -] : \mathbf{Set} \longrightarrow \mathbf{Set}$ is finitary
b. the set X_i is finite for each $i \in I$

c. there is a sequence $(P(n))_{n \in \mathbb{N}}$ *of sets such that*

$$\coprod_{i \in I}[X_i, -] \cong \coprod_{n \in \mathbb{N}} P(n) \times (-)^n.$$

Proof

(a) $\Rightarrow$ **(b)** For each $i \in I$ we have a pullback square

$$
\begin{array}{ccc}
[X_i, -] & \xrightarrow{\ \mathrm{copr}_i\ } & \coprod_{i \in I}[X_i, -] \\
\downarrow & & \downarrow{\scriptstyle \mathrm{pr}} \\
\Delta 1 & \xrightarrow[\ \mathrm{copr}_i\ ]{} & \Delta I
\end{array}
$$

in [**Set**, **Set**], where $\Delta K : \textbf{Set} \longrightarrow \textbf{Set}$ denotes the constant functor
with value K. We know that all four functors except perhaps $[X_i, -]$ are
finitary; since pullbacks commute with filtered colimits in **Set** it follows
that $[X_i, -]$ is finitary too. This in turn implies that X_i is finite, as is
well-known (Adámek and Rosický, 1994, p. 9).

(b) $\Rightarrow$ **(c)** For each $n \in \mathbb{N}$, put

$$P(n) = \{i \in I \mid \mathrm{cardinality}(X_i) = n\}.$$

By choosing for each $i \in I$ a bijection $X_i \xrightarrow{\ \sim\ } \{1, \ldots, n\}$, we obtain
isomorphisms

$$\coprod_{i \in I}[X_i, X] \cong \coprod_{n \in \mathbb{N}} \coprod_{i \in P(n)} [X_i, X] \cong \coprod_{n \in \mathbb{N}} \coprod_{i \in P(n)} X^n \cong \coprod_{n \in \mathbb{N}} P(n) \times X^n$$

natural in $X \in \textbf{Set}$.

(c) $\Rightarrow$ **(a)** If (c) holds then $\coprod_{i \in I}[X_i, -]$ is a colimit of endofunctors of **Set** of
the form $(-)^n$ $(n \in \mathbb{N})$; each of these is finitary, so the whole functor is
finitary. $\qquad\qquad\square$

So an endofunctor of **Set** is finitary and familially representable if and only if
it is, in the sense of (c), operadic. The story for *monads* is quite different, as we
see below (C.2.8). A hint of the subtlety is that the isomorphism constructed in
the proof of (b) $\Rightarrow$ (c) is not canonical: it involves an arbitrary choice of a total
ordering of X_i, for each $i \in I$.

Let us turn, then, to monads on **Set**. A **familially representable monad**
(T, μ, η) on **Set** is one whose functor part T is familially representable and
whose natural transformation parts μ and η are cartesian. (For example, any
operadic monad on **Set** is familially representable.) To see what μ and η look

like in terms of the representing family of T, we need to consider composition of familially representable functors.

So, let

$$S \cong \coprod_{h \in H} [W_h, -], \qquad T \cong \coprod_{i \in I} [X_i, -]$$

be familially representable endofunctors of **Set**. Conditions (a) and (b) of Proposition C.2.4 make it clear that $T \circ S$ must be familially representable, but what is the representing family? For any set X we have

$$TSX \cong \coprod_{i \in I} \left[X_i, \coprod_{h \in H} [W_h, X] \right]$$

$$\cong \coprod_{i \in I} \coprod_{g \in [X_i, H]} \left[\coprod_{x \in X_i} W_{g(x)}, X \right].$$

Hence

$$T \circ S \cong \coprod_{j \in J} [Y_j, -]$$

where

$$J = \coprod_{i \in I} [X_i, H] = TH, \qquad Y_{i,g} = \coprod_{x \in X_i} W_{g(x)}$$

for $i \in I$ and $g \in [X_i, H]$ (that is, $(i, g) \in J$). In particular, the case $S = T$ gives

$$T^2 \cong \coprod_{j \in J} [Y_j, -]$$

where

$$J = \coprod_{i \in I} [X_i, I] = TI, \qquad Y_{i,g} = \coprod_{x \in X_i} X_{g(x)}.$$

The familial representation of the identity functor is

$$\mathrm{id} \cong \coprod_{i \in I} [1, -].$$

A familially representable monad on **Set** therefore consists of

- a family $(X_i)_{i \in I}$ of sets, inducing the functor $T = \coprod_{i \in I} [X_i, -]$

- a function

$$m : \{(i, g) \mid i \in I, g \in [X_i, I]\} \longrightarrow I$$

and for each $i \in I$ and $g \in [X_i, I]$, a bijection

$$X_{m(i,g)} \xrightarrow{\sim} \coprod_{x \in X_i} X_{g(x)},$$

inducing the natural transformation $\mu : T^2 \longrightarrow T$ whose component μ_X at a set X is the composite

$$\coprod_{i \in I, g \in [X_i, I]} \left[\coprod_{x \in X_i} X_{g(x)}, X \right] \xrightarrow{\sim} \coprod_{i \in I, g \in [X_i, I]} [X_{m(i,g)}, X]$$

$$\xrightarrow{\text{canonical}} \coprod_{k \in I} [X_k, X]$$

- an element $e \in I$ such that X_e is a one-element set, inducing the natural transformation $\eta : 1 \longrightarrow T$ whose component η_X at a set X is the composite

$$X \xrightarrow{\sim} [X_e, X] \hookrightarrow \coprod_{i \in I} [X_i, X],$$

such that μ and η obey associativity and unit laws. The monad is finitary just when all the X_is are finite.

Example C.2.7 Let (T, μ, η) be the free monoid monad on **Set**. Then

$$TX = \coprod_{i \in \mathbb{N}} [i, X],$$

so in this case $I = \mathbb{N}$ and $X_i = i$. (We use i to denote an i-element set, say $\{1, \ldots, i\}$.) The multiplication function

$$m : \{(i, g) \mid i \in \mathbb{N}, g \in [i, \mathbb{N}]\} \longrightarrow \mathbb{N}$$

is given by

$$m(i, g) = g(1) + \cdots + g(i),$$

and there is an obvious choice of bijection

$$X_{g(1)+\cdots+g(i)} \xrightarrow{\sim} X_{g(1)} + \cdots + X_{g(i)}.$$

The unit element e is $1 \in \mathbb{N}$.

This explicit form for familially representable monads on **Set** will enable us to prove, for instance, that a commutative monoid is naturally an algebra for any finitary familially representable monad on **Set** (Example C.4.3).

We have now proved almost all of the equalities and inclusions in Fig. C-A. All that remains is to prove that the first inclusion is proper: not every finitary familially representable monad on **Set** is operadic (strongly regular). At this point we come to an error in Carboni and Johnstone (1995). In their Proposition 3.2, they prove that every strongly regular monad on **Set** is familially representable; but they also claim the converse, which is false. The error is in the final sentence of the proof. Having started with a familially representable monad (T, μ, η), they construct a strongly regular monad – let us call it (T', μ', η') – and claim that the monads (T, μ, η) and (T', μ', η') are isomorphic; and while there is indeed a (non-canonical) isomorphism between the functors T and T', there is not in general an isomorphism that commutes with the monad structures. The following counterexample is from Carboni and Johnstone's corrigenda (2002).

Example C.2.8 The monad (T, μ, η) on **Set** corresponding to the theory of monoids with an anti-involution (4.1.8, 4.2.10) is familially representable and finitary, but not operadic.

The free monoid with anti-involution on a set X is the set of expressions $x_1^{\sigma_1} x_2^{\sigma_2} \cdots x_n^{\sigma_n}$ where $n \geq 0$, $x_i \in X$, and $\sigma_i \in \{+1, -1\}$: so

$$TX \cong \coprod_{n \in \mathbb{N}} (\{+1, -1\} \times X)^n \cong \coprod_{i \in I} [X_i, X]$$

where $I = \coprod_{n \in \mathbb{N}} \{+1, -1\}^n$ and $X_{(\sigma_1, \ldots, \sigma_n)} = n$. The unit map at X is

$$
\begin{array}{rccc}
\eta_X : & X & \longrightarrow & TX \\
& x & \longmapsto & (+1, x) \in (\{+1, -1\} \times X)^1.
\end{array}
$$

So far this is the same as for monoids with *involution* (4.1.7, 4.2.9), but the multiplication is different: its component μ_X at X is the map

$$\coprod_{n \in \mathbb{N}} \left(\{+1, -1\} \times \coprod_{k \in \mathbb{N}} (\{+1, -1\} \times X)^k \right)^n \longrightarrow \coprod_{m \in \mathbb{N}} (\{+1, -1\} \times X)^m$$

given by

$$
\begin{aligned}
&\mu_X \left((\sigma_1, (\sigma_1^1, x_1^1, \ldots, \sigma_1^{k_1}, x_1^{k_1})), \ldots, (\sigma_n, (\sigma_n^1, x_n^1, \ldots, \sigma_n^{k_n}, x_n^{k_n})) \right) \\
&= (\hat{\sigma}_1^1, \hat{x}_1^1, \ldots, \hat{\sigma}_1^{k_1}, \hat{x}_1^{k_1}, \ldots, \hat{\sigma}_n^1, \hat{x}_n^1, \ldots, \hat{\sigma}_n^{k_n}, \hat{x}_n^{k_n})
\end{aligned}
$$

where for $i \in \{1, \ldots, n\}$,

$$(\hat{\sigma}_i^1, \hat{x}_i^1, \ldots, \hat{\sigma}_i^{k_i}, \hat{x}_i^{k_i}) = \begin{cases} (\sigma_i^1, x_i^1, \ldots, \sigma_i^{k_i}, x_i^{k_i}) & \text{if } \sigma_i = +1 \\ (-\sigma_i^{k_i}, x_i^{k_i}, \ldots, -\sigma_i^1, x_i^1) & \text{if } \sigma_i = -1. \end{cases}$$

The functor T is plainly familially representable, and finitary by C.2.6. The unit and multiplication are cartesian by the observations on p. 366. So the monad is familially representable and finitary, as claimed.

On the other hand, the following argument – reminiscent of a standard proof of Brouwer's fixed point theorem – shows that it is not operadic. For suppose it were. Then by Proposition 6.2.1, there is a cartesian natural transformation ψ from T to the free monoid monad S, commuting with the monad structures. There is also a cartesian natural transformation $\theta : S \longrightarrow T$ commuting with the monad structures: in the notation of p. 366, it is given by the function

$$\phi : \quad \mathbb{N} \quad \longrightarrow \quad \{+1, -1\}^n$$
$$n \quad \longmapsto \quad (+1, \ldots, +1)$$

together with the identity map $f_n : n \longrightarrow n$ for each $n \in \mathbb{N}$; on categories of algebras, it induces the functor $\mathbf{Set}^T \longrightarrow \mathbf{Set}^S$ forgetting anti-involutions. So we have a commutative triangle

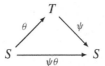

of cartesian natural transformations, each commuting with the monad structures. The members of the representing family of S all have different cardinalities, so any cartesian natural endomorphism of S is actually an automorphism. Hence in the induced triangle of functors on categories of algebras,

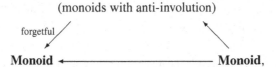

the functor along the bottom is an isomorphism. This implies that the forgetful functor is surjective on objects. So every monoid admits an anti-involution, and in particular, every monoid is isomorphic to its opposite (the same monoid with the order of multiplication reversed); but, for instance, the monoid of endomorphisms of a two-element set does not have this property. This is the desired contradiction.

C.3 Familially representable monads on presheaf categories

The cartesian monads arising in this book are typically monads on presheaf categories. Very often they are familially representable (in a sense defined shortly), and this section provides some of the theory of such monads. After some general preliminaries, we concentrate on the special case of presheaf categories $[\mathbb{B}^{op}, \mathbf{Set}]$ where $\mathbb{B}$ is discrete, arriving eventually at an explicit description of finitary familially representable monads on such categories. This will be used in the next section to provide a link between symmetric and generalized multicategories.

Let $\mathcal{A}$ be a category and $\mathbb{B}$ a small category. What should it mean for a functor $T : \mathcal{A} \longrightarrow [\mathbb{B}^{op}, \mathbf{Set}]$ to be familially representable? If $\mathbb{B}$ is discrete then the answer is clear: for each $b \in \mathbb{B}$ the b-component

$$T_b = T(-)(b) : \mathcal{A} \longrightarrow \mathbf{Set}$$

should be familially representable. If $\mathbb{B}$ is not discrete then the answer is not quite so obvious: we certainly do want each T_b to be familially representable, but we also want the representing family to vary functorially as b varies in $\mathbb{B}$. Thus:

Definition C.3.1 Let $\mathcal{A}$ be a category and $\mathbb{B}$ a small category. A functor $T : \mathcal{A} \longrightarrow [\mathbb{B}^{op}, \mathbf{Set}]$ is **familially representable** if there exists a functor R making the diagram

$$
\begin{array}{ccc}
\mathbb{B}^{op} & \overset{R}{\dashrightarrow} & \mathbf{Fam}(\mathcal{A}) \\
 & {\scriptstyle \overline{T}} \searrow & \downarrow {\scriptstyle \mathbf{y}} \\
 & & [\mathcal{A}, \mathbf{Set}]
\end{array}
$$

commute up to natural isomorphism. Here $\overline{T}$ is the transpose of T and $\mathbf{y}$ is the 'Yoneda' functor of (C.5) (p. 366).

Note that when $\mathbb{B}$ is the terminal category, this is compatible with the definition of familial representability for set-valued functors. Note also that since $\mathbf{y}$ is full and faithful, the functor R is determined uniquely up to isomorphism, if it exists.

This definition is morally the correct one but in practical terms needlessly elaborate: if each T_b is familially representable then the representing family *automatically* varies functorially in b, as the equivalence (c) ⇔ (d) in the following result shows.

Proposition C.3.2 *Let $\mathcal{A}$ be a category with the adjoint functor property and a terminal object. Let $\mathbb{B}$ be a small category. The following conditions on a functor $T : \mathcal{A} \longrightarrow [\mathbb{B}^{\mathrm{op}}, \mathbf{Set}]$ are equivalent:*

a. *T preserves connected limits*
b. *T preserves wide pullbacks*
c. *for each $b \in \mathbb{B}$, the functor $T_b : \mathcal{A} \longrightarrow \mathbf{Set}$ is familially representable*
d. *T is familially representable.*

In our applications of this result, $\mathcal{A}$ will be a presheaf category. Such an $\mathcal{A}$ does have the adjoint functor property: for by Borceux (1994a, 4.7.2), it satisfies the hypotheses of the Special Adjoint Functor Theorem.

Proof

(a) $\Leftrightarrow$ (b) $\Leftrightarrow$ (c) Since limits in a presheaf category are computed pointwise, T preserves limits of a given shape if and only if T_b preserves them for each $b \in \mathbb{B}$. So these implications follow from Proposition C.2.4.

(c) $\Rightarrow$ (d) Choose for each $b \in \mathbb{B}$ a representing family $(X_{b,i})_{i \in I(b)}$ and an isomorphism

$$\alpha_b : T_b \xrightarrow{\ \sim\ } \mathbf{y}((X_{b,i})_{i \in I(b)}) = \coprod_{i \in I(b)} \mathcal{A}(X_{b,i}, -).$$

Since $\mathbf{y}$ is full and faithful, the assignment $b \longmapsto R(b) = (X_{b,i})_{i \in I(b)}$ extends uniquely to a functor R such that α is a natural isomorphism $\overline{T} \xrightarrow{\ \sim\ } \mathbf{y} \circ R$.

(d) $\Rightarrow$ (c) Trivial. $\square$

A familially representable functor $\mathcal{A} \longrightarrow [\mathbb{B}^{\mathrm{op}}, \mathbf{Set}]$ is determined by a functor $R : \mathbb{B}^{\mathrm{op}} \longrightarrow \mathbf{Fam}(\mathcal{A})$. Explicitly, such a functor R consists of

- a functor $I : \mathbb{B}^{\mathrm{op}} \longrightarrow \mathbf{Set}$
- for each $b \in \mathbb{B}$, a family $(X_{b,i})_{i \in I(b)}$ of objects of $\mathcal{A}$
- for each $b' \xrightarrow{\ g\ } b$ in $\mathbb{B}$ and each $i \in I(b)$, a map

$$X_{g,i} : X_{b',(Ig)(i)} \longrightarrow X_{b,i}$$

in $\mathcal{A}$

such that

- if $b'' \xrightarrow{\ g'\ } b' \xrightarrow{\ g\ } b$ in $\mathbb{B}$ and $i \in I(b)$ then $X_{g' \circ g,i} = X_{g,i} \circ X_{g',(Ig)(i)}$
- if $b \in \mathbb{B}$ then $X_{1_b,i} = 1_{X_{b,i}}$.

The resulting functor $T : \mathcal{A} \longrightarrow [\mathbb{B}^{\text{op}}, \mathbf{Set}]$ is given at $X \in \mathcal{A}$ and $b \in \mathbb{B}$ by

$$(TX)(b) = \coprod_{i \in I(b)} \mathcal{A}(X_{b,i}, X).$$

A good example of a familially representable functor into a presheaf category is the free strict ω-category functor T discussed in Chapter 8 and Appendix F. Here is the 1-dimensional version.

Example C.3.3 Let $\mathbb{B} = (0 \overset{\sigma}{\underset{\tau}{\rightrightarrows}} 1)$, so that $[\mathbb{B}^{\text{op}}, \mathbf{Set}]$ is the category **DGph** of directed graphs; let $\mathcal{A} = \mathbf{DGph}$ too. Then the free category functor $T :$ **DGph** $\longrightarrow$ **DGph** is familially representable. The directed graph I is defined by $I(0) = 1$ and $I(1) = \mathbb{N}$. The families $(X_{b,i})_{i \in I(b)}$ ($b \in \mathbb{B}$) are defined as follows. For $b = 1$ and $i \in \mathbb{N}$, let $X_{1,i}$ be the graph

$$\underset{0}{\bullet} \longrightarrow \underset{1}{\bullet} \longrightarrow \quad \cdots \quad \longrightarrow \underset{i}{\bullet}$$

with $(i + 1)$ vertices and i edges. For $b = 0$, writing $1 = \{j\}$, let

$$X_{0,j} = \bullet = X_{1,0}.$$

The maps $0 \overset{\sigma}{\underset{\tau}{\rightrightarrows}} 1$ in $\mathbb{B}$ induce, for each $i \in \mathbb{N}$, the graph maps

$$X_{0,j} \rightrightarrows X_{1,i}$$

sending $X_{0,j}$ to the first and last vertex of $X_{1,i}$, respectively. This data does represent T: if X is a directed graph then

$$\coprod_{i \in I(1)} \mathbf{DGph}(X_{1,i}, X) = \coprod_{i \in \mathbb{N}} \{\text{strings of } i \text{ arrows in } X\} = (TX)_1$$

and

$$\coprod_{j \in I(0)} \mathbf{DGph}(X_{0,j}, X) = X_0 = (TX)_0,$$

as required.

A short digression: the **Artin gluing** of a functor $T : \mathcal{A} \longrightarrow \mathcal{B}$ is the category $\mathcal{B} \!\downarrow\! T$ in which an object is a triple (X, Y, π) with $X \in \mathcal{A}$, $Y \in \mathcal{B}$, and $\pi : Y \longrightarrow TX$ in $\mathcal{B}$, and a map $(X, Y, \pi) \longrightarrow (X', Y', \pi')$ is a pair of maps $(X \longrightarrow X', Y \longrightarrow Y')$ making the evident square commute.

Proposition C.3.4 (Carboni–Johnstone) *Let $\mathcal{A}$ and $\mathcal{B}$ be presheaf categories and $T : \mathcal{A} \longrightarrow \mathcal{B}$ a functor. Then $\mathcal{B} \!\downarrow\! T$ is a presheaf category if and only if T preserves wide pullbacks.*

Proof See Carboni and Johnstone (1995, 4.4(v)). Their proof of 'if' is a little roundabout, so the following sketch of a direct version may be of interest.

By Proposition C.3.2, T is familially representable. Suppose that $\mathcal{A} = [\mathbb{A}^{op}, \mathbf{Set}]$ and $\mathcal{B} = [\mathbb{B}^{op}, \mathbf{Set}]$, and represent T by the families $(X_{b,i})_{i \in I(b)}$, as above. An object of $\mathcal{B} \downarrow T$ consists of functors $X : \mathbb{A}^{op} \longrightarrow \mathbf{Set}$ and $Y : \mathbb{B}^{op} \longrightarrow \mathbf{Set}$ and a map

$$\pi_b : Yb \longrightarrow \coprod_{i \in I(b)} [\mathbb{A}^{op}, \mathbf{Set}](X_{b,i}, X)$$

for each $b \in \mathbb{B}$, satisfying naturality axioms. Equivalently, it consists of

a. a functor $X : \mathbb{A}^{op} \longrightarrow \mathbf{Set}$,
b. a family $(Y(b, i))_{b \in \mathbb{B}, i \in I(b)}$ of sets, functorial in b, and
c. for each $b \in \mathbb{B}$, $i \in I(b)$ and $y \in Y(b, i)$, a natural transformation $X_{b,i} \longrightarrow X$,

satisfying axioms; indeed, (c) can equivalently be replaced by

c'. for each $b \in \mathbb{B}, i \in I(b), a \in \mathbb{A}$ and $x \in X_{b,i}(a)$, a function $Y(b, i) \longrightarrow X(a)$.

Equivalently, it is a functor $\mathbb{C}^{op} \longrightarrow \mathbf{Set}$, where $\mathbb{C}$ is the category whose object-set $\mathbb{C}_0$ is the disjoint union

$$\mathbb{C}_0 = \mathbb{A}_0 + \{(b, i) \mid b \in \mathbb{B}_0, i \in I(b)\},$$

whose maps are given by

$$\mathbb{C}(a', a) = \mathbb{A}(a', a),$$
$$\mathbb{C}((b, i), (b', i')) = \{g \in \mathbb{B}(b, b') \mid (Ig)(i') = i\},$$
$$\mathbb{C}(a, (b, i)) = X_{b,i}(a),$$
$$\mathbb{C}((b, i), a) = \emptyset$$

$(a, a' \in \mathbb{A}_0, b, b' \in \mathbb{B}_0, i \in I(b), i' \in I(b'))$, and whose composition and identities are the evident ones. $\qquad \square$

Returning to the main story of familially representable functors into presheaf categories, let us restrict our attention to functors of the form

$$T : \mathbf{Set}^C \longrightarrow \mathbf{Set}^B$$

where C and B are sets. This makes the calculations much easier but still provides a wide enough context for our applications.

A familially representable functor $T : \mathbf{Set}^C \longrightarrow \mathbf{Set}^B$ consists of

- a family $(I(b))_{b \in B}$ of sets
- for each $b \in B$, a family $(X_{b,i})_{i \in I(b)}$ of objects of $\mathbf{Set}^C$,

and the actual functor T is then given by

$$(TX)(b) \cong \coprod_{i \in I(b)} \mathbf{Set}^C(X_{b,i}, X) \cong \coprod_{i \in I(b)} \prod_{c \in C} [X_{b,i}(c), X(c)]$$

for each $X \in \mathbf{Set}^C$ and $b \in B$.

Let $T' : \mathbf{Set}^C \longrightarrow \mathbf{Set}^B$ be another such functor, with representing families $(X_{b,i'})_{i' \in I'(b)}$ $(b \in B)$. A natural transformation

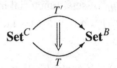

consists merely of a natural transformation

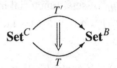

for each $b \in B$. So by the results of the previous section, a transformation $T' \longrightarrow T$ is described by

- for each $b \in B$, a function $\phi_b : I'(b) \longrightarrow I(b)$
- for each $b \in B$ and $i' \in I'(b)$, a map $f_{b,i'} : X_{b,\phi_b(i')} \longrightarrow X'_{b,i'}$ in $\mathbf{Set}^C$,

and the induced map $(T'X)(b) \longrightarrow (TX)(b)$ is the composite

$$\coprod_{i' \in I'(b)} \mathbf{Set}^C(X'_{b,i'}, X)$$
$$\xrightarrow[\text{canonical}]{\coprod_{i' \in I'(b)} f^*_{b,i'}} \coprod_{i' \in I'(b)} \mathbf{Set}^C(X_{b,\phi_b(i')}, X)$$
$$\coprod_{i \in I(b)} \mathbf{Set}^C(X_{b,i}, X)$$

$(X \in \mathbf{Set}^C, b \in B)$. The transformation is cartesian if and only if each map $f_{b,i'}$ is an isomorphism.

Just as for functors $\mathbf{Set} \longrightarrow \mathbf{Set}$, the theory tells us that the class of familially representable functors between presheaf categories is closed under composition. Working out the representing family of a composite seems extremely complicated for presheaf categories in general, but is manageable in our restricted context of direct powers of **Set**. So, take sets D, C and B and familially representable functors

$$\mathbf{Set}^D \xrightarrow{\ S\ } \mathbf{Set}^C \xrightarrow{\ T\ } \mathbf{Set}^B$$

given by

$$(SW)(c) \cong \coprod_{h\in H(c)} \mathbf{Set}^D(W_{c,h}, W), \qquad (TX)(b) \cong \coprod_{i\in I(b)} \mathbf{Set}^C(X_{b,i}, X)$$

($W \in \mathbf{Set}^D$, $X \in \mathbf{Set}^C$, $c \in C$, $b \in B$). Then for $W \in \mathbf{Set}^D$ and $b \in B$,

$$
\begin{aligned}
(TSW)(b) &\cong \coprod_{i\in I(b)} \mathbf{Set}^C(X_{b,i}, SW) \\
&\cong \coprod_{i\in I(b)} \prod_{c\in C} [X_{b,i}(c), \coprod_{h\in H(c)} \mathbf{Set}^D(W_{c,h}, W)] \\
&\cong \coprod_{i\in I(b)} \prod_{c\in C} \coprod_{g\in[X_{b,i}(c),H(c)]} \prod_{x\in X_{b,i}(c)} \mathbf{Set}^D(W_{c,g(x)}, W) \\
&\cong \coprod_{i\in I(b)} \coprod_{\gamma\in\mathbf{Set}^C(X_{b,i},H)} \prod_{c\in C} \prod_{x\in X_{b,i}(c)} \mathbf{Set}^D(W_{c,\gamma_c(x)}, W) \\
&\cong \coprod_{(i,\gamma)\in J(b)} \mathbf{Set}^D(Y_{b,i,\gamma}, W),
\end{aligned}
$$

where

$$
\begin{aligned}
J(b) &= \coprod_{i\in I(b)} \mathbf{Set}^C(X_{b,i}, H) \cong (TH)(b), \\
Y_{b,i,\gamma} &= \coprod_{c\in C, x\in X_{b,i}(c)} W_{c,\gamma_c(x)}.
\end{aligned}
$$

The identity functor on $\mathbf{Set}^B$ is also familially representable: for $X \in \mathbf{Set}^B$,

$$X(b) \cong \coprod_{i\in 1} \mathbf{Set}^B(\delta_b, X)$$

where, treating B as a discrete category, $\delta_b = B(-, b) \in \mathbf{Set}^B$.

Assembling these descriptions gives a description of monads on $\mathbf{Set}^B$ whose functor parts are familially representable and whose natural transformation parts are cartesian – **familially representable monads** on $\mathbf{Set}^B$, as we call them. Such a monad consists of:

- for each $b \in B$, a set $I(b)$ and a family $(X_{b,i})_{i\in I(b)}$ of objects of $\mathbf{Set}^B$, inducing the functor

$$
\begin{aligned}
T : \ &\mathbf{Set}^B \longrightarrow \mathbf{Set}^B, \\
&(TX)(b) = \coprod_{i\in I(b)} \mathbf{Set}^B(X_{b,i}, X)
\end{aligned}
$$

- for each $b \in B$, a function

$$m_b : \{(i, \gamma) \mid i \in I(b), \gamma \in \mathbf{Set}^B(X_{b,i}, X)\} \longrightarrow I(b),$$

and for each $b \in B$, $i \in I(b)$ and $\gamma \in \mathbf{Set}^B(X_{b,i}, X)$, an isomorphism

$$X_{b,m_b(i,\gamma)} \overset{\sim}{\longrightarrow} \coprod_{c \in B, x \in X_{b,i}(c)} X_{c,\gamma_c(x)},$$

inducing the natural transformation $\mu : T^2 \longrightarrow T$ whose component

$$\mu_{X,b} : (T^2 X)(b) \longrightarrow (TX)(b)$$

is the composite

$$\coprod_{i \in I(b), \gamma \in \mathbf{Set}^B(X_{b,i}, I)} \mathbf{Set}^B (\coprod_{c \in B, x \in X_{b,i}(c)} X_{c,\gamma_c(x)}, X)$$

$$\overset{\sim}{\underset{\text{canonical}}{\longrightarrow}} \quad \coprod_{i \in I(b), \gamma \in \mathbf{Set}^B(X_{b,i}, I)} \mathbf{Set}^B (X_{b,m_b(i,\gamma)}, X)$$

$$\coprod_{k \in I(b)} \mathbf{Set}^B (X_{b,k}, X)$$

- for each $b \in B$, an element $e_b \in I(b)$ such that $X_{b,e_b} \cong \delta_b$, inducing the natural transformation $\eta : 1 \longrightarrow T$ whose component

$$\eta_{X,b} : X(b) \longrightarrow (TX)(b)$$

is the composite

$$X(b) \overset{\sim}{\longrightarrow} \mathbf{Set}^B (X_{b,e_b}, X) \hookrightarrow \coprod_{i \in I(b)} \mathbf{Set}^B (X_{b,i}, X),$$

such that μ and η obey associativity and unit laws. This is complicated, but rest assured that it gets no worse.

The aim of this section was to describe finitary familially representable monads on categories of the form $\mathbf{Set}^B$, and we are nearly there. All that remains is 'finitary'.

Proposition C.3.5 *Let C and B be sets. Let $T : \mathbf{Set}^C \longrightarrow \mathbf{Set}^B$ be a famil-ially representable functor with representing families $(X_{b,i})_{i \in I(b)}$ ($b \in B$). The following are equivalent:*

a. T is a finitary functor

b. $\coprod_{c \in C} X_{b,i}(c)$ is a finite set for each $b \in B$ and $i \in I(b)$.

Proof T is finitary if and only if the functor

$$T_b = \coprod_{i \in I(b)} \mathbf{Set}^C (X_{b,i}, -) : \mathbf{Set}^C \longrightarrow \mathbf{Set}$$

is finitary for each $b \in B$. By the arguments in the proof of Proposition C.2.6, T_b is finitary if and only if the functor

$$\mathbf{Set}^C(X_{b,i}, -) : \mathbf{Set}^C \longrightarrow \mathbf{Set}$$

is finitary for each $i \in I(b)$. But $\mathbf{Set}^C(X_{b,i}, -)$ is finitary if and only if $\coprod_{c \in C} X_{b,i}(c)$ is a finite set (Adámek and Rosický, 1994, p. 9). $\square$

C.4 Cartesian structures from symmetric structures

A commutative monoid is a structure in which every finite family of elements has a well-defined sum. So if $T = \coprod_{i \in I}[X_i, -]$ is any finitary familially representable monad on **Set** then any commutative monoid A is naturally a T-algebra via the map

$$TA = \coprod_{i \in I}[X_i, A] \longrightarrow A$$

whose i-component sends a family $(a_x)_{x \in X_i}$ to $\sum_{x \in X_i} a_x$.

This is the simplest case of the theme of this section: how symmetric structures give rise to cartesian structures. We prove two main results. The first is that if B is any set, T any finitary familially representable monad on $\mathbf{Set}^B$, and A any commutative monoid, then the constant family $(A)_{b \in B}$ is naturally a T-algebra. (Just now we looked at the case $B = 1$.) It follows that any commutative monoid is naturally a T_n-operad, where T_n is the nth opetopic monad (Section 7.1). We can also extract a rigorous definition of the set-with-multiplicities of n-opetopes underlying a given n-dimensional opetopic pasting diagram.

The second main result is that any symmetric multicategory A gives rise to a T-multicategory. Here T is, again, any finitary familially representable monad on $\mathbf{Set}^B$ for some set B, and the object-of-objects of the induced T-multicategory is $(A_0)_{b \in B}$. So this is like the result for commutative monoids but one level up; it states that symmetric multicategories play some kind of universal role for generalized multicategories, despite (apparently) not being generalized multicategories themselves. A corollary is that any symmetric multicategory is naturally a T_n-multicategory for each n, previously stated as Theorem 7.1.3 and pictorially very plausible – see p. 224.

We start with commutative monoids. The first main result mentioned above is reasonably clear informally. I shall, however, prove it with some care in preparation for the proof of the second main result, which involves many of the same thoughts in a more complex setting.

Notation: for any set B there are adjoint functors

$$\mathbf{Set}^B \; \underset{\Delta}{\overset{\Sigma}{\rightleftarrows}} \; \mathbf{Set},$$

where $\Sigma X = \coprod_{b \in B} Xb$ and $(\Delta A)(b) = A$.

Theorem C.4.1 *Let B be a set and T a finitary familially representable monad on $\mathbf{Set}^B$. Then there is a canonical functor*

$$\mathbf{CommMon} \longrightarrow (\mathbf{Set}^B)^T$$

making the diagram

$$
\begin{array}{ccc}
\mathbf{CommMon} & \longrightarrow & (\mathbf{Set}^B)^T \\
{\scriptstyle \text{forgetful}} \downarrow & & \downarrow {\scriptstyle \text{forgetful}} \\
\mathbf{Set} & \underset{\Delta}{\longrightarrow} & \mathbf{Set}^B
\end{array}
$$

commute.

(The standard notation is awkward here: $(\mathbf{Set}^B)^T$ is the category of algebras for the monad T on the category $\mathbf{Set}^B$.)

Proof To make the link between commutative monoids and familial representability we use the fat commutative monoids of Section A.1, which come equipped with explicit operations for summing arbitrary finite families (not just finite sequences) of elements. By Theorem A.1.3, it suffices to prove the present theorem with '**CommMon**' replaced by '**FatCommMon**'.

Represent the functor T by families $(X_{b,i})_{i \in I(b)}$ $(b \in B)$, the unit η of the monad by e, and the multiplication μ by m and a nameless isomorphism, as in the description at the end of Section C.3.

For any set A and any $b \in B$ we have

$$(T \Delta A)(b) \cong \coprod_{i \in I(b)} (X_{b,i}, \Delta A) \cong \coprod_{i \in I(b)} [\Sigma X_{b,i}, A],$$

and by Proposition C.3.5, the set $\Sigma X_{b,i}$ is finite. So given a fat commutative monoid A, we have a map

$$\theta_B^A : (T \Delta A)(b) \longrightarrow (\Delta A)(b) = A$$

whose component at $i \in I(b)$ is summation

$$\sum_{\Sigma X_{b,i}} : [\Sigma X_{b,i}, A] \longrightarrow A.$$

If we can show that θ^A is a T-algebra structure on ΔA then we are done: the functoriality is trivial.

For the multiplication axiom we have to show that the square

$$
\begin{array}{ccc}
(T^2 \Delta A)(b) & \xrightarrow{\mu_{\Delta A, b}} & (T \Delta A)(b) \\
{\scriptstyle (T\theta^A)_b} \downarrow & & \downarrow {\scriptstyle \theta^A_b} \\
(T \Delta A)(b) & \xrightarrow[\theta^A_b]{} & A
\end{array}
$$

commutes, for each $b \in B$. We have

$$
(T^2 \Delta A)(b) \cong \coprod_{(i,\gamma) \in J(b)} [\Sigma Y_{b,i,\gamma}, A]
$$

where

$$
J(b) = \{ (i, \gamma) \mid i \in I(b), \gamma \in \mathbf{Set}^B(X_{b,i}, I) \},
$$
$$
Y_{b,i,\gamma} = \coprod_{c \in B, x \in X_{b,i}(c)} X_{c, \gamma_c(x)}.
$$

Let $i \in I(b)$ and $\gamma \in \mathbf{Set}^B(X_{b,i}, I)$. Write $k = m_b(i, \gamma) \in I$; then part of the data for the monad is an isomorphism $X_{b,k} \xrightarrow{\sim} Y_{b,i,\gamma}$. The clockwise route around the square has (i, γ)-component

$$
[\Sigma Y_{b,i,\gamma}, A] \xrightarrow{\sim} [\Sigma X_{b,k}, A]
$$
$$
\downarrow {\scriptstyle \Sigma_{\Sigma X_{b,k}}}
$$
$$
A,
$$

and by Lemma A.1.2, this is just $\Sigma_{\Sigma Y_{b,i,\gamma}}$. For the anticlockwise route, an element of $[\Sigma Y_{b,i,\gamma}, A]$ is a family $(a_{c,x,d,w})_{c,x,d,w}$ indexed over

$$
c \in B, x \in X_{b,i}(c), d \in B, w \in X_{c, \gamma_c(x)}(d).
$$

For each $c \in B$ and $x \in X_{b,i}(c)$ we therefore have a family

$$
(a_{c,x,d,w})_{d,w} \in [\Sigma X_{x, \gamma_c(x)}, A],
$$

and

$$
\theta^A_c((a_{c,x,d,w})_{d,w}) = \sum_{d,w} a_{c,x,d,w} \in A,
$$

so

$$(T\theta^A)_b(a_{c,x,d,w})_{c,x,d,w} = \left(\sum_{d,w} a_{c,x,d,w}\right)_{c,x} \in [\Sigma X_{b,i}, A].$$

So the anticlockwise route sends $(a_{c,x,d,w})_{c,x,d,w}$ to $\sum_{c,x} \sum_{d,w} a_{c,x,d,w}$, and the clockwise route sends it to $\sum_{c,x,d,w} a_{c,x,d,w}$; the two are equal.

For the unit axiom we have to show that

$$(\Delta A)(b) \xrightarrow{\eta_{\Delta A,b}} (T\Delta A)(b)$$

commutes, for each $b \in B$. Write $k = e_b \in I$; we know that $X_{b,k} \cong \delta_b$, and so $\Sigma X_{b,k}$ is a one-element set. So if $a \in A$ then

$$\theta_b^A(\eta_{\Delta A,b}(a)) = \theta_b^A((a)_{x \in \Sigma X_{b,k}}) = \sum_{x \in \Sigma X_{b,k}} a = a,$$

as required. □

Example C.4.2 Let $n \in \mathbb{N}$; take B to be the set O_{n+1} of $(n+1)$-opetopes and T to be the $(n+1)$th opetopic monad T_{n+1} on $\mathbf{Set}/O_{n+1} \simeq \mathbf{Set}^{O_{n+1}}$. By Proposition 6.5.6 and induction, T_{n+1} is finitary and preserves wide pullbacks, so is familially representable by Proposition C.3.2. Theorem C.4.1 then gives a canonical functor

$$\mathbf{CommMon} \longrightarrow (\mathbf{Set}/O_{n+1})^{T_{n+1}} \simeq T_n\text{-}\mathbf{Operad},$$

proving Theorem 7.1.2.

Example C.4.3 For any finitary familially representable monad T on $\mathbf{Set}$, there is a canonical functor $\mathbf{CommMon} \longrightarrow \mathbf{Set}^T$ preserving underlying sets. Most such Ts that we have met have been operadic, in which case something stronger is true: there is a canonical functor from the category of not-necessarily-commutative monoids to $\mathbf{Set}^T$, as noted after Proposition 6.2.1. But, for instance, if T is the monad corresponding to the theory of monoids with an anti-involution (Example C.2.8) then the commutativity is necessary. Concretely, Theorem C.4.1 produces the functor

$$\mathbf{CommMon} \longrightarrow \text{(monoids with an anti-involution)}$$
$$(A, \cdot, 1) \longmapsto (A, \cdot, 1, (\ )^\circ)$$

defined by taking the anti-involution ( )° to be the identity; that this does define an anti-involution is exactly commutativity.

Write M for the free commutative monoid monad on **Set**. Lemma 6.1.1 tells us that Theorem C.4.1 is equivalent to:

Corollary C.4.4 *Let B be a set and T a finitary familially representable monad on* **Set**^{B}. *Then there is a canonical natural transformation*

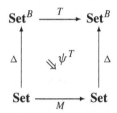

with the property that $(\Delta,\ \psi^T)$ *is a lax map of monads* (**Set**, M) $\longrightarrow$ (**Set**^{B}, T).   □

When $B = 1$, this corollary provides a natural transformation $\psi^T : T \longrightarrow M$ commuting with the monad structures. It is straightforward to check that if $\alpha : T' \longrightarrow T$ is a cartesian natural transformation commuting with the monad structures then $\psi^T \circ \alpha = \psi^{T'}$. So M is the vertex (codomain) of a cone on the inclusion functor

    (finitary familially representable monads on **Set**
    + cartesian transformations commuting with the monad structures)
↪ (monads on **Set**
    + transformations commuting with the monad structures),

in which the coprojections are the transformations ψ^T. In fact, it is a colimit cone: the theory of commutative monoids plays a universal role for finitary familially representable monads on **Set**, despite not being familially representable itself. Since this fact will not be used, I will not prove it; the main tactic is to consider the free algebraic theory on a single n-ary operation.

As explained on p. 186, we can translate between lax and colax maps of monads using mates. Applying this to Corollary C.4.4 gives:

Corollary C.4.5 *Let B be a set and T a finitary familially representable monad on* $\mathbf{Set}^B$. *Then there is a canonical natural transformation*

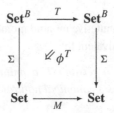

with the property that (Σ, ϕ^T) *is a colax map of monads* $(\mathbf{Set}^B, T) \longrightarrow$ $(\mathbf{Set}, M)$.

Proof Take ϕ^T to be the mate of ψ^T under the adjunction $\Sigma \dashv \Delta$. □

Example C.4.6 Let $n \in \mathbb{N}$ and take $B = O_n$ and $T = T_n$, as in Example C.4.2 but with the indexing shifted. An object X of $\mathbf{Set}^{O_n} \simeq \mathbf{Set}/O_n$ is thought of as a set of labelled n-opetopes. An element of $T_n X$ (or rather, of its underlying set $\Sigma T_n X$) is then an X-labelled n-pasting diagram; on the other hand, an element of $M \Sigma X$ is a finite set-with-multiplicities of labels (disregarding shapes completely). So there ought to be a forgetful function $\Sigma T_n X \longrightarrow M \Sigma X$, and there is: $\phi_X^{T_n}$.

When X is the terminal object of $\mathbf{Set}^{O_n}$ we have $\Sigma X = O_n$ and $\Sigma T_n X = O_{n+1}$, so $\phi_X^{T_n}$ is a map $O_{n+1} \longrightarrow M O_n$. This sends an n-pasting diagram to the set-with-multiplicities of its constituent n-opetopes: for example, the 2-pasting diagram (7.2) on p. 221 (stripped of its labels) is sent to the set-with-multiplicities

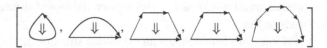

of 2-opetopes.

We now consider symmetric multicategories. Much of what follows is similar to what we did for commutative monoids but at a higher level of complexity.

Theorem C.4.7 *Let B be a set and T a finitary familially representable monad on* $\mathbf{Set}^B$. *Then there is a canonical functor*

$$\mathbf{FatSymMulticat} \longrightarrow T\text{-}\mathbf{Multicat}$$

making the diagram

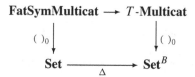

commute, where in both cases ( )$_0$ is the functor assigning to a multicategory its object of objects.

Before we prove the Theorem let us gather a corollary and some examples.

Corollary C.4.8 *Theorem C.4.7 holds with* '**FatSymMulticat**' *replaced by* '**SymMulticat**' *and 'canonical' by 'canonical up to isomorphism'.*

Proof Follows from Theorem A.2.4. □

Example C.4.9 For any $n \in \mathbb{N}$, let B be the set O_n of n-opetopes and let T_n be the nth opetopic monad. Then, as observed in C.4.2, T_n is finitary and familially representable. So the Corollary produces a functor

$$\textbf{SymMulticat} \longrightarrow T_n\textbf{-Multicat},$$

proving Theorem 7.1.3.

Example C.4.10 Taking $B = 1$, any symmetric multicategory is naturally a T-multicategory for any finitary familially representable monad T on **Set**. If T is operadic then we do not need the symmetries: the canonical natural transformation from T to the free monoid monad induces a functor

$$\textbf{Multicat} \longrightarrow T\textbf{-Multicat}.$$

This is analogous to the situation for commutative monoids described in C.4.3.

We finish with a proof of Theorem C.4.7. Undeniably it is complicated, but there is no great conceptual difficulty; the main struggle is against drowning in notation. It may help to keep Example C.4.9 in mind.

Proof of Theorem C.4.7 Let T be a finitary familially representable monad on $\textbf{Set}^B$. As in the description just before Proposition C.3.5, represent T by families $(X_{b,i})_{i \in I(b)}$ ($b \in B$), the unit η by e, and the multiplication μ by m and a bijection

$$s_{b,i,\gamma} : X_{b,m_b(i,\gamma)} \overset{\sim}{\longrightarrow} \coprod_{c \in B, x \in X_{b,i}(c)} X_{c,\gamma_c(x)}$$

for each $b \in B$, $i \in I$, and $\gamma \in \mathbf{Set}^B(X_{b,i}, I)$. Given such b, i, and γ, and given $d \in B$, write

$$s_{b,i,\gamma,d}(v) = (c(v), x(v), w(v))$$

for any $v \in X_{b,m_b(i,\gamma)}(d)$; so here $c(v) \in B$, $x(v) \in X_{b,i}(c)$, and $w \in X_{c,\gamma_c(x)}(d)$.

Since T is finitary, Proposition C.3.5 tells us that the set $\Sigma X_{b,i}$ is finite for each $b \in B$ and $i \in I(b)$.

Let P be a fat symmetric multicategory. Our main task is to define from P a T-multicategory C.

Define $C_0 = \Delta P_0$ (as we must). Then for each $b \in B$ we have

- $C_0(b) = P_0$
- $(TC_0)(b) = \coprod_{i \in I(b)} [\Sigma X_{b,i}, P_0]$, so an element of $(TC_0)(b)$ consists of an element $i \in I(b)$ together with a family $(q_{c,x})_{c,x}$ of objects of P indexed over $c \in B$ and $x \in X_{b,i}(c)$.

Define the T-graph

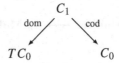

by declaring an element of $C_1(b)$ with domain $(i, (q_{c,x})_{c,x}) \in (TC_0)(b)$ and codomain $r \in (C_0)(b)$ to be a map

$$(q_{c,x})_{c,x} \xrightarrow{\theta} r$$

in P.

To define comp : $C_1 \circ C_1 \longrightarrow C_1$ we first have to compute $C_1 \circ C_1$. With a little effort we find that an element of $(C_1 \circ C_1)(b)$ consists of the following data (Fig. C-B):

- an element $i \in I(b)$ and a function $\gamma : X_{b,i} \longrightarrow I$
- a map $(q_{c,x})_{c,x} \xrightarrow{\theta} r$ in P
- a family of maps $((p_{c,x,d,w})_{d,w} \xrightarrow{\phi_{c,x}} q_{c,x})_{c,x}$ in P, where the inner indexing is over $d \in B$ and $w \in X_{c,\gamma_c(x)}(d)$.

Going down the left-hand slope of the diagram in Fig. C-B, the image of this element of $(C_1 \circ C_1)(b)$ in $(TC_0)(b)$ is the element $m_b(i, \gamma)$ of $I(b)$ together with the family $(p_{c(v),x(v),d,w(v)})_{d,v}$ indexed over $d \in B$ and $v \in X_{b,m_b(i,\gamma)}(d)$. So to define composition, we must derive from our data a map

$$(p_{c(v),x(v),d,w(v)})_{d,v} \longrightarrow r \qquad \text{(C.6)}$$

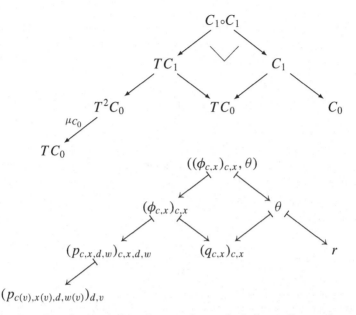

Fig. C-B. Data for composition in C. Each entry in the lower diagram is an element of the b-component of the corresponding object in the upper diagram.

in P. But we have the composite

$$\theta \circ (\phi_{c,x})_{c,x} : (p_{c,x,d,w})_{c,x,d,w} \longrightarrow r$$

in P and the bijection

$$\Sigma s_{b,i,\gamma} : \Sigma X_{b,m_b(i,\gamma)} \overset{\sim}{\longrightarrow} \Sigma \left(\coprod_{c \in B, x \in X_{b,i}(c)} X_{c,\gamma_c(x)} \right)$$

$$(d, v) \longmapsto (c(v), x(v), d, w(v)),$$

so $(\theta \circ (\phi_{c,x})_{c,x}) \cdot (\Sigma s_{b,i,\gamma})$ is a map of the form (C.6); and this is what we define the composite to be.

To define $\mathrm{ids} : C_0 \longrightarrow C_1$, first note that if $p \in C_0(b) = P_0$ then $\eta_{C_0,b}(p) \in (TC_0)(b)$ consists of the element $e_b \in I(b)$ together with the family $(p)_{c,u}$ indexed over $c \in B$ and $u \in U = X_{b,e_b}(c)$, as in the diagrams

$$\begin{array}{ccc} & C_0 & \\ {}^{\eta_{C_0}}\!\nearrow & & \searrow^{1} \\ TC_0 & & C_0 \end{array} \qquad \begin{array}{ccc} & p & \\ \nearrow & & \searrow \\ (p)_{c,u} & & p. \end{array}$$

Since U is a one-element set, we may define $\mathrm{ids}(p) \in C_1(b)$ to be the identity map

$$1_p^U : (p)_{c,u} \longrightarrow p$$

in P, and this has the correct domain and codomain.

We have now defined all the data for the T-multicategory C. It only remains to check the associativity and identity axioms, and these follow from the associativity and identity axioms on the fat symmetric multicategory P. So we have defined the desired functor

$$\textbf{FatSymMulticat} \longrightarrow T\textbf{-Multicat}$$

on objects.

The rest is trivial. If $f : P \longrightarrow P'$ is a map of fat symmetric multicategories and C and C' are the corresponding T-multicategories then we have to define a map $h : C \longrightarrow C'$. We take $h_0 = \Delta f_0$ (as we must) and define h to act on maps as f does. Since f preserves all the structure, so does h. Functoriality is immediate. $\square$

Notes

The main result of Section C.1 – that the theories described by plain operads are exactly the strongly regular ones – must exist as a subconscious principle, at least, in the mind of anyone who has worked with operads. Nevertheless, this is as far as I know the first proof. A brief sketch proof was given in my (1998a, 4.6).

The theory of familially representable monads on presheaf categories presented here is clearly unsatisfactory. The proof of Theorem C.4.7 is so complicated that it is at the limits of tolerability, for this author at least. Moreover, we have not even attempted to describe explicitly a cartesian monad structure on a familially representable endofunctor on a general presheaf category $[\mathbb{B}^{\mathrm{op}}, \textbf{Set}]$ (where $\mathbb{B}$ need not be discrete), nor to describe explicitly algebras for such monads. So the theory works, but only just.

Other ideas on the relation between cartesian and symmetric structures can be found in Weber (2001) and Batanin (2002b).

Appendix D

Free multicategories

In this appendix we define what it means for a monad or a category to be 'suitable' and prove in outline the results on free multicategories stated in Section 6.5.

First we need some terminology. Let $\mathcal{E}$ be a cartesian category, $\mathbb{I}$ a small category, $D : \mathbb{I} \longrightarrow \mathcal{E}$ a functor for which a colimit exists, and $(D(I) \xrightarrow{p_I} L)_{I \in \mathbb{I}}$ a colimit cone. Any map $L' \longrightarrow L$ gives rise to a new functor $D' : \mathbb{I} \longrightarrow \mathcal{E}$ and a new cone $(D'(I) \xrightarrow{p'_I} L')_{I \in \mathbb{I}}$ by pullback: if ΔL denotes the functor $\mathbb{I} \longrightarrow \mathcal{E}$ constant at L then

$$
\begin{array}{ccc}
D' & \longrightarrow & D \\
{\scriptstyle p'}\downarrow & & \downarrow{\scriptstyle p} \\
\Delta L' & \longrightarrow & \Delta L
\end{array}
$$

is a pullback square in the functor category $[\mathbb{I}, \mathcal{E}]$. We say that the colimit $(D(I) \xrightarrow{p_I} L)_{I \in \mathbb{I}}$ is **stable under pullback** if for any map $L' \longrightarrow L$ in $\mathcal{E}$, the resulting cone $(D'(I) \xrightarrow{p'_I} L')_{I \in \mathbb{I}}$ is also a colimit.

The maps p_I in a colimit cone $(D(I) \xrightarrow{p_I} L)_{I \in \mathbb{I}}$ are called the **coprojections** of the colimit, so we say that the colimit of D 'has monic coprojections' if each p_I is monic.

A category is said to have **disjoint finite coproducts** if it has finite coproducts, these coproducts have monic coprojections, and for any pair A, B of objects, the square

$$
\begin{array}{ccc}
0 & \longrightarrow & B \\
\downarrow & & \downarrow \\
A & \longrightarrow & A + B
\end{array}
$$

is a pullback.

Let ω be the natural numbers with their usual ordering. A **nested sequence** in a category $\mathcal{E}$ is a functor $\omega \longrightarrow \mathcal{E}$ in which the image of every map in ω is monic; in other words, it is a diagram

$$A_0 \rightarrowtail A_1 \rightarrowtail \cdots$$

in $\mathcal{E}$, where $\rightarrowtail$ indicates a monic. A functor that preserves pullbacks also preserves monics, so it makes sense for such a functor to 'preserve colimits of nested sequences'.

Let $\mathbb{I}$ and $\mathbb{J}$ be small categories and $\mathcal{E}$ a category with all limits of shape $\mathbb{I}$ and colimits of shape $\mathbb{J}$. We say that limits of shape $\mathbb{I}$ and colimits of shape $\mathbb{J}$ **commute** in $\mathcal{E}$ if for each functor $P : \mathbb{I} \times \mathbb{J} \longrightarrow \mathcal{E}$, the canonical map

$$\varinjlim_{\to \mathbb{J}} \varprojlim_{\leftarrow \mathbb{I}} P \longrightarrow \varprojlim_{\leftarrow \mathbb{I}} \varinjlim_{\to \mathbb{J}} P \qquad (D.1)$$

is an isomorphism. In particular, let $\mathbb{I}$ be the three-object category such that limits over $\mathbb{I}$ are pullbacks, and let $\mathbb{J} = \omega$; we say that pullbacks and colimits of nested sequences commute in $\mathcal{E}$ if this canonical map is an isomorphism for all functors P such that $P(I, -) : \omega \longrightarrow \mathcal{E}$ is a nested sequence for each $I \in \mathbb{I}$.

A category $\mathcal{E}$ is **suitable** if

- $\mathcal{E}$ is cartesian
- $\mathcal{E}$ has disjoint finite coproducts, and these are stable under pullback
- $\mathcal{E}$ has colimits of nested sequences; these commute with pullbacks and have monic coprojections.

A monad (T, μ, η) is **suitable** if

- (T, μ, η) is cartesian
- T preserves colimits of nested sequences.

D.1 Proofs

We sketch proofs of each of the results stated in 6.5.

Theorem 6.5.1 *Any presheaf category is suitable. Any finitary cartesian monad on a cartesian category is suitable.*

Proof First note that the category ω is filtered. The suitability of **Set** then reduces to a collection of standard facts. Presheaf categories are also suitable, as limits and colimits in them are computed pointwise. The second sentence is trivial. $\qquad\qquad\square$

Before we embark on the proofs of the main theorems, here is the main idea. A T-multicategory with object-of-objects E is a monoid in the monoidal category $\mathcal{E}/(TE \times E)$ (p. 199), so a free T-multicategory is a free monoid of sorts. The usual formula for the free monoid on an object X of a monoidal category is $\coprod_{n\in\mathbb{N}} X^{\otimes n}$. But this only works if the tensor product preserves countable coproducts on each side, and this is only true in our context if T preserves countable coproducts, which is often not the case – consider plain multicategories, for instance. So we need a more subtle construction. What we actually do corresponds to taking the colimit of the sequence

$$\coprod_{k=0}^{0} X^{\otimes k} \rightarrowtail \coprod_{k=0}^{1} X^{\otimes k} \rightarrowtail \coprod_{k=0}^{2} X^{\otimes k} \rightarrowtail \cdots$$

in the case that the monoidal category *does* have coproducts preserved by the tensor; in the general case we replace $\coprod_{k=0}^{n} X^{\otimes k}$ by $X^{(n)}$, defined recursively by

$$X^{(0)} = I, \qquad X^{(n+1)} = I + (X \otimes X^{(n)}).$$

Theorem 6.5.2 *Let T be a suitable monad on a suitable category $\mathcal{E}$. Then the forgetful functor*

$$(\mathcal{E}, T)\text{-}\mathbf{Multicat} \longrightarrow \mathcal{E}^{+} = (\mathcal{E}, T)\text{-}\mathbf{Graph}$$

has a left adjoint, the adjunction is monadic, and if T^{+} is the induced monad on $\mathcal{E}^{+}$ then both T^{+} and $\mathcal{E}^{+}$ are suitable.

Proof We proceed in four steps:

a. construct a functor $F : \mathcal{E}^{+} \longrightarrow (\mathcal{E}, T)\text{-}\mathbf{Multicat}$
b. construct an adjunction between F and the forgetful functor U
c. check that $\mathcal{E}^{+}$ and T^{+} are suitable
d. check that the adjunction is monadic.

Each step goes as follows.

a. *Construct a functor $F : \mathcal{E}^{+} \longrightarrow (\mathcal{E}, T)\text{-}\mathbf{Multicat}$.* Let X be a T-graph. For each $n \in \mathbb{N}$, define a T-graph by

- $X_1^{(0)} = X_0$, $d_0 = \eta_{X_0}$, and $c_0 = 1$
- $X_1^{(n+1)} = X_0 + X_1 \circ X_1^{(n)}$ (where $\circ$ is 1-cell composition in the bicategory $\mathcal{E}_{(T)}$), with the obvious choices of d_{n+1} and c_{n+1}.

For each $n \in \mathbb{N}$, define a map $i_n : X_1^{(n)} \longrightarrow X_1^{(n+1)}$ by taking

- $i_0 : X_0 \longrightarrow X_0 + X_1 \circ X_0$ to be first coprojection
- $i_{n+1} = 1_{X_0} + (1_{X_1} * i_n)$.

Then each i_n is monic, and by taking X_1^* to be the colimit of

$$X_1^{(0)} \stackrel{i_0}{\rightarrowtail} X_1^{(1)} \stackrel{i_1}{\rightarrowtail} \cdots$$

we obtain a T-graph

$$FX = \left(\begin{array}{ccc} & X_1^* & \\ {}^{d}\swarrow & & \searrow^{c} \\ TX_0 & & X_0 \end{array} \right).$$

This T-graph FX naturally has the structure of a T-multicategory. The identities map $X_0 \longrightarrow X_1^*$ is just the coprojection $X_1^{(0)} \rightarrowtail X_1^*$. Composition comes from canonical maps $X_1^{(m)} \circ X_1^{(n)} \longrightarrow X_1^{(m+n)}$ (defined by induction on m for each fixed n), which piece together to give a map $X_1^* \circ X_1^* \longrightarrow X_1^*$. It is the definition of composition that needs most of the suitability axioms.

We have now described what F does to objects, and extension to morphisms is straightforward.

(The colimit of the nested sequence of $X_1^{(n)}$s appears, in light disguise, as the recursive description of the free plain multicategory monad in Section 2.3: $X_1^{(n)}$ is the set of formal expressions that can be obtained from the first clause on p. 85 and up to n applications of the second clause.)

b. *Construct an adjunction between F and U.* We do this by constructing unit and counit transformations and verifying the triangle identities. Both transformations are the identity on the object of objects ('X_0'), so we only need define them on the object of arrows. For the unit, η^+, if $X \in \mathcal{E}^+$ is a T-graph then $\eta_X^+ : X_1 \longrightarrow X_1^*$ is the composite

$$X_1 \stackrel{\sim}{\longrightarrow} X_1 \circ X_0 \stackrel{\mathrm{copr}_2}{\rightarrowtail} X_0 + X_1 \circ X_0 = X_1^{(1)} \rightarrowtail X_1^*.$$

For the counit, ε^+, let C be a T-multicategory; write $X = UC$ and use the notation $X_1^{(n)}$ and X_1^* as in part (a). We need to define a map $\varepsilon_C^+ : X_1^* \longrightarrow C$. To do this, define for each $n \in \mathbb{N}$ a map $\varepsilon_{C,n}^+ : X_1^{(n)} \longrightarrow C_1$ by

- $\varepsilon_{C,0}^+ = (C_0 \xrightarrow{\text{ids}} C_1)$
- $\varepsilon_{C,n+1}^+ = (C_0 + C_1 \circ X_1^{(n)} \xrightarrow{1+(1*\varepsilon_{C,n}^+)} C_0 + C_1 \circ C_1 \xrightarrow{q} C_1)$, where q is ids on the first summand and comp on the second,

and then take ε_C^+ to be the induced map on the colimit. Verification of the triangle identities is straightforward.

c. *Check that $\mathcal{E}^+$ and T^+ are suitable.* The forgetful functor

$$\begin{array}{ccc} \mathcal{E}^+ & \longrightarrow & \mathcal{E} \times \mathcal{E} \\ X & \longmapsto & (X_0, X_1) \end{array}$$

creates pullbacks and colimits. This implies that $\mathcal{E}^+$ possesses pullbacks, finite coproducts and colimits of nested sequences, and that they behave as well as they do in $\mathcal{E}$. So $\mathcal{E}^+$ is suitable, and it is now straightforward to check that T^+ is suitable too.

d. *Check that the adjunction is monadic.* We apply the Monadicity Theorem (Mac Lane, 1971, VI.7) by checking that U creates coequalizers for U-absolute-coequalizer pairs. This is a completely routine procedure and works for any cartesian (not necessarily suitable) $\mathcal{E}$ and T. □

The fixed-object version of the theorem is now easy to deduce:

Theorem 6.5.4 *Let T be a suitable monad on a suitable category $\mathcal{E}$, and let $E \in \mathcal{E}$. Then the forgetful functor*

$$(\mathcal{E}, T)\text{-}\mathbf{Multicat}_E \longrightarrow \mathcal{E}_E^+ = \mathcal{E}/(TE \times E)$$

has a left adjoint, the adjunction is monadic, and if T_E^+ is the induced monad on $\mathcal{E}_E^+$ then both T_E^+ and $\mathcal{E}_E^+$ are suitable.

Proof In the adjunction $(F, U, \eta^+, \varepsilon^+)$ constructed in the proof of 6.5.2, each of F, U, η^+ and ε^+ leaves the object-of-objects unchanged. The adjunction therefore restricts to an adjunction between the subcategories $\mathcal{E}_E^+$ and $(\mathcal{E}, T)\text{-}\mathbf{Multicat}_E$, and the restricted adjunction is also monadic.

All we need to check, then, is that $\mathcal{E}_E^+$ and T_E^+ are suitable. For $\mathcal{E}_E^+$, it is enough to know that the slice of a suitable category is suitable, and to prove this we need only note that for any $E' \in \mathcal{E}$, the forgetful functor $\mathcal{E}/E' \longrightarrow \mathcal{E}$ creates both pullbacks and colimits. Suitability of T_E^+ follows from suitability of T^+ since the inclusion $\mathcal{E}_E^+ \hookrightarrow \mathcal{E}^+$ preserves and reflects both pullbacks and colimits of nested sequences. □

Finally, we prove the result stating that if $\mathcal{E}$ and T have certain special properties beyond suitability then those properties are inherited by $\mathcal{E}^+$ and T^+, or, in the fixed-object case, by $\mathcal{E}_E^+$ and T_E^+.

Proposition 6.5.6 *If $\mathcal{E}$ is a presheaf category and the functor T preserves wide pullbacks then the same is true of $\mathcal{E}^+$ and T^+ in Theorem 6.5.2, and of $\mathcal{E}_E^+$ and T_E^+ in Theorem 6.5.4. Moreover, if T is finitary then so are T^+ and T_E^+.*

(Wide pullbacks were defined on p. 364.)

Proof First we show that $\mathcal{E}^+$ and $\mathcal{E}_E^+$ are presheaf categories. Let $G : \mathcal{E} \longrightarrow \mathcal{E}$ be the functor defined by $G(E) = T(E) \times E$: then in the terminology of p. 374, $\mathcal{E}^+$ is the Artin gluing $\mathcal{E} \downarrow G$. Since T preserves wide pullbacks, so too does G, and Proposition C.3.4 then implies that $\mathcal{E}^+$ is a presheaf category. The fixed-object case is easier: the slice of a presheaf category is a presheaf category (1.1.7).

Next we show that T^+ and T_E^+ preserve wide pullbacks. Recall that in the proof of 6.5.2, free T-multicategories were constructed using coproducts, colimits of nested sequences, pullbacks, and the functor T. (The last two of these were hidden in the notation '$X \circ X_1^{(n)}$'.) It is therefore enough to show that all four of these entities commute with wide pullbacks. The last two are immediate, and Lemma C.2.2 implies that coproducts commute with wide pullbacks in **Set** (and so in any presheaf category); all that remains is to prove that colimits of nested sequences commute with wide pullbacks in **Set**. This is actually not true in general, but a slightly weaker statement is true and suffices. Specifically, let $\mathbb{I}$ be a category of the form $\mathbb{P}_K$ (p. 364), so that a limit over $\mathbb{I}$ is a wide pullback; let $\mathbb{J} = \omega$; and let $P : \mathbb{I} \times \mathbb{J} \longrightarrow$ **Set** be a functor such that

$$
\begin{array}{ccc}
P(I, J) & \longrightarrow & P(I, J') \\
\downarrow & & \downarrow \\
P(I', J) & \longrightarrow & P(I', J')
\end{array}
$$

is a pullback square for each pair of maps $(I \longrightarrow I', J \longrightarrow J')$. Then the canonical map (D.1) (p. 390) is an isomorphism. In the case at hand these squares are of the form

$$
\begin{array}{ccc}
X_1^{(n)} & \xrightarrow{\ i_n\ } & X_1^{(n+1)} \\
{\scriptstyle f_1^{(n)}}\downarrow & & \downarrow{\scriptstyle f_1^{(n+1)}} \\
Y_1^{(n)} & \xrightarrow[\ i_n\ ]{} & Y_1^{(n+1)}
\end{array}
$$

where $f : X \longrightarrow Y$ is some map of T-graphs, and it is easily checked that such squares are pullbacks.

For 'moreover' we have to show that T^+ and T_E^+ preserve filtered colimits if T does. Just as in the previous paragraph, this reduces to the statement that filtered colimits commute with pullbacks in **Set**, whose truth is well-known (Mac Lane, 1971, IX.2). □

Notes

These proofs first appeared in my (1999a). They have much in common with the free monoid construction of Baues, Jibladze and Tonks (1997).

Appendix E
Definitions of tree

I met this guy
and he looked like he might have been a hat-check clerk
at an ice rink
which in fact
he turned out to be

Laurie Anderson (1982)

Trees appear everywhere in higher-dimensional algebra. In this text they were defined in a purely abstract way (2.3.3): **tr** is the free plain operad on the terminal object of $\mathbf{Set}^{\mathbb{N}}$, and an n-leafed tree is an element of $\mathbf{tr}(n)$. But for the reasons laid out at the beginning of Section 7.3, I give here a 'concrete', graph-theoretic, definition of (finite, rooted, planar) tree and sketch a proof that it is equivalent to the abstract definition.

E.1 The equivalence

The main subtlety is that the trees we use are not quite finite graphs in the usual sense: some of the edges have a vertex at only one of their ends. (Recall from 2.3.3 that in a tree, an edge with a free end is not the same thing as an edge ending in a vertex.) This suggests the following definitions.

Definition E.1.1 A (planar) **input-output graph** (Fig. E-A(a)) consists of

- a finite set V (the **vertices**)
- a finite set E (the **edges**), a subset $I \subseteq E$ (the **input edges**), and an element $o \in E$ (the **output edge**)
- a function $s : E \backslash I \longrightarrow V$ (**source**) and a function $t : E \backslash \{o\} \longrightarrow V$ (**target**)
- for each $v \in V$, a total order $\leq$ on $t^{-1}\{v\}$.

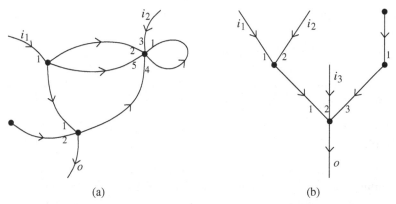

(a)   (b)

Fig. E-A. (a) Input-output graph with four vertices and two input edges i_1, i_2, (b) combinatorial tree with four vertices and three input edges i_1, i_2, i_3. In both, the numbers indicate the order on the edges arriving at each vertex.

We write $v \xrightarrow{e}$ to mean that e is a non-input edge with $s(e) = v$, and similarly $\xrightarrow{e} v'$ to mean that e is a non-output edge with $t(e) = v'$, and of course $v \xrightarrow{e} v'$ to mean that e is a non-input, non-output edge with $s(e) = v$ and $t(e) = v'$.

A tree is roughly speaking a connected, simply connected graph, and the following notion of path allows us to express this.

Definition E.1.2 A **path** from a vertex v to an edge e in an input-output graph is a diagram

$$v = v_1 \xrightarrow{e_1} v_2 \xrightarrow{e_2} \cdots \xrightarrow{e_{l-1}} v_l \xrightarrow{e_l = e}$$

in the graph. That is, a path from v to e consists of

- an integer $l \geq 1$
- a sequence $(v_1, v_2, \ldots, v_l)$ of vertices with $v_1 = v$
- a sequence $(e_1, \ldots, e_{l-1}, e_l)$ of edges with $e_l = e$

such that

$$v_1 = s(e_1), \quad t(e_1) = v_2 = s(e_2), \quad \ldots, \quad t(e_{l-1}) = v_l = s(e_l)$$

and all of these sources and targets are defined.

Definition E.1.3 A **combinatorial tree** is an input-output graph such that for every vertex v, there is precisely one path from v to the output edge.

Fig. E-A(b) shows a combinatorial tree. The ordering of the edges arriving at each vertex encodes the planar embedding. 'Tree' is an abbreviation for 'finite, rooted, planar tree'. If we were doing symmetric operads then we would use non-planar trees, if we were doing cyclic operads then we would use non-rooted trees, and so on.

A combinatorial tree is essentially the same thing as a tree in our sense, where 'essentially the same thing' refers to the obvious notion of isomorphism between combinatorial trees. We write **combtr**(n) for the set of isomorphism classes of combinatorial trees with n input edges.

Proposition E.1.4 *For each $n \in \mathbb{N}$, there is a canonical bijection* **tr**$(n) \cong$ **combtr**(n).

With a little more work we could define an operad structure on $(\textbf{combtr}(n))_{n \in \mathbb{N}}$ and turn the proposition into the stronger statement that the operads **tr** and **combtr** are isomorphic. With more work still we could define maps between combinatorial trees and so define a **Cat**-operad **COMBTR** in which **COMBTR**(n) is the category of n-leafed combinatorial trees; then we could prove that this **Cat**-operad is equivalent to the **Cat**-operad **TR**.

Sketch proof The strategy is to define functions

$$\textbf{tr}(n) \; \underset{\Psi}{\overset{\Phi}{\rightleftarrows}} \; \{\text{combinatorial trees with } n \text{ input edges}\}$$

for each $n \in \mathbb{N}$, such that if $G \cong G'$ then $\Psi(G) = \Psi(G')$, and such that $\Psi(\Phi(\tau)) = \tau$ and $\Phi(\Psi(G)) \cong G$ for each tree τ and combinatorial tree G. The proposition follows. The definition of $\Phi(\tau)$ is by induction on the structure of the tree τ, and the definition of $\Psi(G)$ is by induction on the number of vertices of the combinatorial tree G. All the details are straightforward. $\square$

Notes

Some pointers to the literature on trees can be found in the Notes to Chapter 7.

Appendix F

Free strict n-categories

Here we prove that the forgetful functor

$$(\text{strict } n\text{-categories}) \longrightarrow (n\text{-globular sets})$$

is monadic and that the induced monad is cartesian and finitary. We also prove the analogous results for ω-categories. We used monadicity and that the monad is cartesian in Part III, in order to be able to define and understand globular operads and weak n- and ω-categories. We use the fact that the monad is finitary for technical purposes in Appendix G.

It is frustrating to have to prove this theorem, for two separate reasons. The first is that the proof can *almost* be made trivial: an adjoint functor theorem tells us that the forgetful functor has a left adjoint, a monadicity theorem tells us that it is monadic, and a routine calculation tells us that it is finitary. However, we have no result of the form 'given an adjunction, its induced monad is cartesian if the right adjoint satisfies certain conditions', and so in order to prove that the monad is cartesian we are forced to actually construct the whole adjunction explicitly. The second is that there ought to be some way of simply looking at the theory of strict ω-categories, presented as an algebraic theory on the presheaf category of globular sets (as in 1.4.8), and applying some general principle to deduce the theorem immediately; see p. 362. But again, we currently have no way of doing this.

Despite these frustrations, the proof is quite easy. By exploiting the definition of strict n- and ω-categories by iterated enrichment (1.4.1), we can reduce it to the proof of some straightforward statements about enriched categories.

Here is the idea. Suppose we know how to construct the free strict 2-category on a 2-globular set. Then we can construct the free strict 3-category on a 3-globular set X in two steps:

- For each $x, x' \in X(0)$, take the 2-globular set $X(x, x')$ of cells whose 0-dimensional source is x and whose 0-dimensional target is x', replace it by the free strict 2-category on $X(x, x')$, and reassemble to obtain a new 3-globular set Y. The 0- and 1-cells of Y are the same as those of X, and a typical 3-cell of Y looks like

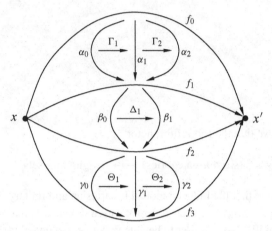

  where x and x' are 0-cells of X, the f_is are 1-cells, the α_is, β_is and γ_is are 2-cells, and the Γ_is, Δ_is and Θ_is are 3-cells.
- Write $\mathcal{A}$ for the category of 2-globular sets, think of Y as a family $(Y(x, x'))_{x,x' \in X(0)}$ of objects of $\mathcal{A}$, and let Z be the free $\mathcal{A}$-enriched category on Y. The 0-cells of Z are the same as those of X and Y, and a typical 3-cell of Z looks like

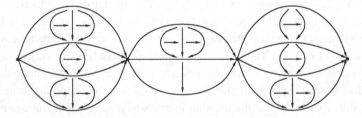

  where the cells making up this diagram are cells of X. So Z is the free strict 3-category on X.

We therefore construct the free strict n-category functor inductively on n, and pass to the limit to reach the infinite-dimensional version.

In the first section, F.1, we prove the necessary results on free enriched categories. In the second, F.2, we apply them to establish the existence and properties of the free strict n- and ω-category functors. Since the construction is explicit, we are then able to verify the intuitively plausible formula (8.11) (p. 270) for the free functor in terms of pasting diagrams.

F.1 Free enriched categories

Recall from Section 1.3 that for every category $\mathcal{A}$ there is a category $\mathcal{A}$-**Gph** of $\mathcal{A}$-graphs, and that if $\mathcal{A}$ has finite products then there is also a category $\mathcal{A}$-**Cat** of $\mathcal{A}$-enriched categories and a forgetful functor $\mathcal{A}$-**Cat** $\longrightarrow$ $\mathcal{A}$-**Gph**. Any finite-product-preserving functor $Q : \mathcal{B} \longrightarrow \mathcal{A}$ between categories with finite products induces a (strictly) commutative square

$$\begin{array}{ccc} \mathcal{B}\text{-}\mathbf{Cat} & \longrightarrow & \mathcal{A}\text{-}\mathbf{Cat} \\ \downarrow & & \downarrow \\ \mathcal{B}\text{-}\mathbf{Gph} & \longrightarrow & \mathcal{A}\text{-}\mathbf{Gph,} \end{array}$$

and in particular induces an unambiguous functor $\mathcal{B}$-**Cat** $\longrightarrow$ $\mathcal{A}$-**Gph**. So if T is a monad on a category $\mathcal{A}$ with finite products then the forgetful functor $\mathcal{A}^T \longrightarrow \mathcal{A}$ induces a forgetful functor $\mathcal{A}^T$-**Cat** $\longrightarrow$ $\mathcal{A}$-**Gph**. This functor will play an important role.

A monad will be called **coproduct-preserving** if its functor part preserves all (small) coproducts.

Proposition F.1.1 *Let $\mathcal{A}$ be a presheaf category.*

a. *The forgetful functor*

$$\mathcal{A}\text{-}\mathbf{Cat} \longrightarrow \mathcal{A}\text{-}\mathbf{Gph}$$

is monadic, and the induced monad is cartesian, finitary and coproduct-preserving.

b. *For any monad T on $\mathcal{A}$, the forgetful functor*

$$\mathcal{A}^T\text{-}\mathbf{Gph} \longrightarrow \mathcal{A}\text{-}\mathbf{Gph}$$

is monadic, and the induced monad is cartesian (respectively, finitary or coproduct-preserving) if T is.

c. *For any coproduct-preserving monad T on $\mathcal{A}$, the forgetful functor*

$$\mathcal{A}^T\text{-}\mathbf{Cat} \longrightarrow \mathcal{A}\text{-}\mathbf{Gph}$$

is monadic; the induced monad $T^\sharp$ is coproduct-preserving, and is cartesian (respectively, finitary) if T is.

Moreover, $T^\sharp$ is given on $\mathcal{A}$-graphs X by $(T^\sharp X)_0 = X_0$ and

$$(T^\sharp X)(x, x') = \coprod_{x=x_0, x_1, \dots, x_{r-1}, x_r = x'} T(X(x_0, x_1)) \times \cdots \times T(X(x_{r-1}, x_r))$$

$(x, x' \in X_0)$, where the coproduct is over all $r \in \mathbb{N}$ and sequences $x_0, \dots, x_r$ of elements of X_0 satisfying $x_0 = x$ and $x_r = x'$.

The hypothesis that $\mathcal{A}$ is a presheaf category is excessive, but makes the proof easier and serves our purpose.

Proof Part (a) is simple: free $\mathcal{A}$-enriched categories are constructed just as free ordinary categories are. The induced monad $\mathbf{fc}_\mathcal{A}$ on $\mathcal{A}$-**Gph** is given on $\mathcal{A}$-graphs X by $(\mathbf{fc}_\mathcal{A} X)_0 = X_0$ and

$$(\mathbf{fc}_\mathcal{A} X)(x, x') = \coprod_{x=x_0, x_1, \dots, x_{r-1}, x_r = x'} X(x_0, x_1) \times \cdots \times X(x_{r-1}, x_r)$$

$(x, x' \in X_0)$. Everything works as in the familiar case $\mathcal{A} = \mathbf{Set}$ because $\mathcal{A}$ is a **Set**-valued functor category.

Part (b) is also straightforward: $(\ )$-**Gph** defines a strict map $\mathbf{CAT} \longrightarrow \mathbf{CAT}$ of strict 2-categories (ignoring set-theoretic worries), and so turns the adjunction between $\mathcal{A}^T$ and $\mathcal{A}$ into an adjunction between the corresponding categories of graphs. Explicitly, the induced monad T_* on $\mathcal{A}$-**Gph** is given on $\mathcal{A}$-graphs X by $(T_*X)_0 = X_0$ and $(T_*X)(x, x') = T(X(x, x'))$. That T_* inherits the properties of T is easily checked.

To prove (c) and 'moreover' it suffices to put a monad structure on the composite functor $\mathbf{fc}_\mathcal{A} \circ T_*$, to construct an isomorphism of categories

$$(\mathcal{A}\text{-}\mathbf{Gph})^{\mathbf{fc}_\mathcal{A} \circ T_*} \cong \mathcal{A}^T\text{-}\mathbf{Cat}$$

commuting with the forgetful functors to $\mathcal{A}$-**Gph**, and to check that the monad $\mathbf{fc}_\mathcal{A} \circ T_*$ is cartesian if T is. (Recall the two-step strategy of the introduction.) The monad structure on $\mathbf{fc}_\mathcal{A} \circ T_*$ consists of the monad structures on T_* and $\mathbf{fc}_\mathcal{A}$ glued together by a distributive law (6.1.2)

$$\lambda : T_* \circ \mathbf{fc}_\mathcal{A} \longrightarrow \mathbf{fc}_\mathcal{A} \circ T_*.$$

The functor $\mathbf{fc}_\mathcal{A} \circ T_*$ is given by the formulas for $T^\sharp$ in 'moreover'; the composite the other way round is given by $((T_* \circ \mathbf{fc}_\mathcal{A})(X))_0 = X_0$ and

$$((T_* \circ \mathbf{fc}_\mathcal{A})X)(x, x') = \coprod_{x=x_0, x_1, \dots, x_{r-1}, x_r = x'} T(X(x_0, x_1) \times \cdots \times X(x_{r-1}, x_r)).$$

So we can define λ_X by the evident natural maps. It is straightforward to check that λ is indeed a distributive law and that λ is cartesian if T is, so the monad

$\mathbf{fc}_{\mathcal{A}} \circ T_*$ is cartesian if T is (6.1.3). All that remains is the isomorphism. That follows from 6.1.5 once we know that the monad on $(\mathcal{A}\text{-}\mathbf{Gph})^{T_*} \cong \mathcal{A}^T\text{-}\mathbf{Gph}$ corresponding to the distributive law λ is $\mathbf{fc}_{\mathcal{A}^T}$, and that too is easily checked.

$\square$

F.2 Free n- and ω-categories

Since we are using the definition of strict n- and ω-categories by iterated enrichment, it is convenient to replace the category of n-globular sets by the equivalent category $n\text{-}\mathbf{Gph}$ of n-graphs introduced in 1.4.9. There is a forgetful functor $U_n : n\text{-}\mathbf{Cat} \longrightarrow n\text{-}\mathbf{Gph}$ defined by taking U_0 to be the identity and U_{n+1} to be the diagonal of the commutative square

$$
\begin{array}{ccc}
(n\text{-}\mathbf{Cat})\text{-}\mathbf{Cat} & \longrightarrow & (n\text{-}\mathbf{Gph})\text{-}\mathbf{Cat} \\
\downarrow & & \downarrow \\
(n\text{-}\mathbf{Cat})\text{-}\mathbf{Gph} & \longrightarrow & (n\text{-}\mathbf{Gph})\text{-}\mathbf{Gph}.
\end{array}
\qquad \text{(F.1)}
$$

(We abbreviate our previous notation for the category of strict n-categories, $\mathbf{Str}\text{-}n\text{-}\mathbf{Cat}$, to $n\text{-}\mathbf{Cat}$.) There is also a restriction functor $R_n : (n+1)\text{-}\mathbf{Gph} \longrightarrow n\text{-}\mathbf{Gph}$ for each $n \in \mathbb{N}$, defined by $R_0(X) = X_0$ and $R_{n+1} = R_n\text{-}\mathbf{Gph}$. These functors fit together into a strictly commutative diagram

$$
\begin{array}{ccccccc}
\cdots (n+1)\text{-}\mathbf{Cat} & \xrightarrow{S_n} & n\text{-}\mathbf{Cat} & \xrightarrow{S_{n-1}} & \cdots & \xrightarrow{S_0} & 0\text{-}\mathbf{Cat} \\
\downarrow{\scriptstyle U_{n+1}} & & \downarrow{\scriptstyle U_n} & & & & \downarrow{\scriptstyle U_0} \\
\cdots (n+1)\text{-}\mathbf{Gph} & \xrightarrow[R_n]{} & n\text{-}\mathbf{Gph} & \xrightarrow[R_{n-1}]{} & \cdots & \xrightarrow[R_0]{} & 0\text{-}\mathbf{Gph},
\end{array}
$$

where the functors S_n are the usual ones (p. 47). Passing to the limit gives a category $\omega\text{-}\mathbf{Gph}$, the ω-**graphs**, and a forgetful functor $U : \omega\text{-}\mathbf{Cat} \longrightarrow \omega\text{-}\mathbf{Gph}$.

Analogous functors can, of course, be defined with $[\mathbb{G}_n^{\mathrm{op}}, \mathbf{Set}]$ in place of $n\text{-}\mathbf{Gph}$ and $[\mathbb{G}^{\mathrm{op}}, \mathbf{Set}]$ in place of $\omega\text{-}\mathbf{Gph}$. It is straightforward to check that there are equivalences of categories

$$
n\text{-}\mathbf{Gph} \simeq [\mathbb{G}_n^{\mathrm{op}}, \mathbf{Set}], \qquad \omega\text{-}\mathbf{Gph} \simeq [\mathbb{G}^{\mathrm{op}}, \mathbf{Set}]
$$

commuting with all these functors. We are therefore at liberty to use graphs in place of globular sets.

Theorem F.2.1 *For each* $n \in \mathbb{N}$, *the forgetful functor* $n\text{-}\mathbf{Cat} \longrightarrow [\mathbb{G}_n^{\mathrm{op}}, \mathbf{Set}]$ *is monadic and the induced monad is cartesian, finitary, and coproduct-preserving.*

Proof We replace this forgetful functor by U_n and use induction. U_0 is the identity. Given $n \in \mathbb{N}$, write T_n for the monad induced by U_n and its left adjoint (which exists by inductive hypothesis). Then U_{n+1} is the diagonal of the square (F.1), and under the equivalence $n\text{-}\mathbf{Cat} \simeq (n\text{-}\mathbf{Gph})^{T_n}$ becomes the forgetful functor

$$(n\text{-}\mathbf{Gph})^{T_n}\text{-}\mathbf{Cat} \longrightarrow (n\text{-}\mathbf{Gph})\text{-}\mathbf{Gph}.$$

The result now follows from Proposition F.1.1(c). $\qquad\square$

We want to deduce the same result for ω-dimensional structures, and morally this should be immediate from their definition by limits. The only problem is that $\omega\text{-}\mathbf{Cat}$ and $\omega\text{-}\mathbf{Gph}$ are defined as *strict*, or 1-categorical, limits in the 2-category $\mathbf{CAT}$, and properties such as adjointness are 2-categorical. So, for instance, the fact that each U_n has a left adjoint F_n does not *a priori* guarantee that U has a left adjoint, since the squares

$$
\begin{array}{ccc}
(n+1)\text{-}\mathbf{Cat} & \xrightarrow{\;S_n\;} & n\text{-}\mathbf{Cat} \\
{\scriptstyle F_{n+1}}\big\uparrow & & \big\uparrow{\scriptstyle F_n} \\
(n+1)\text{-}\mathbf{Gph} & \underset{R_n}{\xrightarrow{\quad}} & n\text{-}\mathbf{Gph}
\end{array}
\tag{F.2}
$$

are only known to commute up to (canonical) isomorphism.

A satisfactory resolution would involve the theory of weak limits in a 2-category. Here, however, we use a short and nasty method, exploiting some special features of the situation.

The key is that each of the functors S_n has the following (easily proved) isomorphism-lifting property: if $C \in (n+1)\text{-}\mathbf{Cat}$ and $j : S_n(C) \xrightarrow{\;\sim\;} D$ is an isomorphism in $n\text{-}\mathbf{Cat}$, then there exists an isomorphism $i : C \xrightarrow{\;\sim\;} C'$ in $(n+1)\text{-}\mathbf{Cat}$ such that $S_n C' = D$ and $S_n i = j$. This allows us to choose left adjoints $F_0, F_1, \ldots$ successively so that the squares (F.2) are strictly commutative.

Observe also that the categories $n\text{-}\mathbf{Cat}$ have all (small) limits and colimits and the functors S_n preserve them, as follows by induction using standard facts about enriched categories. Together with the isomorphism-lifting property, this implies that $\omega\text{-}\mathbf{Cat}$ has all limits and colimits and that a (co)cone in the category $\omega\text{-}\mathbf{Cat}$ is a (co)limit if and only if its image in each of the categories $n\text{-}\mathbf{Cat}$ is a (co)limit. The same is true with $\mathbf{Gph}$ in place of $\mathbf{Cat}$, easily.

Theorem F.2.2 *The forgetful functor* ω-**Cat** $\longrightarrow$ $[\mathbb{G}^{\text{op}}, \textbf{Set}]$ *is monadic and the induced monad is cartesian, finitary, and coproduct-preserving.*

Proof It is equivalent to prove the same properties of $U : \omega\text{-}\textbf{Cat} \longrightarrow \omega\text{-}\textbf{Gph}$. Choose left adjoints F_n to the U_ns so that the squares (F.2) commute, and let F be the induced functor $\omega\text{-}\textbf{Gph} \longrightarrow \omega\text{-}\textbf{Cat}$; this is left adjoint to U. With the aid of the Monadicity Theorem we see that all the properties of U remaining to be proved concern limits and colimits in $\omega\text{-}\textbf{Cat}$ and $\omega\text{-}\textbf{Gph}$, and by the observations above, they are implied by the corresponding properties of U_n (Theorem F.2.1). $\square$

We can now read off an explicit formula for the free strict ω-category monad T. Define the globular set **pd** by taking $\textbf{pd}(0)$ to be a one-element set and $\textbf{pd}(n+1)$ to be the free monoid on $\textbf{pd}(n)$, and define for each pasting diagram π a globular set $\widehat{\pi}$, as in Section 8.1. Write $T_{(n)}$ for the free strict n-category monad on the category $[\mathbb{G}_n^{\text{op}}, \textbf{Set}]$ of n-globular sets, and $X_{(n)}$ for the n-globular set obtained by forgetting all the cells of a globular set X above dimension n.

Proposition F.2.3 *For globular sets X and $n \in \mathbb{N}$, there is an isomorphism*

$$(TX)(n) \cong \coprod_{\pi \in \textbf{pd}(n)} [\mathbb{G}^{\text{op}}, \textbf{Set}](\widehat{\pi}, X)$$

natural in X.

Proof If $\pi \in \textbf{pd}(n)$ then the globular set $\widehat{\pi}$ is empty above dimension n. Also

$$(TX)(n) = (TX)_{(n)}(n) \cong (T_{(n)}X_{(n)})(n)$$

by construction of $T_{(n)}$ and T. So the claimed isomorphism is equivalent to

$$(T_{(n)}X_{(n)})(n) \cong \coprod_{\pi \in \textbf{pd}(n)} [\mathbb{G}_n^{\text{op}}, \textbf{Set}](\widehat{\pi}_{(n)}, X_{(n)}).$$

This is a statement about n-globular sets; let us translate it into one about n-graphs.

First, if $n \in \mathbb{N}$ and Z is an n-graph then the corresponding n-globular set has a set of n-cells, which we write as $Z(n)$. If $n = 0$ then this is given by $Z(0) = Z$, and then inductively,

$$Z(n+1) \cong \coprod_{z,z' \in Z_0} (Z(z, z'))(n).$$

Second, if $n \in \mathbb{N}$ and $\pi \in \textbf{pd}(n)$ then there is an n-graph $\widetilde{\pi}$ corresponding to the n-globular set $\widehat{\pi}_{(n)}$. If $n = 0$ and π is the unique element of $\textbf{pd}(0)$ then $\widetilde{\pi} =$

1. If $\pi \in \mathbf{pd}(n+1)$ then $\pi = (\pi_1, \ldots, \pi_r)$ for some $r \in \mathbb{N}$ and $\pi_i \in \mathbf{pd}(n)$, and in this case $\widetilde{\pi}$ is given by

$$(\widetilde{\pi})_0 = \{0, \ldots, r\}$$

and

$$\widetilde{\pi}(i, j) = \begin{cases} \widetilde{\pi}_j & \text{if } j = i + 1 \\ 0 & \text{otherwise.} \end{cases}$$

The claimed isomorphism is therefore equivalent to

$$(T_{(n)}Y)(n) \cong \coprod_{\pi \in \mathbf{pd}(n)} n\text{-}\mathbf{Gph}(\widetilde{\pi}, Y)$$

for $n \in \mathbb{N}$ and n-graphs Y.

For $n = 0$ this is trivial. Then inductively, using 'moreover' of Proposition F.1.1 for the second step,

$$(T_{(n+1)}Y)(n+1) = \coprod_{y, y' \in Y_0} ((T_{(n+1)}Y)(y, y'))(n)$$

$$\cong \coprod_{r \in \mathbb{N}, y_0, \ldots, y_r \in Y_0} \left(T_{(n)}(Y(y_0, y_1)) \times \cdots \times T_{(n)}(Y(y_{r-1}, y_r)) \right)(n)$$

$$\cong \coprod_{r \in \mathbb{N}, y_0, \ldots, y_r \in Y_0} (T_{(n)}(Y(y_0, y_1)))(n) \times \cdots \times (T_{(n)}(Y(y_{r-1}, y_r)))(n)$$

$$\cong \coprod_{\substack{r \in \mathbb{N}, y_0, \ldots, y_r \in Y_0, \\ \pi_1, \ldots, \pi_r \in \mathbf{pd}(n)}} n\text{-}\mathbf{Gph}(\widetilde{\pi}_1, Y(y_0, y_1)) \times \cdots \times n\text{-}\mathbf{Gph}(\widetilde{\pi}_r, Y(y_{r-1}, y_r))$$

$$\cong \coprod_{r \in \mathbb{N}, \pi_1, \ldots, \pi_r \in \mathbf{pd}(n)} (n+1)\text{-}\mathbf{Gph}(\widetilde{(\pi_1, \ldots, \pi_r)}, Y)$$

$$\cong \coprod_{\pi \in \mathbf{pd}(n+1)} (n+1)\text{-}\mathbf{Gph}(\widetilde{\pi}, Y),$$

as required. $\square$

Notes

The material here first appeared in my thesis (2000b, App. C).

Appendix G

Initial operad-with-contraction

There existed another ending to the story of O.
Réage (1954)

We prove Proposition 9.2.2: the category **OC** of operads-with-contraction has an initial object. This was needed in Chapter 9 for the definition of weak ω-category.

The explanation in Section 9.1 suggests an explicit construction of the initial operad-with-contraction: ascend through the dimensions, at each stage freely adding in elements obtained by contraction and then freely adding in elements obtained by operadic composition. Here we take a different approach, exploiting a known existence theorem.

G.1 The proof

The following result appears to be due to Kelly (1980, 27.1).

Theorem G.1.1 *Let*

$$
\begin{array}{ccc}
\mathcal{D} & \longrightarrow & \mathcal{C} \\
\downarrow & & \downarrow {\scriptstyle V} \\
\mathcal{B} & \xrightarrow{\; U \;} & \mathcal{A}
\end{array}
$$

be a (strict) pullback diagram in **CAT**. *If* $\mathcal{A}$ *is locally finitely presentable and each of* U *and* V *is finitary and monadic then the functor* $\mathcal{D} \longrightarrow \mathcal{A}$ *is also monadic.* $\qquad\square$

Actually, all we need is:

Corollary G.1.2 *In the situation of Theorem G.1.1,* $\mathcal{D}$ *has an initial object.*

407

Proof A locally finitely presentable category is by definition cocomplete, so $\mathcal{A}$ has an initial object. The functor $\mathcal{D} \longrightarrow \mathcal{A}$ has a left adjoint (being monadic), which applied to the initial object of $\mathcal{A}$ gives an initial object of $\mathcal{D}$. $\qquad \square$

Let T be the free strict ω-category monad on the category $[\mathbb{G}^{\mathrm{op}}, \mathbf{Set}]$ of globular sets, as in Chapters 8 and 9. Write **Coll** for the category $[\mathbb{G}^{\mathrm{op}}, \mathbf{Set}]/\mathbf{pd}$ of collections (p. 271), **CC** for the category of collections-with-contraction (defined by replacing 'operad' by 'collection' throughout Definition 9.2.1), and **Operad** for the category of globular operads. Then there is a (strict) pullback diagram

$$
\begin{array}{ccc}
\mathbf{OC} & \longrightarrow & \mathbf{Operad} \\
\downarrow & & \downarrow{\scriptstyle V} \\
\mathbf{CC} & \underset{U}{\longrightarrow} & \mathbf{Coll}
\end{array}
$$

in **CAT**, made up of forgetful functors.

To prove that **OC** has an initial object, we verify the hypotheses of Theorem G.1.1. The only non-routine part is showing that we can freely add a contraction to any collection.

Coll is locally finitely presentable

Since **Coll** is a slice of a presheaf category, it is itself a presheaf category (1.1.7) and so locally finitely presentable (Borceux, 1994b, Example 5.2.2(b)).

U is finitary and monadic

It is straightforward to calculate that U creates filtered colimits; and since **Coll** possesses all filtered colimits, U preserves them too. It is also easy to calculate that U creates coequalizers for U-split coequalizer pairs. So we have only to show that U has a left adjoint.

Let P be a collection. We construct a new collection FP, a contraction κ^P on FP, and a map $\alpha_P : P \longrightarrow FP$, together having the appropriate universal property; so the functor $P \longmapsto (FP, \kappa^P)$ is left adjoint to U, with α as unit. The definitions of FP and α_P are by induction on dimension:

- if π is the unique element of $\mathbf{pd}(0)$ then $(FP)(\pi) = P(\pi)$
- if $n \geq 1$ and $\pi \in \mathbf{pd}(n)$ then $(FP)(\pi) = P(\pi) + \mathrm{Par}_{FP}(\pi)$
- $\alpha_{P,\pi} : P(\pi) \hookrightarrow (FP)(\pi)$ is inclusion as the first summand, for all π
- if $n \geq 1$ and $\pi \in \mathbf{pd}(n)$ then the source map $s : (FP)(\pi) \longrightarrow (FP)(\partial\pi)$

is defined on the first summand of $(FP)(\pi)$ as the composite

$$P(\pi) \xrightarrow{\ s\ } P(\partial\pi) \xrightarrow{\ \alpha_{P,\partial\pi}\ } (FP)(\partial\pi)$$

and on the second summand as first projection; the target map is defined dually.

The globularity equations hold, so FP forms a collection. Clearly $\alpha_P :$ $P \longrightarrow FP$ is a map of collections. The contraction κ^P on FP is defined by taking

$$\kappa^P_\pi : \mathrm{Par}_{FP}(\pi) \longrightarrow (FP)(\pi)$$

$(n \geq 1, \pi \in \mathbf{pd}(n))$ to be inclusion as the second summand. It is easy to check that FP, κ^P and α^P have the requisite universal property: so U has a left adjoint.

V is finitary and monadic

The functor T is finitary, by F.2.2. This implies by 6.5.1 that the monad T is suitable, and so by 6.5.4 that V is monadic. It also implies by the 'moreover' of 6.5.6 that the monad induced by V and its left adjoint is finitary. If a category has colimits of a certain shape and a monad on it preserves colimits of that shape, then so too does the forgetful algebra functor; hence V is finitary, as required.

Notes

I thank Steve Lack and John Power for telling me that the result I needed, G.1.1, was in Kelly (1980), and Sjoerd Crans for telling me exactly where.

References

Adámek, J., Rosický, J. (1994), *Locally Presentable and Accessible Categories*, London Mathematical Society Lecture Note Series 189, Cambridge University Press.

Adams, J. F. (1978), *Infinite Loop Spaces*, Princeton University Press.

Anderson, D. W. (1971), Spectra and Γ-sets, in *Algebraic Topology (Proceedings of Symposia in Pure Mathematics 22)*, AMS.

Anderson, L. (1982), Let $X = X$, on *Big Science*, Warner Brothers.

The Athenian (1980), That wretched kilt-wearer: the paintings of Theofilos, May.

Baez, J. (1997), An introduction to n-categories, e-print q-alg/9705009; also in *Category Theory and Computer Science (Santa Margherita Ligure, 1997)*, Lecture Notes in Computer Science 1290, Springer.

Baez, J., Dolan, J. (1995), Higher-dimensional algebra and topological quantum field theory, *Journal of Mathematical Physics* **36** (11), 6073–6105; also e-print q-alg/9503002 (without diagrams).

(1997), Higher-dimensional algebra III: n-categories and the algebra of opetopes, e-print q-alg/9702014; also *Advances in Mathematics* **135** (1998), no. 2, 145–206.

Batanin, M. (1998a), Monoidal globular categories as a natural environment for the theory of weak n-categories, *Advances in Mathematics* **136**, no. 1, 39–103.

(1998b), Computads for finitary monads on globular sets, in *Higher Category Theory (Evanston, IL, 1997)*, Contemporary Mathematics 230, AMS.

(2002a), On the Penon method of weakening algebraic structures, *Journal of Pure and Applied Algebra* **172**, no. 1, 1–23.

(2002b), Computads and slices of operads, e-print math.CT/0209035.

Batanin, M., Street, R. (2000), The universal property of the multitude of trees, *Journal of Pure and Applied Algebra* **154**, no. 1–3, 3–13.

Baues, H., Jibladze, M., Tonks, A. (1997), Cohomology of monoids in monoidal categories, in J.-L. Loday, J. Stasheff, A. Voronov (eds.), *Operads: Proceedings of Renaissance Conferences*, Contemporary Mathematics 202, AMS.

Beck, J. (1969), Distributive laws, in *Seminar on Triples and Categorical Homology Theory (ETH, Zürich, 1966/67)*, Springer.

Beilinson, A., Drinfeld, V. (*c.*1997), Chiral algebras, preprint,

Bénabou, J. (1967), Introduction to bicategories, in J. Bénabou *et al.* (eds.), *Reports of the Midwest Category Seminar*, Lecture Notes in Mathematics 47, Springer.

411

Berger, C. (2002), A cellular nerve for higher categories, *Advances in Mathematics* **169**, no. 1, 118–175.

Blackwell, R., Kelly, G. M., Power, A. J. (1989), Two-dimensional monad theory, *Journal of Pure and Applied Algebra* **59**, no. 1, 1–41.

Boardman, J. M., Vogt, R. (1973), *Homotopy Invariant Algebraic Structures on Topological Spaces*, Lecture Notes in Mathematics 347, Springer.

Borceux, F. (1994a), *Handbook of Categorical Algebra 1: Basic Category Theory*, Encyclopedia of Mathematics and its Applications 50, Cambridge University Press.

(1994b), *Handbook of Categorical Algebra 2: Categories and Structures*, Encyclopedia of Mathematics and its Applications 51, Cambridge University Press.

Borcherds, R. (1997), Vertex algebras, e-print q-alg/9706008; also in *Topological Field Theory, Primitive Forms and Related Topics (Kyoto, 1996)*, Progress in Mathematics 160, Birkhäuser Boston, 1998.

Bright, S. (1999), *Full Exposure*, HarperCollins.

Brown, J. R. (1999), *Philosophy of Mathematics: an Introduction to the World of Proofs and Pictures*, Routledge.

Burroni, A. (1971), *T*-catégories (catégories dans un triple), *Cahiers de Topologie et Géométrie Différentielle* **12**, 215–321.

(1991), Higher-dimensional word problem, in *Category Theory and Computer Science (Paris, 1991)*, Lecture Notes in Computer Science 530, Springer.

(1993), Higher-dimensional word problems with applications to equational logic, *Theoretical Computer Science* **115**, no. 1, 43–62.

Carboni, A., Johnstone, P. (1995), Connected limits, familial representability and Artin glueing, *Mathematical Structures in Computer Science* **5**, 441–459.

(2002), Corrigenda to 'Connected limits, familial representability and Artin glueing', to appear.

Carboni, A., Kasangian, S., Walters, R. (1987), An axiomatics for bicategories of modules, *Journal of Pure and Applied Algebra* **45**, 127–141.

Cheng, E. (2003a), Weak *n*-categories: opetopic and multitopic foundations, e-print math.CT/0304277; also *Journal of Pure and Applied Algebra*, to appear.

(2003b), Weak *n*-categories: comparing opetopic foundations, e-print math.CT/0304279; also *Journal of Pure and Applied Algebra*, to appear.

(2003c), The category of opetopes and the category of opetopic sets, e-print math.CT/0304284; also *Theory and Applications of Categories* **11** (2003), no. 16, 353–374.

(2003d), Opetopic bicategories: comparison with the classical theory, e-print math.CT/0304285.

(2003e), An alternative characterisation of universal cells in opetopic *n*-categories, e-print math.CT/0304286.

Crans, S. (1999), A tensor product for **Gray**-categories, *Theory and Applications of Categories* **5**, no. 2, 12–69.

(2000), On braidings, syllepses and symmetries, *Cahiers de Topologie et Géométrie Différentielle Catégorique* **41**, no. 1, 2–74.

Duskin, J. (2001), A simplicial-matrix approach to higher dimensional category theory II: bicategory morphisms and simplicial maps, preprint.

(2002), Simplicial matrices and the nerves of weak *n*-categories I: nerves of bicategories, *Theory and Applications of Categories* **9**, no. 10, 198–308.

Dwyer, W., Kan, D., Smith, J. (1989), Homotopy commutative diagrams and their real-izations, *Journal of Pure and Applied Algebra* **57**, no. 1, 5–24.

Eckmann, B., Hilton, P. (1962), Group-like structures in general categories, I: multiplications and comultiplications, *Mathematische Annalen* **145**, 227–255.

Eco, U. (1995), *The Search for the Perfect Language*, Blackwell.

Ehresmann, C. (1965), *Catégories et Structures*, Dunod.

Gabriel, P., Zisman, M. (1967), *Calculus of Fractions and Homotopy Theory*, Springer.

Gan, W. L. (2002), Koszul duality for dioperads, e-print math.QA/0201074.

Getzler, E., Kapranov, M. (1994), Modular operads, e-print dg-ga/9408003; also *Compositio Mathematica* **110** (1998), no. 1, 65–126.

(1995), Cyclic operads and cyclic homology, in *Geometry, Topology, & Physics*, Conference Proceedings and Lecture Notes in Geometry and Topology IV, International Press.

Ginzburg, V., Kapranov, M. (1994), Koszul duality for operads, *Duke Mathematical Journal* **76**, no. 1, 203–272.

Gordon, R., Power, A. J., Street, R. (1995), *Coherence for Tricategories*, Memoirs of the AMS **117**, no. 558.

Grandis, M., Paré, R. (1999), Limits in double categories, *Cahiers de Topologie et Géométrie Différentielle Catégorique* **40**, no. 3, 162–220.

(2003), Adjoints for double categories, preprint.

Grothendieck, A. (1983), Pursuing stacks, manuscript.

Hermida, C. (2000), Representable multicategories, *Advances in Mathematics* **151**, no. 2, 164–225.

Hermida, C., Makkai, M., Power, J. (1998), Higher-dimensional multigraphs, in *Thirteenth Annual IEEE Symposium on Logic in Computer Science (Indianapolis, IN, 1998)*, IEEE Computer Society.

(2000), On weak higher dimensional categories I.1, *Journal of Pure and Applied Algebra* **154**, no. 1–3, 221–246.

(2001), On weak higher-dimensional categories I.2, *Journal of Pure and Applied Algebra* **157**, no. 2–3, 247–277.

(2002), On weak higher-dimensional categories I.3, *Journal of Pure and Applied Algebra* **166**, no. 1–2, 83–104.

Hutchinson, J. (1981), Fractals and self-similarity, *Indiana University Mathematics Journal* **30**, no. 5, 713–747.

Joyal, A. (1986), Foncteurs analytiques et espèces de structures, in *Combinatoire Énumérative (Montreal, Quebec, 1985)*, Lecture Notes in Mathematics 1234, Springer.

(1997), Disks, duality and Θ-categories, preprint.

(2002), Quasi-categories and Kan complexes, *Journal of Pure and Applied Algebra* **175**, no. 1–3, 207–222.

Joyal, A., Street, R. (1993), Braided tensor categories, *Advances in Mathematics* **102**, no. 1, 20–78.

Kapranov, M., Voevodsky, V. (1994), 2-categories and Zamolodchikov tetrahedra equations, in *Algebraic Groups and their Generalizations: Quantum and Infinite-Dimensional Methods*, Proceedings of Symposia in Pure Mathematics 56, Part 2, AMS.

Kelly, G. M. (1972), Many-variable functorial calculus I, in *Coherence in Categories*, Lecture Notes in Mathematics 281, Springer.

(1974), On clubs and doctrines, in *Category Seminar*, Lecture Notes in Mathematics 420, Springer.

(1980), A unified treatment of transfinite constructions for free algebras, free monoids, colimits, associated sheaves, and so on, *Bulletin of the Australian Mathematical Society* **22**, 1–83.

(1982), *Basic Concepts of Enriched Category Theory*, London Mathematical Society Lecture Note Series 64, Cambridge University Press.

(1992), On clubs and data-type constructors, in *Applications of Categories in Computer Science (Durham, 1991)*, London Mathematical Society Lecture Note Series 177, Cambridge University Press.

Kelly, G. M., Street, R. (1974), Review of the elements of 2-categories, in *Category Seminar (Sydney, 1972/1973)*, Lecture Notes in Mathematics 420, Springer.

Khovanov, M. (2001), A functor-valued invariant of tangles, e-print math.QA/0103190; also *Algebraic and Geometric Topology* **2** (2002), 665–741.

Kontsevich, M. (1999), Operads and motives in deformation quantization, e-print math.QA/9904055; also *Letters in Mathematical Physics* **48**, no. 1, 35–72.

Kontsevich, M., Manin, Yu. (1994), Gromov–Witten classes, quantum cohomology and enumerative geometry, e-print hep-th/9402147; also *Communications in Mathematical Physics* **164** (1994), no. 3, 525–562.

Koslowski, J. (2003), A monadic approach to polycategories, *Electronic Notes in Theoretical Computer Science* **69**.

Lambek, J. (1969), Deductive systems and categories II: standard constructions and closed categories, in P. Hilton (ed.), *Category Theory, Homology Theory and their Applications, I (Battelle Institute Conference, Seattle, 1968, Vol. One)*, Lecture Notes in Mathematics 86, Springer.

Lawrence, R. (1996), An introduction to topological field theory, in *The Interface of Knots and Physics*, Proceedings of Symposia in Applied Mathematics 51, AMS, 89–128.

Lazard, M. (1955), Lois de groupes et analyseurs, *Annales Scientifiques de l'École Normale Supérieure (3)* **72**, 299–400.

Leinster, T. (1998a), General operads and multicategories, e-print math.CT/9810053.

(1998b), Basic bicategories, e-print math.CT/9810017.

(1999a), Generalized enrichment for categories and multicategories, e-print math.CT/9901139.

(1999b), Up-to-homotopy monoids, e-print math.QA/9912084.

(2000a), Homotopy algebras for operads, e-print math.QA/0002180.

(2000b), Operads in higher-dimensional category theory, e-print math.CT/0011106; also *Theory and Applications of Categories*, to appear.

(2001a), Structures in higher-dimensional category theory, e-print math.CT/0109021.

(2001b), A survey of definitions of *n*-category, e-print math.CT/0107188; also *Theory and Applications of Categories* **10** (2002), no. 1, 1–70.

(2002), Generalized enrichment of categories, e-print math.CT/0204279; also *Journal of Pure and Applied Algebra* **168** (2002), no. 2–3, 391–406.

(2003), Fibrations of multicategories, to appear.

Lewis, G. (1972), Coherence for a closed functor, in G. M. Kelly, M. Laplaza, G. Lewis, S. Mac Lane (eds.), *Coherence in Categories*, Lecture Notes in Mathematics 281, Springer.

Loday, J.-L., Stasheff, J., Voronov, A. (eds.) (1997), *Operads: Proceedings of Renaissance Conferences*, AMS.

Mac Lane, S. (1963), Natural associativity and commutativity, *Rice University Studies* **49**, no. 4, 28–46.

(1971), *Categories for the Working Mathematician*, Graduate Texts in Mathematics 5, Springer, revised edition 1998.

Mac Lane, S., Moerdijk, I. (1992), *Sheaves in Geometry and Logic*, Springer.

Makkai, M. (1996), Avoiding the axiom of choice in general category theory, *Journal of Pure and Applied Algebra* **108**, no. 2, 109–173.

(1999), The multitopic ω-category of all multitopic ω-categories, preprint.

Makkai, M., Zawadowski, M. (2001), Duality for simple ω-categories and disks, *Theory and Applications of Categories* **8**, no. 7, 114–243.

Manes, E. (1976), *Algebraic Theories*, Graduate Texts in Mathematics 26, Springer.

Marcus, B. (2002), *Notable American Women*, Vintage.

Markl, M., Shnider, S., Stasheff, J. (2002), *Operads in Algebra, Topology and Physics*, Mathematical Surveys and Monographs 96, AMS.

May, J. P. (1972), *The Geometry of Iterated Loop Spaces*, Lectures Notes in Mathematics 271, Springer.

(2001), Operadic categories, A_∞-categories and n-categories, preprint.

Moore, G., Seiberg, N. (1989), Classical and quantum conformal field theory, *Communications in Mathematical Physics* **123**, no. 2, 177–254.

Penon, J. (1999), Approche polygraphique des ∞-catégories non strictes, *Cahiers de Topologie et Géométrie Différentielle Catégorique* **40**, no. 1, 31–80.

Réage, P. (1954), *Story of O*.

Ruelle, D. (1999), Conversations on mathematics with a visitor from outer space, in V. Arnold, M. Atiyah, P. Lax, B. Mazur (eds.), *Mathematics: Frontiers and Perspectives*, International Mathematical Union/AMS.

Salvatore, P. (2000), The universal operad of iterated loop spaces, talk at *Workshop on Operads*, Osnabrück.

Schwänzl, R., Vogt, R. (1992), Homotopy homomorphisms and the hammock localization, *Boletín de la Sociedad Matemática Mexicana (2)* **37**, no. 1–2, 431–448.

Segal, G. (1968), Classifying spaces and spectral sequences, *Institut des Hautes Études Scientifiques Publications Mathématiques* **34**, 105–112.

(1974), Categories and cohomology theories, *Topology* **13**, 293–312.

Shakespeare, W. (1596), *King John*.

Shum, M. C. (1994), Tortile tensor categories, *Journal of Pure and Applied Algebra* **93**, no. 1, 57–110.

Simpson, C. (1997), A closed model structure for n-categories, internal $\underline{Hom}$, n-stacks and generalized Seifert–Van Kampen, e-print alg-geom/9704006.

(1998), On the Breen–Baez–Dolan stabilization hypothesis for Tamsamani's weak n-categories, e-print math.CT/9810058.

Snydal, C. (1999a), Equivalence of Borcherds G-vertex algebras and axiomatic vertex algebras, e-print math.QA/9904104.

(1999b), Relaxed multicategory structure of a global category of rings and modules, e-print math.CT/9912075; also *Journal of Pure and Applied Algebra* **168** (2002), no. 2–3, 407–423.

Soibelman, Y. (1997), Meromorphic tensor categories, e-print q-alg/9709030.

(1999), The meromorphic braided category arising in quantum affine algebras, e-print math.QA/9901003; also *International Mathematics Research Notices* 1999, no. 19, 1067–1079.

Sokal, A., Bricmont, J. (1998), *Fashionable Nonsense*, Picador.

Stasheff, J. (1963a), Homotopy associativity of H-spaces I, *Transactions of the AMS* **108**, 275–292.

(1963b), Homotopy associativity of H-spaces II, *Transactions of the AMS* **108**, 293–312.

Street, R. (1972), The formal theory of monads, *Journal of Pure and Applied Algebra* **2**, 149–168.

(1976), Limits indexed by category-valued 2-functors, *Journal of Pure and Applied Algebra* **8**, no. 2, 149–181.

(1987), The algebra of oriented simplexes, *Journal of Pure and Applied Algebra* **49**, no. 3, 283–335.

(1988), Fillers for nerves, in *Categorical Algebra and its Applications (Louvain-la-Neuve, 1987)*, Lecture Notes in Mathematics 1348, Springer.

(1996), Categorical structures, in M. Hazewinkel (ed.), *Handbook of Algebra Vol. 1*, North-Holland.

(1998), The role of Michael Batanin's monoidal globular categories, in *Higher Category Theory (Evanston, IL, 1997)*, Contemporary Mathematics 230, AMS.

(2003), Weak ω-categories, in *Diagrammatic Morphisms and Applications*, Contemporary Mathematics 318, AMS.

Szabo, M. (1975), Polycategories, *Communications in Algebra* **3**, no. 8, 663–689.

Tamsamani, Z. (1995), Sur des notions de n-catégorie et n-groupoïde non strictes via des ensembles multi-simpliciaux, e-print alg-geom/9512006; also *K-Theory* **16** (1999), no. 1, 51–99.

Tillmann, U. (1998), S-structures for k-linear categories and the definition of a modular functor, e-print math.GT/9802089; also *Journal of the London Mathematical Society (2)* **58** (1998), no. 1, 208–228.

Toën, B., Vezzosi, G. (2002), Segal topoi and stacks over Segal categories, e-print math.AG/0212330.

Vonnegut, K. (1963), *Cat's Cradle*, Gollancz.

Voronov, A. (1998), The Swiss-cheese operad, e-print math.QA/9807037; also in *Homotopy Invariant Algebraic Structures (Baltimore, MD, 1998)*, Contemporary Mathematics 239, AMS, 1999.

Walters, R. (1981), Sheaves and Cauchy-complete categories, *Cahiers de Topologie et Géométrie Différentielle Catégorique* **22**, no. 3, 283–286.

Weber, M. (2001), Symmetric operads for globular sets, Ph.D. thesis, Macquarie University.

Worth, S. (1975), Pictures can't say 'ain't', *Versus: Quaderni di studi semiotici* **12**, 85–105.

Yetter, D. (2001), *Functorial Knot Theory*, Series on Knots and Everything 26, World Scientific Publishing Co.

Index of notation

'It's a revealing thing, an author's index of his own work,' she informed me. 'It's a shameless exhibition – to the *trained* eye.'

Vonnegut (1963)

Page numbers indicate where the term is defined; multiple page numbers mean that the term is used in multiple related senses. Boxes (□) stand for some or all of 'lax', 'colax', 'wk', and 'str'.

Latin letters

Ab	category of abelian groups	35
Alg( )	category of algebras for a multicategory	66, 153
Alg$_\square$( )	category of algebras for a **Cat**-operad	103–104
B	initial operad with coherence and system of compositions	313
Bicat$_\square$	category of bicategories	53
( )-Bicat$_\square$	category of 'bicategories' according to some theory	125
Braid	category of braids	16
CartMnd$_\square$	2-category of cartesian monads and lax maps	207, 208
Cat	(2-)category of small categories	24, 41
Cat$_2$	weak double category of categories	171
Cat( )	category of internal categories	47
( )-Cat	category of enriched categories	39
CatAlg( )	category of categorical algebras	125, 316
Cat-Operad	(2-)category of **Cat**-operads	103, 108
( )^{co}	dual of bicategory, reversing 2-cells	51
cod	codomain map of (multi)category	30, 140
CommMon	category of commutative monoids	223
comp	composition map in (multi)category	30, 39, 140
Cone( )	cone on space	244
ctr	operad of classical trees	87
D	discrete category	24, 204
$\mathbb{D}$	augmented simplex category	31

417

Greek letters

Other symbols

Index

423